Acid Sulfate Weathering
SSSA Special Publication Number 10

Proceedings of a symposium sponsored
by Divisions S-9, S-2, S-5, and S-6 of the
Soil Science Society of America in
Fort Collins, Colorado, 5–10 Aug. 1979.

Editorial Committee
J. A. Kittrick
D. S. Fanning
L. R. Hossner

Managing Editor
David M. Kral

Assistant Editor
Sherri Hawkins

1982
Published by the
SOIL SCIENCE SOCIETY OF AMERICA
677 South Segoe Road
Madison, Wisconsin 53711

Copyright 1982 by the Soil Science Society of America, Inc.
ALL RIGHTS RESERVED UNDER THE U.S. COPYRIGHT LAW OF 1978 (P.L. 94-553). Any and all uses beyond the limitations of the "fair use" provisions of the law require written permission from the publisher(s) and/or the author(s): not applicable to contributions prepared by officers or employees of the U.S. Government as part of their official duties.

Soil Science Society of America
677 South Segoe Road, Madison, Wisconsin 53711 USA

Library of Congress Catalog Card Number: 82-80099
Standard Book Number: 0-89118-770-7

Printed in the United States of America

Contents

	Page
Foreword	v
Preface	vii

PART I
SULFIDE ACCUMULATION IN SEDIMENTS

Physiography of Coastal Sediments and Development of Potential Soil Acidity
    L. J. Pons, N. Van Breemen, and P. M. Driessen 1

Controls and Consequences of Sulfate Reduction in Recent Marine Sediments
    M. B. Goldhaber and I. R. Kaplan 19

PART II
CHEMISTRY, MICROBIOLOGY AND MINERALOGY OF ACID SULFATE WEATHERING

Aqueous Pyrite Oxidation and the Consequent Formation of Secondary Iron Minerals
    Darrell Kirk Nordstrom 37

Microbiological Transformations of Iron and Sulfur and their Applications to Acid Sulfate Soils and Tidal Marshes
    K. C. Ivarson, G. J. Ross, and N. M. Miles 57

Microbial Formation of Basic Ferric Sulfates in Laboratory Systems and in Soils
    G. J. Ross, K. C. Ivarson, and N. M. Miles 77

PART III
SOILS WITH ACID SULFATE WEATHERING FEATURES OR WITH GYPSUM OF OTHER ORIGIN

Genesis, Morphology and Classification of Acid Sulfate Soils in Coastal Plains
    N. Van Breemen.. 95

Morphological and Mineralogical Features Related to Sulfide Oxidation under Natural and Disturbed Land Surfaces in Maryland
    D. P. Wagner, D. S. Fanning, J. E. Foss, M. S. Patterson, and P. A. Snow... 109

Alfisols and Ultisols with Acid Sulfate Weathering Features in Texas
    C. D. Carson, D. S. Fanning, and J. B. Dixon 127

Gypsiferous Soils in the Western United States
   W. D. Nettleton, R. E. Nelson, B. R. Brasher, and
   P. S. Derr 147

PART IV
EARTH SURFACE MANIPULATION AND MINESPOIL RECLAMATION

Mineralogical Properties of Lignite Overburden as they Relate
to Mine Spoil Reclamation
   J. B. Dixon, L. R. Hossner, A. L. Senkayi, and
   K. Egashira... 169

Relation of Pyritic Sandstone Weathering to Soil and Minesoil
Properties
   R. N. Singh, W. E. Grube, Jr., R. M. Smith, and
   R. F. Keefer 193

Mineralogical Alterations that Affect Pedogenesis in Minesoils
from Bituminous Coal Overburdens
   W. E. Grube, Jr., R. M. Smith, and J. T. Ammons 209

Characteristics and Reclamation of "Acid Sulfate" Mine Spoils
   R. I. Barnhisel, J. L. Powell, G. W. Akin, and
   M. W. Ebelhar 225

Foreword

Acid sulfate weathering is a subject of increased interest both nationally and internationally. Acid sulfate soils, in general, result from processes that release sulfuric acid into the soil system as the soil forms. This term is in turn applied to soils in which sulfuric acids have been, are being, or will be produced in amounts that have a lasting effect on principal soil characteristics. Such soils occur in all climatic zones of the earth with the majority of them being located in relatively recent coastal marine sediments. However, sulphidic materials which produce acid sulfates on oxidation are not limited to coastal regions. They are often associated with pyritic materials such as lignite. When such materials are brough to the soil surface through mining, construction, or other activities that disturb the soil, sulfuric acid may form making revegetation of the soil very difficult and releasing pollutants into surface and subsurface waters.

In depth understanding of the nature and properties of acid sulfate soils is necessary if they are to be reclaimed as a resource to be used in crop production. This reclamation is of increasing importance because of the expanding areas of potentially acid sulfate soils associated with expanded mining activities. While reclamation of these soils is important, it is far from simple because of the complexity of their chemical, microbiological, and mineralogical relationships. Increased understanding of these relationships must be developed so that these lands may be made productive, and that they not be sources of pollutants to the environment.

A symposium was held during the 1979 meetings of ASA and SSSA to bring together those working on acid sulfate soils so that the present state of knowledge could be shared. The intent was to broaden our understanding of the nature and properties of acid sulfate soils, so that reclamation attempts would be based on the combined knowledge of these individuals. This publication is the result of that symposium. It includes contributions of leading scientists in the area of acid sulfate soils. The Society is indebted to these authors as well as the organizers of the symposium and the editors of this special publication.

R. G. Gast, SSSA President, 1982

Preface

An international symposium held at Wageningen, The Netherlands, in 1972 (Dost, 1973) had a large impact in bringing the existence and nature of acid sulfate soils to the attention of scientists around the world. This 1972 symposium and a second international acid sulfate soils symposium, held in Thailand and Malaysia in 1981, have been concerned primarily with acid sulfate soil development and reclamation in modern sediments near sea coasts.

This publication, which contains papers presented at a symposium held at the Soil Science Society of America (SSSA) meetings at Colorado State University on 6 Aug. 1979, is not intended to replace publications resulting from the international symposia, but rather to emphasize that acid sulfate weathering is not confined to coastal soils. This publication shows the wide application of the principles of acid sulfate weathering in understanding and managing soils and geologic columns—particularly those that are subject to major manipulation in construction, in recovering mineral resources, etc. At the same time the publication further elucidates the pedogeochemistry of acid sulfate weathering and implies that many soils have been affected by this weathering.

More specifically the papers presented here seek a) to explain how sulfide bearing sediments accumulate (for further explanation of this subject readers should also consult publications such as the extensive paper by D. T. Rickard (1973); b) to describe the physical chemistry, microbiology, and mineralogy of acid sulfate weathering; c) to illustrate effects of acid sulfate weathering and associated pedogeochemical changes upon young and old soils and associated substrata; and d) to present examples of how this knowledge is being applied in man's manipulations of the earth's surface and of undesirable situations that may develop after such manipulations when this knowledge is ignored. An additional paper on gypsiferous soils reminds us that not all soils containing large amounts of sulfate minerals have been directly influenced by acid sulfate weathering.

The symposium that led to this publication was organized by Division S-9, Soil Mineralogy, of SSSA, and was co-sponsored by Divisions S-2 (Soil Chemistry), S-5 (Soil morphology, Genesis, and Classification), and S-6 (Soil and Water Management and Conservation). The SSSA further contributed to the symposium by supporting the travel expenses of Dr. Nico van Breemen of the Netherlands to the symposium.

Editorial Committee
J. A. Kittrick
D. S. Fanning
L. R. Hossner

REFERENCE

Rickard, D. T. 1973. Acid sulfate soils. *In* H. Dost (ed.) ILRI Publ. 18, Vols. I and II. Int. Institute for Land Reclamation and Improvement. P.O. Box 45, Wageningen, The Netherlands.

Chapter 1

Physiography of Coastal Sediments and Development of Potential Soil Acidity[1]

L. J. PONS, N. VAN BREEMEN, AND P. M. DRIESSEN[2]

ABSTRACT

Potentially acid sedimentary material contains pyrite in excess of acid-neutralizing substances. Formation of such material requires (1) ingredients for pyrite formation (sulfate, sulfate reducers, organic matter, iron, and anaeroby alternating with limited aeration), (2) low contents of acid-neutralizing substances, and (3) removal of dissolved alkalinity formed during sulfate reduction. Intertidal environments with mangroves or reeds are particularly favorable for pyritization of ferric iron. Highest pyrite contents build up where tidal flushing is strong. Rapid rises in relative sea level, as after the last glaciation, caused deposition of extensive, thick and highly pyritic sediments (examples: interior parts of the Chao Phraya, Mekong and Orinoco deltas, parts of Sumatra, old sea clay of Holland). After stabilization of the sea level, some 5,000 years B.P., pyrite contents remained low where high rates of sedimentation and coastal accretion caused a rapid shift of the intertidal zone (Irrawaddy and Mekong deltas, Guyana coast). High pyrite contents in the most recent sediments are associated with low sedimentation rates (e.g., along the Saigon, Niger, and Gambia rivers), or with a high density of tidal creeks. In humid climates very low sedimentation rates result in the formation of pyritic peaty material on top of older pyritic clay (Niger delta, western Netherlands).

---

[1] Contribution from the Dep. of Soil Science and Geology, Agricultural Univ., Wageningen, the Netherlands.
[2] Professor of regional soil science, and soil scientists, respectively, Dep. of Soil Science and Geology, P.O. Box 37, Wageningen, the Netherlands.

Copyright © 1982 Soil Science Society of America, 677 S. Segoe Rd., Madison, WI 53711. *Acid Sulfate Weathering.*

INTRODUCTION

An estimated 10 million hectares of land in recent coastal plains, mainly in the tropics, consist of highly pyritic material that will acidify upon aeration, or has already done so. Over an even larger area pyritic coastal sediments are covered with nonpyritic fluvial sediments or with peat. In addition, Pleistocene, Tertiary or still older potentially acid pyritic sediments, often originally deposited in tidal environments, are common in many inland areas.

Whereas synsedimentary pyrite formation in seabottom sediments has been studied in considerable detail by geochemists and oceanographers (Berner, 1970, Goldhaber and Kaplan, 1974), little is known about the formation of pyrite in tidal swamps and marshes, the source areas for most acid sulfate soils. Highly relevant in this regard are the questions of why many tidal swamp sediments accumulate pyrite to higher concentrations than most seabottom sediments, and why some tidal swamp sediments have very high pyrite contents and are potentially acid, whereas others, though seemingly similar, are not. In this paper we will present data supporting the hypothesis that the formation of potentially acid sediments proceeds rapidly under tropical mangrove swamp conditions where tidal flushing is strong, and we will discuss various chemical, physiographic and sedimentary factors that presumably influence pyrite accumulation in tidal swamps.

FORMATION OF POTENTIAL SOIL ACIDITY: MECHANISM

Acid sulfate soils form where the quantity of sulfuric acid, formed by oxidation of reduced S-compounds, exceeds the acid-neutralizing capacity of adsorbed bases and easily weatherable minerals to the extent that the pH drops below 4. Pyrite is the dominant sulfur mineral in tidal swamps; contents of Fe(II)-monosulfide and elemental S are generally very low.

Factors influencing the pyrite content and the acid-neutralizing capacity of pyritic sediments will be discussed in general before relating them to specific environmental aspects of recent and former tidal swamps.

Formation of Pyrite

Formation of pyrite (cubic FeS_2) requires (1) reduction of sulfate to sulfide under the influence of dissimilatory sulfate-reducing bacteria in an anaerobic environment; (2) partial oxidation of sulfide to polysulfide or elemental sulfur; and (3) either formation of Fe(II)-monosulfide (from Fe(III)-oxides or Fe-containing silicates and dissolved sulfide) followed by combination of elemental sulfur and Fe(II)-monosulfide to pyrite, or direct precipitation of pyrite from dissolved Fe(II) iron and polysulfide ions (Roberts et al., 1969, Goldhaber and Kaplan, 1974). Whatever the mechanism in operation, formation of pyrite with any Fe(III)-oxide as the source of iron will take place according to the following overall reaction equation (CH_2O stands for organic matter):

$$Fe_2O_{3(s)} + 4SO_{4(aq)}^{2-} + 8CH_2O + \tfrac{1}{2}O_{2(g)}$$

$$\rightarrow 2FeS_{2(s)} + 8HCO_3^-{}_{(aq)} + 4H_2O_{(l)} \qquad [1]$$

This overall reaction includes reduction of all sulfate to sulfide, followed by oxidation of sulfide (with Fe(III) and O_2 as oxidants) to disulfide (S_2^{2-}). Thus, essential ingredients for the formation of pyrite are: sulfate, iron-containing minerals, metabolizable organic matter, sulfate-reducing bacteria, and anaeroby alternating with limited aeration.

Little is known about the rate of pyrite formation in situ. It is generally agreed that the solid-solid reaction $FeS + S \rightarrow FeS_2$ is a slow process, that takes months or years to produce measurable quantities of pyrite, whereas direct precipitation from dissolved Fe(II) iron and polysulfide may, under favorable conditions, yield pyrite within days (Goldhaber and Kaplan, 1974). Both direct precipitation ($Fe^{2+} + S_2^{2-} \rightarrow FeS_2$) and the solid-solid reaction ($FeS + S \rightarrow FeS_2$) are kinetically favored by a low pH (Rickard, 1975; Goldhaber and Kaplan, 1974). The latter authors suggest that when a solution is supersaturated with both Fe(II)- monosulfide and pyrite, FeS will precipitate preferentially, even though the supersaturation with pyrite exceeds that with FeS by far. Once formed, FeS will only slowly be transformed into FeS_2. However, at pH < 6.5, or at low concentrations of dissolved sulfide, the solution may be undersaturated with FeS but still strongly supersaturated with FeS_2, and pyrite could precipitate without competition of FeS. This would enhance pyrite formation under slightly acid conditions. Howarth (1979) observed very rapid pyrite formation (within 48 hours) in a tidal salt marsh, but his method of applying $^{35}SO_4^{2-}$ and $FeSO_4$ may have influenced this result.

Acid-neutralizing Capacity of Soil Material

The acid-neutralizing capacity of soil material is determined by the amount of exchangeable bases and by the contents of carbonates and easily weatherable silicate minerals. Most marine heavy clay soils have appreciable amounts of smectite clay, and their exchange complex, when fully saturated with bases, is capable of inactivating most of the acidity released by the oxidation of up to 0.5% pyrite-S, so that the pH will not drop below 4.0. If the clay fraction is predominantly kaolinitic, or if clay contents are low, less than 0.5% pyrite-S may make the soil potentially acid.

Calcium carbonate contents are low or nill in most marine sediments of the humid tropics but may be appreciable (frequently higher than 10%) in sediments of arid and humid temperate regions. The acidity from 1% (mass fraction) of pyrite-S is approximately balanced by 3% calcium carbonate. If sea water is entrapped in a sediment and all dissolved sulfate is reduced to sulfide, the increase in HCO_3^- would lead to supersaturation with calcium carbonate. Yet, calcium carbonate rarely precipitates, pre-

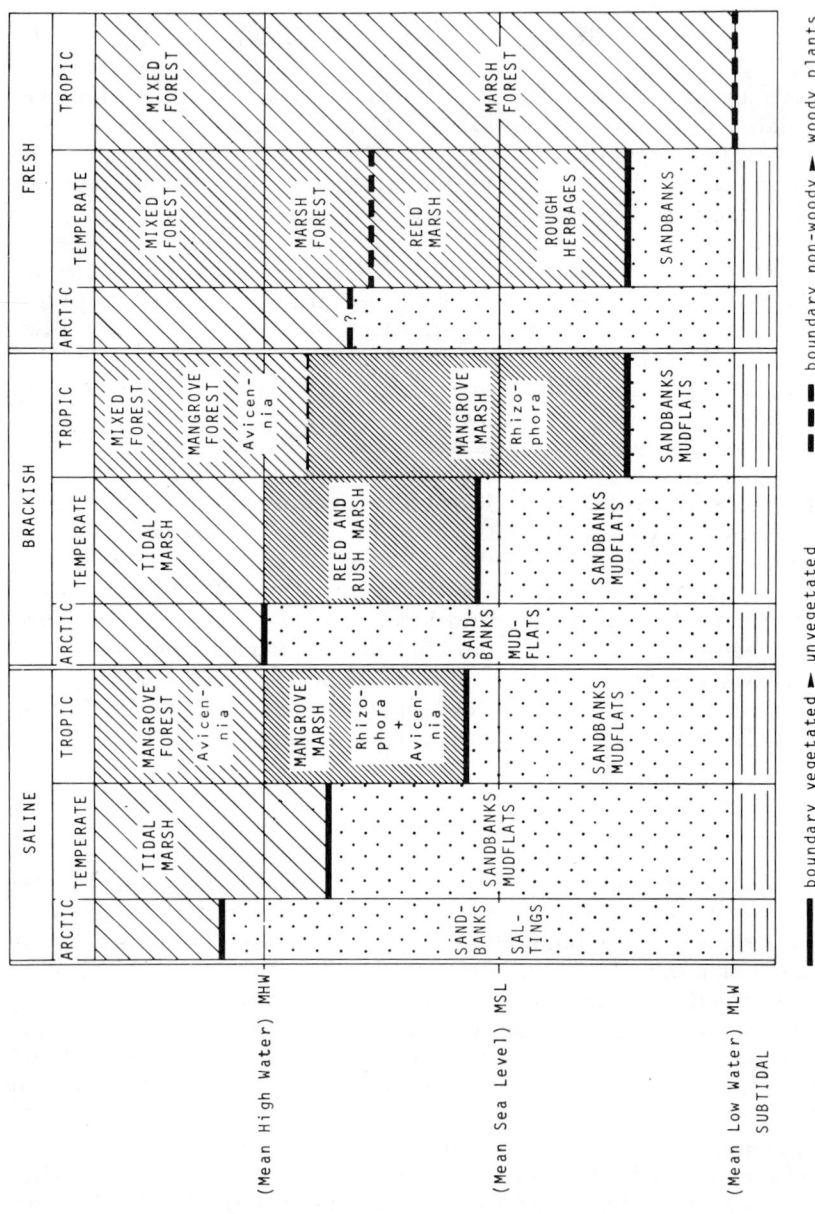

Fig. 1. Tidal vegetation in humid climates along arctic, temperate and tropical coasts.

sumably because of an inhibiting effect of dissolved organic matter (Berner et al., 1970). As a matter of fact, the reverse process, i.e., dissolution of calcium carbonate in tidal sediments in which sulfate reduction takes place, has been frequently reported (Van der Sluijs, 1970; Brummer, 1968; Salomons, 1974). This dissolution is perhaps caused by the combination of strong CO_2-production by decomposing organic matter and leaching by tidal action. Oxidation of some pyrite during low tides would also contribute to removal of calcium carbonate (Kooistra, 1978).

Potential acidity due to excess reduced sulfur can develop only if at least part of the alkalinity (HCO_3^-) formed during sulfate reduction (reaction 1) is removed from the system. Drever (1971) suggests that iron (III) is extracted from smectite during formation of pyrite in anoxic seabottom sediments, and replaced by magnesium from solution. This process could store three-fourths of the alkalinity released during sulfate reduction in the sediment. Van Breemen (1976) found evidence that this process, and its reverse (replacement of smectite-Mg by Fe(III) from oxidizing pyrite), take place in the Bangkok Plain, Thailand, sometimes to the extent that some 1% of pyrite-S is neutralized during oxidation.

FORMATION OF POTENTIAL SOIL ACIDITY: INFLUENCE OF PHYSIOGRAPHY

The ingredients for the formation of potential acidity are most often found in the following land systems:
  (1) saline and brackish tidal flats and tidal swamps,
  (2) bottoms of saline and brackish lagoons, seas and lakes, and
  (3) inland valleys subject to influx of sulfate-rich water.

System 1 comprises bare tidal flats, marshes with a herbaceous vegetation mangrove swamps, with associated tidal creeks. The lowest parts of the system are inundated most of the time and have permanently reduced sediments, the highest parts have been silted up to spring tide levels and have a predominantly aerated surface soil.

Organic carbon contents of the sediments deposited in this system are generally low in the tropics (0.5 to 2%) but may be high in temperate regions (frequently up to 6%). Organic carbon contents in the lower range probably limit pyrite formation in nonvegetated tidal flats and creek bottoms (Pons, 1965). However, vegetated tidal flats commonly receive a high and steady supply of organic matter from mangroves (in the tropics) and reeds and rushes (in temperate regions). Adaption of the vegetation to salinity and prolonged inundation is greater in the tropics than in temperate and arctic zones. Figure 1 suggests that mangroves inhabit nearly all saline and brackish areas between Mean High Water (MHW) and Mean Low Water (MLW) in the humid tropics, while in temperate regions only the land above Mean Sea Level (MSL) harbors plants, provided the water is not too saline. Very high salinities caused by evaporation of seawater in the absence of rain or supply of fresh water by runoff hamper the growth of mangroves. This probably explains the absence of mangroves from arid and semiarid coastal areas (Marius, 1972). Tidal land in arctic regions is devoid of vegetation irrespective of salinity.

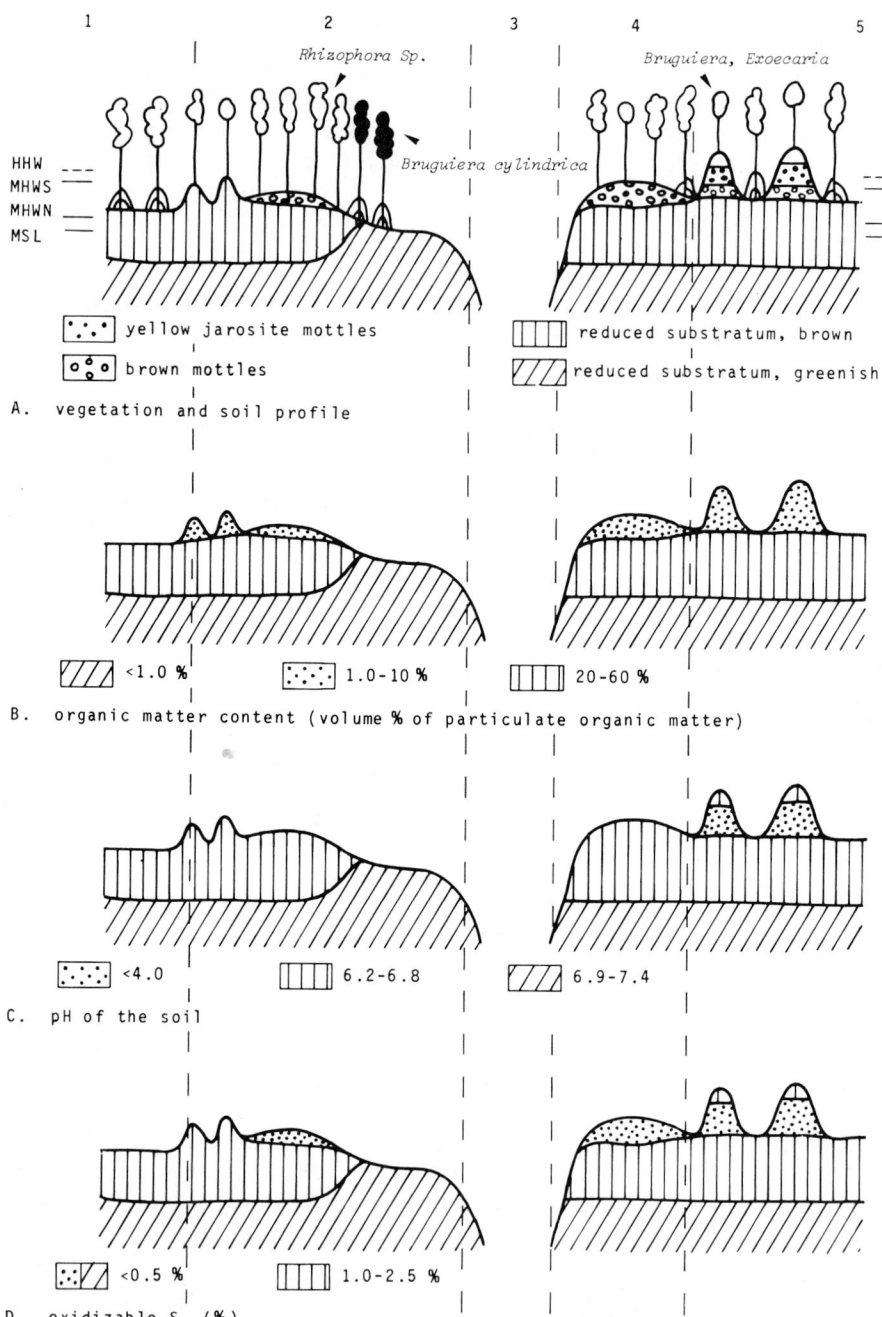

Fig. 2a. Cross section through an estuarine type of mangrove forest in the Merbok area, Malaysia. 1: basin, 2: accreting inner bend of creek, 3: tidal creek, 4: eroding outer bend, 5: high tidal flat with lobster mounds. After Diemont and Van Wijngaarden (1974).

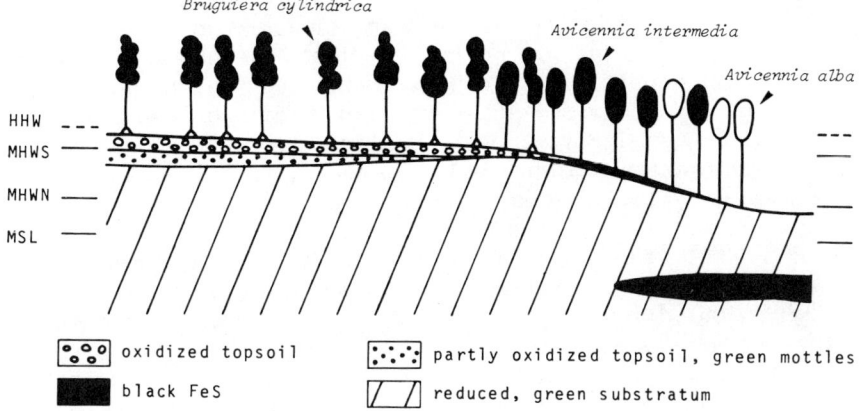
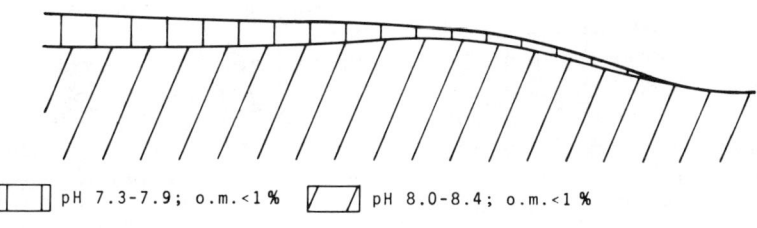

Fig. 2b. Cross section through an accreting coast with mangrove forest, Kuala Selangor, Malaysia. Oxidizable S contents are below 0.5% at all depths. After Diemont and Van Wijngaarden (1974).

Conditions for pyrite formation vary strongly in mangrove swamps. Diemont and Van Wijngaarden (1974) found a number of striking differences in chemistry and pedogenetic development between sediments of an open accreting coastal system and those of an estuarine area dissected by tidal creeks in Malaysia (Fig. 2). The reduced substratum of open coastal swamps had less than 0.5% pyrite-sulfur, was low in organic matter and had a field pH between 8 and 8.4, reflecting high concentrations of HCO_3^- (10 to 26 mol/m^3) in the soil solution. The reduced substrata of estuarine swamps had 1–2.5% pyrite-sulfur, a high content of undecomposed organic matter and a field pH between 6.2 and 6.8 in the upper meter, with interstitial water lower in dissolved HCO_3^- (2 to 10 mol/m^3) than in swamps along an open coast. At greater depth, the reduced substratum of the estuarine swamps was lower in organic matter and pyrite, and higher in pH than its upper meter. Black FeS was found here and there in soils along the open coast, but not in the estuarine sediments. Most of the time concentrations of dissolved sulfide were similar in swamps of both types but dissolved sulfide was lowered to undetectable levels during spring tides in estuarine swamps whereas an almost constant level was observed throughout the monthly tidal cycle in swamps along an open coast (Diemont and Van Wijngaarden, personal communication). The apparent removal of dissolved sulfide and bicarbonate from sediments of estuarine swamps is attributed to more effective tidal flushing as a result of proximity of tidal creeks and higher hydraulic conductivity values due to abundant plant remains.

By lowering the pH, tidal flushing would kinetically favor pyrite formation. Tidal flushing could further speed up pyrite formation by breaking down diffusion-controlled rate-limiting processes, and by supplying limited amounts of dissolved oxygen necessary for complete pyritization of reduced sulfate.

Apart from influencing the pyrite content, tidal flushing enhances removal of sedimentary carbonate and of the bicarbonate formed during sulfate reduction, thereby increasing the potential acidity of the sediment. The elevated parts of the system are generally low in pyrite because of insufficient anaeroby. Their extent varies with the difference between MHW level and MLW level and with the incidence of exceptionally high tides. If sedimentation is slow, deposition of mineral material may be outweighed by accumulation of organic debris and peat growth, and iron may become the limiting factor in the formation of pyrite. Pyritic mangrove peats are known to exist in the Niger delta, in Senegal, in Kenya and in Malaysia and Indonesia; pyritic reed and rush peats are not uncommon in temperate areas, e.g., in the Netherlands. Such pyritic peat layers are normally shallow and witness a transitional phase from a marine environment to a topogenous fresh water swamp without tidal influence.

System 2, the bottoms of saline and brackish lagoons, seas, and lakes nearly always involves clastic sediments brought in by rivers. When the material is high in organic matter as in boreal and arctic regions where decay of organic matter is slow, accumulation of sulfide can be considerable. Often, these sediments contain as much Fe(II)-monosulfide as pyrite-sulfur, indicating an arrested stage of diagenesis probably due to insufficient aeration of the bottom waters. The bottom sediments of the Black Sea and the Littorina sediments of the Baltic and its former extensions are examples of such materials relatively high in FeS (Berner, 1971; Wiklander et al., 1950). Whereas Littorina and Black Sea sediments contain up to 2% reduced sulfur, most bottom sediments have less than 1% sulfide-S. Organic matter usually limits pyrite formation in such environments (Berner, 1970). Isostatic rises of the land after the last glaciation have brought the Littorina sediments above the present sea level, inducing acid sulfate soil formation.

System 3, poorly drained inland valleys with influx of sulfate-rich water, is commonly associated with drainage water from sulfur-containing formations. This situation is comparatively rare. Examples are the pyritic papyrus peats of Uganda (Chenery, 1954), pyritic sands in a few valleys in the eastern Netherlands (Poelman, 1973), and the sulfidic peat soils of Minnesota.

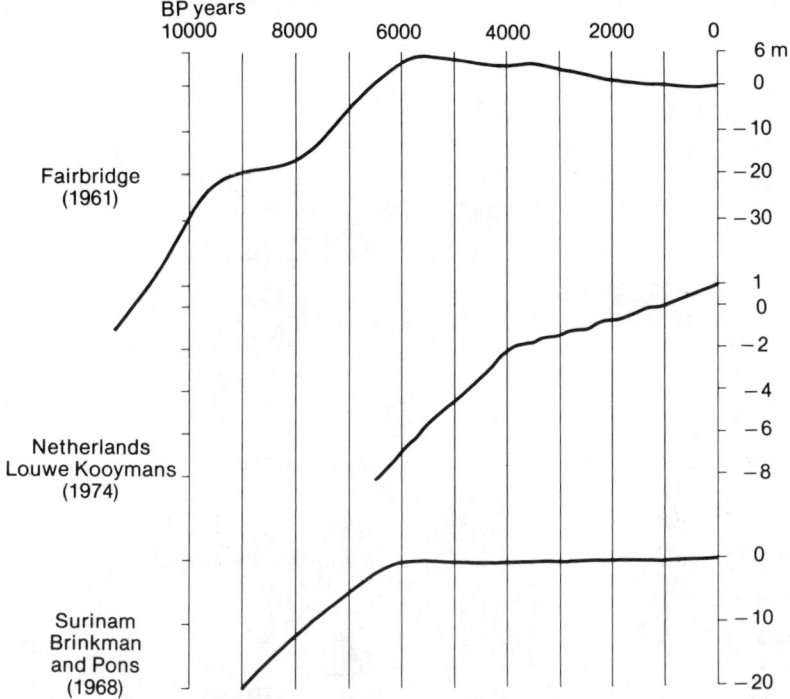

Fig. 3. Changes in relative sea level during the Holocene.

Influence of Relative Sea Level Changes and Coastal Accretion Rates

Both the formation of coastal landforms and the development of potential acidity in their sediments can be affected by relative sea level changes. After the last glaciation the sea level rose by about 3 to 4 m per 1,000 years (Blackwelder et al., 1979) and levelled off until a maximum was reached some 5,500 years B.P. This high sea level remained fairly stable, probably with a slight drop starting some 5,000 years ago. Figure 3 shows reconstructions of the combined effects of land and sea level movements for three regions. The pattern given by Fairbridge (1961) shows the effect of sea level changes only and would be relevant for stable parts of the earth crust.

In the Netherlands, the rise in sea level continued slowly after 5,500 years B.P. because of land subsidence, whereas in Surinam a slight subsidence resulted in a practically constant sea level after 5,500 years B.P.

Figure 4 illustrates how changes in sea level and sediment supply affect coastal areas. The area of estuarine mangrove swamps will increase when the sea level rises and when the sediment supply keeps pace with the rise in relative sea level. If the sediment supply is low relative to the rise in sea level, transgression will take place; if the sediment supply is relatively high the land area will increase.

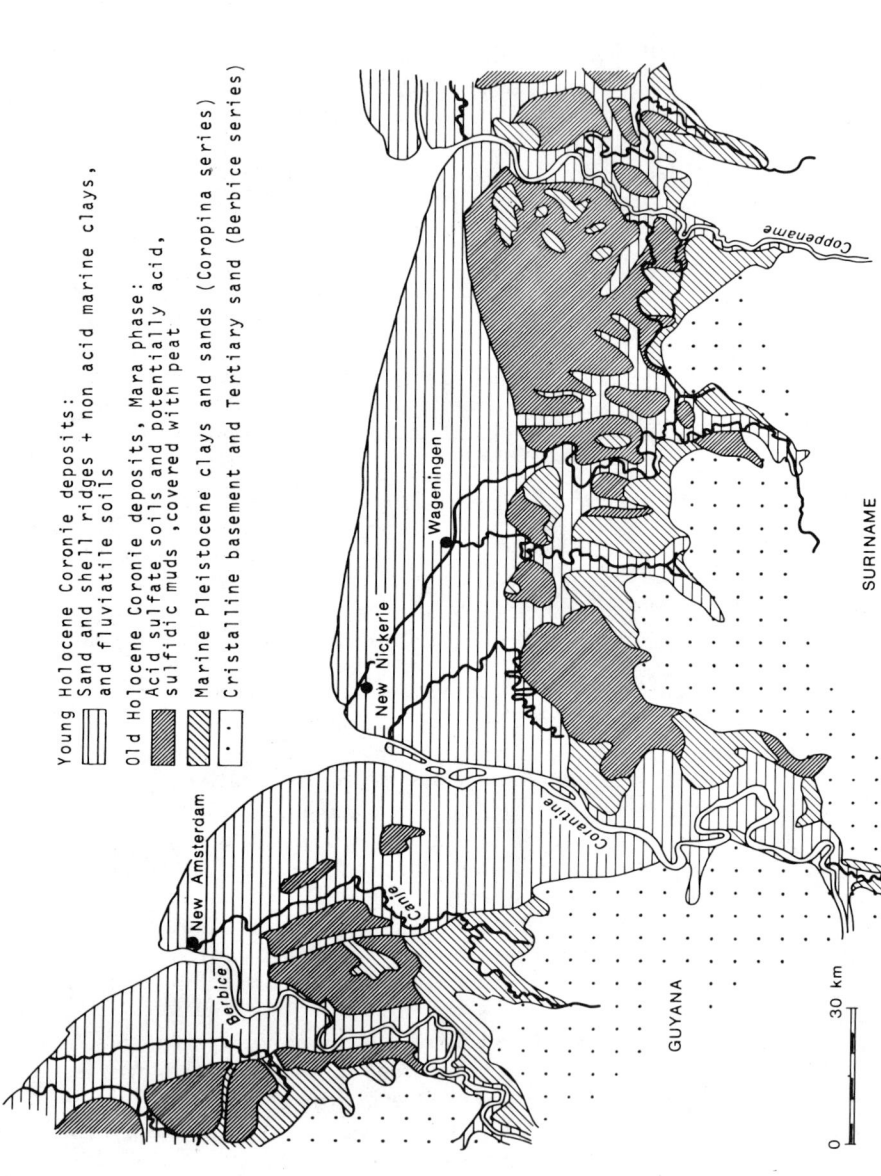

Fig. 4. Effect of the rates of sea level change and sediment supply on the change in the surface area of estuarine mangrove swamps.

After the last glaciation, in many areas the rise in sea level during the early Holocene was approximately balanced by the sediment supply, resulting in a vertical build-up of sediments in an area of stable proportions; lateral coastal accretion started after the late Holocene stabilization of the sea level (Brinkman and Pons, 1968). An additional effect of a relative rise in sea level was that silting up of creeks was retarded resulting in increased tidal flushing of the sediments. As a result mangrove swamps or reed marshes dissected by tidal creeks presisted for long periods of time in those areas, giving rise to thick, highly pyritic sediments. After stabilization of the sea level, highly pyritic sediments could only develop where sedimentation rates were low. However, where the sedimentation rates were high, rapid lateral coastal accretion took place, shifting the intertidal zone with mangroves and reed marshes. This limited the period during which pyrite could be formed, and hence, precluded the build-up of high pyrite contents. In addition to limited time, the less favorable chemical environment for pyrite formation in sediments of rapidly aggrading coasts contributed to the generally low levels of pyrite in such materials. Increased deforestation in watersheds of major rivers during the last 1,000 to 2,000 years caused increased upstream erosion and downstream sedimentation, and has probably contributed to rapid coastal accretion in many areas.

Although these general trends apply to large stretches of coastal land throughout the world, the relationship between sea level changes, coastal accretion rates, and accumulation of pyrite is not always obvious. Secondary transportation of sedimentary material can influence its pyrite content and acid-neutralizing capacity. Detailed soil studies remain indispensable for the characterization of coastal land in terms of potential soil acidification hazard.

COASTAL PHYSIOGRAPHY AND POTENTIAL SOIL ACIDITY; REGIONAL EXAMPLES

The following examples of Holocene coastal sediments in various parts of the world illustrate the theory discussed in the previous sections.

The Americas

The Guyana Coastal Plain. The Guyana coastal plain stretches along the Atlantic Ocean and includes parts of northeast Brazil, French Guyana, Surinam, Guyana, and Venezuela (Fig. 5). Amazon sediments, laid down on tidal flats and in mangrove swamps form the bulk of the Holocene sedimentation sequence of the Demerara series. This series has two main components: the Mara deposits in the interior and the Coronie deposits along the present coast (Brinkman and Pons, 1968). The Mara deposits are at least 20 m thick and consist of highly pyritic (2 to 3% S) potentially acid clay. The shallow Coronie deposits include clayey areas practically free from potential acidity, and coarse textured beach ridges on top of former mudflats. Pollen analyses and ^{14}C-determinations have

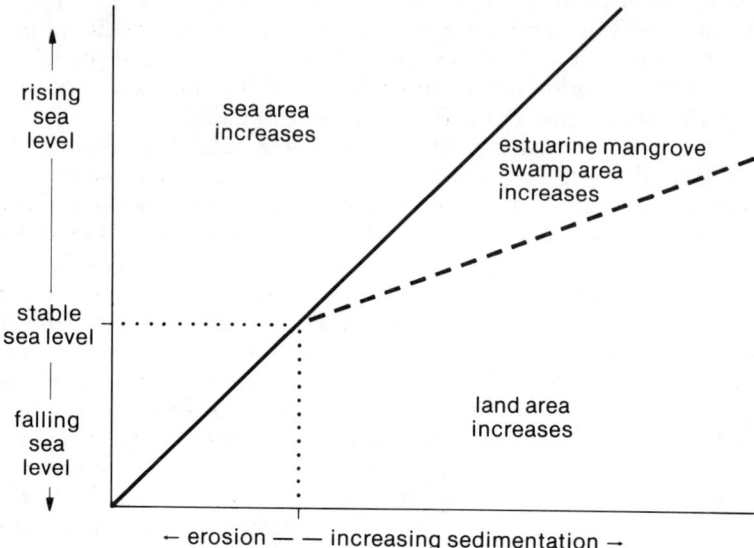

Fig. 5. Holocene sedimentation sequence in the Guyana coastal plain. After Brinkman and Pons (1968).

indicated that the Mara deposits formed before 6,000 years in *Rhizophora* spp. swamps along a stable coastline at a rising sea level; the younger Coronie deposits are characteristic of present accretion areas and consist of narrow strips of tidal land with *Avicennia germinans* (L.) mangroves. The Mara deposits are covered with peat that increases in thickness towards the west. Most of these sediments have not acidified because they have been continuously reduced. Acid sulfate soils and highly pyritic clay occurs locally in creek fillings in the Coronie deposits; stronger and prolonged tidal flushing has probably accelerated pyrite accumulation there.

The Mississippi Deltaic Plain. The Mississippi deltaic plain is characterized by high sedimentation rates and a rapid rise in relative sea level due to subsidence at a rate of about 1 cm per year (DeLaune and Patrick, 1978). This, together with high organic matter contents from a prolific tidal marsh vegetation (mainly *Spartina* sp.) would seem to make conditions ideal for the accumulation of pyrite. However, total S contents are only moderate (an average of 1.4% S was found in 14 peaty samples; Fisk, 1960) and the soil rarely acidifies below pH 5 when aerated (Chabreck, 1972).

Possibly, the low intensity of tidal flushing (due to the small tidal amplitude, which is generally less than 50 cm) and the very rapid vertical accretion limit pyrite accumulation in this area.

Western Europe

The Netherlands. The Holocene marine deposits of the Netherlands have been differentiated into the older Calais deposits, formed until some

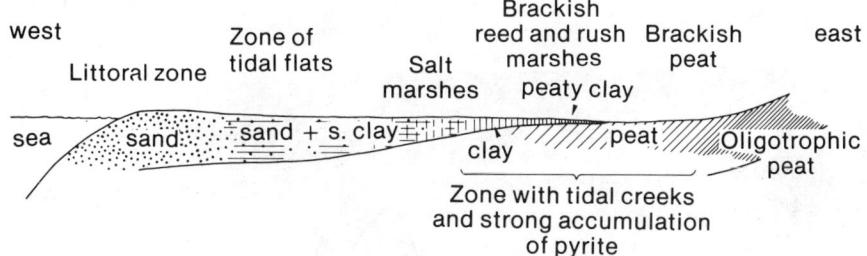

Fig. 6. Schematic cross-section through western Holland during the formation of the Calais IVb deposits.

4,500 years ago under conditions of a rapidly rising sea level, and the younger Duinkerken deposits, associated with slowly rising or nearly stable sea levels (Fig. 3). Each of the four phases identified in the Calais deposits comprises three sedimentation zones: a littoral sandy zone of beach ridges and dunes; a clayey zone of bare tidal flats; and, still further from the sea, a zone of brackish reed marshes with numerous tidal creeks, grading into peat formations (Fig. 6). Strong accumulation of pyrite took place in the reed marshes. Their originally calcareous clay decalcified completely.

During formation of the Duinkerken deposits, a major part of the older Calais deposits was sheltered from marine influence by a completely closed beach barrier, and became overgrown by some 4 m of freshwater peat. Removal of this peat by man and subsequent empoldering of the developed lakes brought the older deposits back to the surface. The present intricate system of calcareous creek deposits and noncalcareous pyrite clays with numerous reed roots bears witness of the former situation in many polders (Fig. 7).

The Duinkerken deposits, especially the youngest ones, are products of the rapid lateral accretion of land associated with strong erosion in the catchment areas of the Rhine and the Meuse in historic times, and rarely contain high contents of pyrite.

Southeast Asia

Burma, Thailand, Vietnam. Major formations of potentially acid sediments occur in this region. Extensive clay plains formed until 5,500 years B.P. by vertical sedimentation under conditions of a rising sea level. Prolonged deposition of sedimentary material under a dense mangrove swamp vegetation built up thick strata with a high pyrite content. Subsequent stabilization of the sea level resulted in accelerated lateral accretion of coastal sediments with lower pyrite content and without potential soil acidity. The acid sulfate soils developed in the young Holocene pyritic sediments are at, or slightly above present mean sea level, depending on a varying degree of local subsidence. They are covered with fluvial deposits of varying thickness and extent. The available evidence suggests that covers of fluvial sediments are most extensive where rivers carry a substantial load of suspended material.

Extensive blankets of these fluvial sediments occur in the delta of the Irrawaddy river which has a high sediment concentration. Here potential

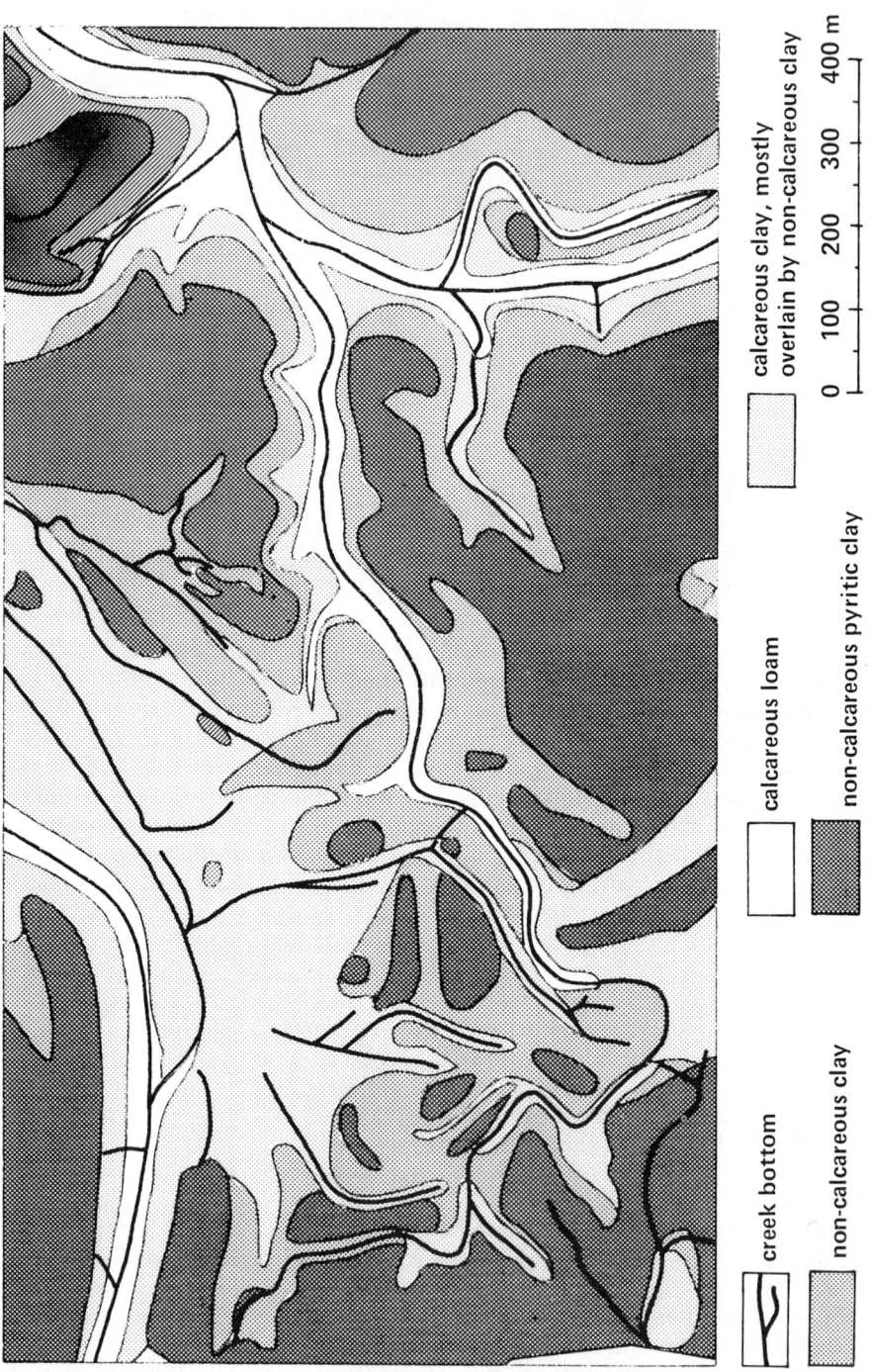

Fig. 7. Soil map of a part of the Eendrachtspolder, the Netherlands. After Edelman, 1950.

acidity is confined to the deeper substrata and poses little danger to agriculture (Ye Goung et al., 1978). Fluvial sedimentation is less widespread where bed loads are smaller e.g., due to less erosion in the hinterland or to a lower carrying capacity of the river itself or where the potentially acid clay plains occur at a relatively high elevation. The deposits of the Chao Phraya river, Thailand, cease at about 100 km from the present shoreline, and highly pyritic and potentially acid marine sediments are exposed in a vast area between the fluvial blanket sediments in the north and a strip of younger, not potentially acid mudflats and swamps adjacent to the sea (Kevie and Yenmanas, 1965). However, marked potential acidity does occur in the vicinity of tidal creeks near river mouths where tidal flushing is strong. The Mekong deltaic system holds an intermediate position (Moormann, 1961). Its rivers are lined with fluvial levee deposits but large areas of potentially acid marine material remain exposed in the backswamps. The examples discussed are schematically depicted in Fig. 8. The Saigon river delta, adjacent to the Mekong delta, is characterized by a lower rate of coastal accretion due to lower contents of suspended material in the river water. The Saigon river delta is dissected by many tidal creeks and its mangrove belt is much wider than that of the Mekong delta; as a result, its sediments are potentially acid (Moormann and Pons, 1974).

Indonesia. Potential soil acidity occurs in several million hectares of coastal lowland along Sumatra's east coast, the southern, western, and northern coasts of Borneo and in the southern lowlands of Irian Jaya (Driessen and Soepraptohardjo, 1974). All belong to land system 1 as described in this paper.

The swamp lands along Sumatra's east coast include the estuaries of the rivers Rokan, Siak, Kampar, Indragiri, Batang Hari, and Musi; the areas in between are filled in with deposits of minor rivers and currents parallel to the coast. All are colonized by a dense mangrove vegetation, mainly *Avicennia* sp., that follows the progressing shoreline in a belt a few kilometers wide. Field observations suggest that potential acidity is most pronounced in the northern part of this coastal strip, and becomes gradually less severe towards the south-east. This difference is attributed to the nature of rock formations in the hinterland which is dominated by a huge fault extending from north Sumatra to Java and the Lesser Sunda islands. This fault is associated with volcanic efflata that are extremely poor in bases in the north (the liparitic tuffs of Riau) but become dacitic and andesitic in South Sumatra and West Java, and mafic further east. The inherent acid-neutralizing capacity of its weathering products changes accordingly.

Severe potential acidity occurs in the older sedimentation areas which formed under conditions of a slowly rising sea level. These are now largely covered by 4 to 10 m of fresh water peat. The more recent sediments close to the coast were deposited at a stable sea level and at a high lateral accretion rate (5 to 9 m per year); they are commonly free from potential acidity. Though generally valid, this picture is complicated by local variations in sedimentation rate (LPT, 1976a, 1976b).

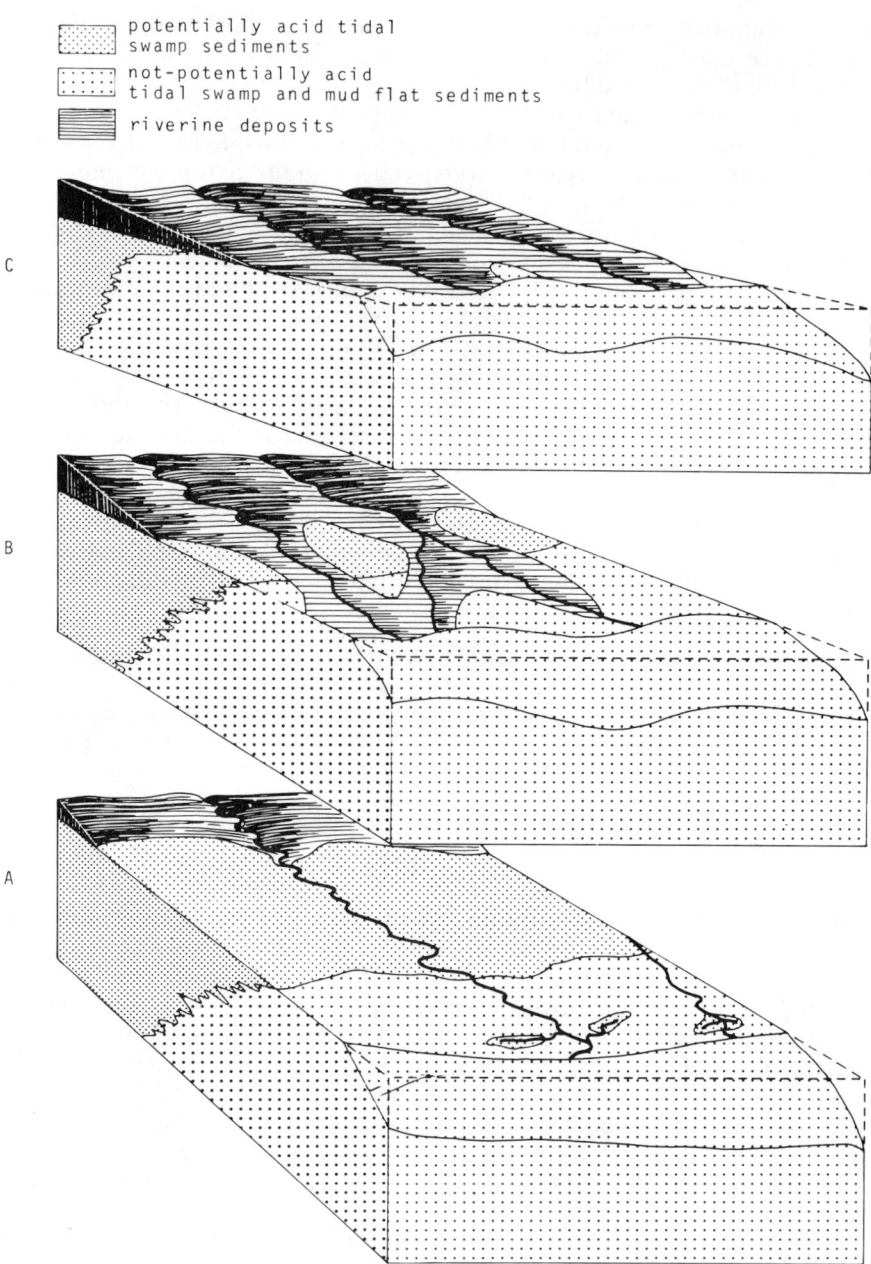

Fig. 8. Block diagrams of Southeast Asian coastal plains with low (A), moderate (B), and high (C) rates of deposition of riverine material. The diagrams schematically represent the situations in the Chao Phraya Delta, Thailand (A), the Mekong Delta, Vietnam (B), and the Jrrawaddy Delta, Burma (C).

The coastal swamps of Borneo show basically the same picture albeit that the recent, not potentially acid sediments adjacent to the coast are considerably less extensive than those in Sumatra. As in Sumatra, conditions of a slowly rising sea level persisted for a long time; some fresh water peats on top of potentially acid sediments in Sarawak started to form at slightly above sea level but have their base well below sea level today. Under this regime, extensive coastal mangrove swamps could form on base-poor weathering products from granite, schists, and shales. These sediments contain high amounts of pyrite and are potentially very acid. The pyritic sediments nearest to the hinterland became subsequently covered with fresh water peats or fluvial material, particularly so in the estuarine areas of the rivers Kapuas and Barito. Potentially acid marine material remained exposed in the downstream basin areas only.

LITERATURE CITED

1. Berner, R. A. 1970. Sedimentary pyrite formation. Am. J. Sci. 268:1–23.
2. ————. 1971. Principles of chemical sedimentology. McGraw-Hill Book Co. 240 p.
3. ————, M. R. Schott, and C. Thomlinson. 1970. Carbonate alkalinity in the pore waters of anoxic marine sediments. Limnol. Oceanogr. 15:544–549.
4. Blackwelder, B. W., O. H. Pilkey, and J. D. Howard. 1979. Late Wisconsinan sea levels on the Southeast U.S. Atlantic shelf based on in-place shore line indicators. Science 204:618–620.
5. Brinkman, R., and L. J. Pons. 1968. A pedo-geomorphological classification and map of the Holocene sediments in the coastal plain of the three Guiana's. Soil Survey Paper 4. Soil Survey Inst. Wageningen, 40 p.
6. Brümmer, G. 1968. Untersuchungen zur Genese der Marschen. Dissertation, Christian-Albrechts-Universität, Kiel. 350 p.
7. Chabreck, R. H. 1972. Vegetation, water and soil characteristics of the Louisiana Coastal Region. Louisiana State Univ. Agric. Exp. Stn. Bull. 664.
8. Chenery, E. M. 1954. Acid sulphate soils in Central Africa. Trans. 5th Int. Congress Soil Sci., Leopoldville, 4:195–198.
9. DeLaune, R. D., and W. H. Patrick, Jr. 1978. Sedimentation rates determined by ^{137}Cs dating in a rapidly accreting salt marsh. Nature 275:532–533.
10. Diemont, W. H., and W. van Wijngaarden. 1974. Sedimentation patterns, soils, mangrove vegetation and land use in tidal areas of West Malaysia. p. 513–528. In G. Walsh et al. (ed.) Proc. Int. Symp. Biology and Management of Mangroves. East-West Center, Hawaii.
11. Drever, J. I. 1971. Magnesium-iron replacement in clay minerals in anoxic marine sediments. Science 172:1334–1336.
12. Driessen, P. M., and M. Soepraptohardjo. 1974. Soils for agricultural expansion in Indonesia. Soil Research Institute, Bull. 1. Bogor, Indonesia.
13. Edelman, C. H. 1950. Soils of the Netherlands. North-Holland Publ. 10, Amsterdam. 177 p.
14. Fairbridge, R. W. 1961. Eustatic changes in sea level. p. 99–185. In Physics and chemistry of the earth. Vol. 4. Pergamon Press, London.
15. Fisk, H. N. 1960. Recent Mississippi river sedimentation. E. van Aelst (ed.) Compte Rendu, 4ème Congrès. L'avancement des études de stratigraphie et de géologie du Carbonifère, Tome 1:187–199. Maestricht.
16. Goldhaber, M. B., and I. R. Kaplan. 1974. The sulfur cycle. p. 527–655. In E. D. Goldberg (ed.) The sea, Vol. 5. Marine chemistry. Wiley-Interscience. New York.
17. Howarth, R. W. 1979. Pyrite: its rapid formation in a salt marsh and its importance in ecosystem metabolism. Science 203:49–51.

18. Kevie, W. van der, and B. Yenmanas. 1965. Report on the soil survey in the Phet Buri irrigation tract. Soil Survey Rep. No. 45, Land Development Department, Bangkok.
19. Kooistra, M. J. 1978. Soil development in recent marine sediments of the intertidal zone in the Oosterschelde—the Netherlands. Soil Survey Paper 14, Soil Survey Institute, Wageningen. 183 p.
20. Louwe Kooymans, L. P. 1974. The Rhine/Meuse delta; four studies on its prehistoric occupation and Holocene geology. Diss. Leiden. 421 p.
21. Lembaga Penelitian Tanah. 1976a. Laporan survey dan pemetaan tanah daerah sungai Rokan (Riau). LPT, No. 5/1976. Bogor, Indonesia.
22. Lembaga Penelitian Tanah. 1976b. Laporan survey dan pemetaan tanah daerah sungai Siak (Riau). LPT, No. 6/1976. Bogor, Indonesia.
23. Marius, C. 1972. Végétation et écologie des mangroves. Bull. de Liaison du Thème C. Numéro 2:22–54, Centre O.R.S.T.O.M. de Dakar.
24. Moormann, F. R. 1961. The soils of the Republic of Vietnam, Min. of Agric. Saigon.
25. ————, and L. J. Pons. 1974. Characteristics of mangrove soils in relation to their agricultural land use and potential. Proc. Int. Symp. on Biol. and Man. of mangroves. Vol. II, p. 529–547.
26. Poelman, J. N. B. 1973. Soil material rich in pyrite in non-coastal areas. p. 197–207. In H. Dost (ed.) Acid sulphate soils. Proc. Int. Symp., Wageningen, Publ. 18, Vol. II, Int. Inst. Land Reclamation Improvement.
27. Pons, L. J. 1965. Pyrites as a factor controlling chemical ripening and formation of cat clay, with special reference to the coastal plain of Surinam. Agric. Exp. Stn., Paramaribo, Bull. No. 82:141–162.
28. Rickard, D. T. 1975. Kinetics and mechanism of pyrite formation at low temperatures. Am. J. Sci. 275:636–652.
29. Roberts, W. M. B., A. L. Walker, and A. S. Buchanan. 1969. The chemistry of pyrite formation in aqueous solution and its relation to the depositional environment. Mineral. Deposita 4:18–29.
30. Salomons, W. 1974. Chemical and isotopic composition of carbonates during an erosion-sedimentation cycle. Diss. Abstr. Int. XXXV:3.
31. Van Breemen, N. 1976. Genesis and solution chemistry of acid sulfate soils in Thailand. Agric. Res. Rep. 848, PUDOC, Wageningen. 263 p.
32. ————. 1980. Magnesium-ferric iron replacement in smectite during aeration of pyritic sediments. Clay Mineral. 15:101–110.
33. Van der Sluijs, P. 1970. Decalcification of marine clay soils connected with decalcification during silting. Geoderma 4:209–227.
34. Wiklander, L., G. Hallgren, N. Brink, and E. Jonsson. 1950. Studies on gyttja soils. II. Some characteristics of two profiles from northern Sweden. Ann. Royal Agric. College Sweden, 17:24–36.
35. Ye, Goung, Khin Win, and Win Thin. 1978. Rice soils of Burma. p. 57–71. In Int. Rice Res. Institute, Soils and Rice. Los Banos, Philippines.

Reproduced after slight alterations with permission from Soil Science, Vol. 119, 1975, p. 42–55.
Copyright © 1975, The Williams and Wilkins Company, Baltimore, MD 21202, USA.

Chapter 2

Controls and Consequences of Sulfate Reduction Rates in Recent Marine Sediments[1]

M. B. GOLDHABER AND I. R. KAPLAN[2]

ABSTRACT

A brief review is given of the process of bacterial sulfate reduction in respect to other processes in ocean sediments. In particular, rates of sulfate reduction are discussed in context of control mechanisms and geochemical consequences. It is concluded that besides temperature and pressure, which are cosmopolitan parameters influencing most biological processes, the rate of sulfate reduction is dependent on 1) total organic carbon preserved in sediment and 2) state of complexing of the organic matter and its availability for biogenic degradation. These two parameters are in turn influenced by 3) the environment of deposition and 4) the rate of sediment accumulation. Correlations are presented showing a direct relationship between rate of sulfate reduction and rate of sediment accumulation. The consequences of different rates of sulfate reduction on pyrite formation and isotope fractionation are discussed.

INTRODUCTION

Bacterial sulfate reduction only occurs in the absence of oxygen. As a consequence of rapid oxygen consumption by aerobic respiration relative to the rate of replenishment, anoxic conditions are frequently established just beneath the surface of marine sediments. The importance of sulfate as an electron acceptor is in part related to the fact that it is the dominant oxygen-containing dissolved ionic species involved in the anaerobic remineralization of organic matter at shallow depths of burial. This results in a number of consequences during early diagenesis.

---

[1] Contribution No. 1374, Institute of Geophysics and Planetary Physics, Univ. of California, Los Angeles, CA.
[2] U.S. Geological Survey, Uranium Thorium Branch, M.S. 916 Denver, CO 80225, and Dep. of Geology and Institute of Geophysics and Planetary Physics, Univ. of California, Los Angeles, CA 90024, respectively.

Copyright © 1982 Soil Science Society of America, 677 S. Segoe Rd., Madison, WI 53711. *Acid Sulfate Weathering.*

The products of organic matter decomposition include the nutrients ammonia and phosphate, which can increase in concentration in the interstitial water of sediments to several orders of magnitude above their concentration in the water column. These nutrients are therefore susceptible to release from the sediments to the overlying water by mixing (biogenic or non-biogenic) and/or diffusion, which can potentially enhance biological productivity in the overlying waters. It is also known that carbonate alkalinity (mainly HCO_3^-) arising from oxidation of organic matter by sulfate reducing bacteria leads to supersaturation and precipitation of alkaline earth carbonates (e.g., Berner, 1966; Presley and Kaplan, 1968). Furthermore, it has been convincingly argued that the pH of pore waters of recent marine sediments is controlled by the progressive addition of protolytic species resulting from sulfate reduction, together with the selective removal of some of these to form authigenic minerals (Thorstenson, 1970; Nissenbaum et al., 1972; Ben-Yaakov, 1973; Gardner, 1973).

Environmental consequences of sulfate reduction result from the production of dissolved sulfide (H_2S aq. and HS^- aq.). Dissolved sulfide is highly toxic to many aerobic organisms (Theede et al., 1969) and in extreme cases may decrease the diversity of benthic organisms inhabiting upper sediment layers. Base and transition metal sulfides which form readily, have exceedingly low solubility products, and such sulfides might be expected to form during early diagenesis. However, with the important exception of iron, these metals are present in only trace amounts in normal marine sediments. The pathways of reaction of sulfide with sedimentary iron have recently been reviewed (Sweeney, 1972; Goldhaber and Kaplan, 1974) and are shown schematically in Fig. 1. An important mechanism involves initial formation of an intermediate metastable iron sulfide such as mackinawite (tetragonal FeS) and/or greigite (cubic Fe_3S_4) and subsequent transformation of these intermediates to pyrite (cubic FeS_2) by addition of elemental sulfur. The intermediate iron sulfides are referred to as acid volatile sulfides because they dissolve in hydrochloric acid, whereas pyrite does not. Pyrite is the major end product of sulfur diagenesis in a sulfur-rich environment such as reducing marine sediments.

A particularly interesting aspect of sulfur diagenesis is the modification of the stable isotope distribution of sulfur. It is well established that the light stable isotope, ^{32}S, reacts more rapidly than does ^{34}S during bacterial reduction (Thode et al., 1961; Jones and Starkey, 1957; Harrison and Thode, 1958; Kaplan and Rittenberg, 1964). This results in dissolved sulfide, acid volatile sulfide, and pyrite sulfur enriched in the light isotope. When expressed as the per mil deviation, $\delta^{34}S$, of the isotopic ratio $^{34}S/^{32}S$ compared to a standard[3], this enrichment may exceed 60°/oo, compared to the starting sea water sulfate (Vinogradov et al., 1962; Kaplan et al., 1963; Hartmann and Nielsen, 1969).

Because the various iron sulfide phases do not commonly occur as detrital minerals (Kaplan et al., 1963), their presence in sediments is evi-

---

[3] $\delta^{34}S \equiv [(R_X - R_{STD})/R_{STD}]\,1,000$ where $R_X = {}^{34}S/{}^{32}S$ sample, $R_{STD} = {}^{34}S/{}^{32}S$ of Canyon Diablo Meteorite troilite.

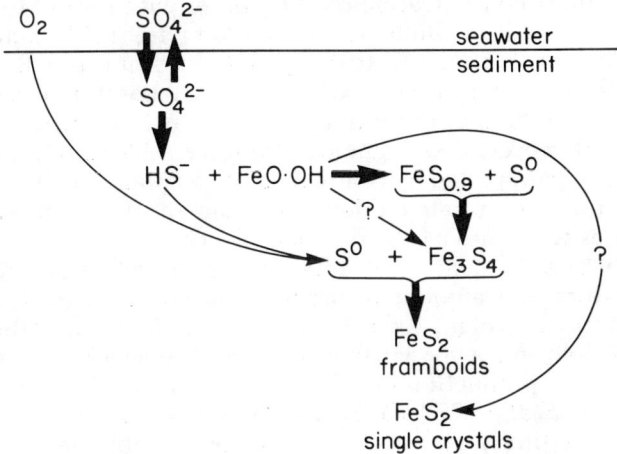

Fig. 1. Pathway of Sedimentary Pyrite formation. After Goldhaber and Kaplan, 1974.

dence for in situ bacterial sulfate reduction. Such evidence is also provided by the progressive depletion in pore water sulfate with depth in sediments, as well as the progressive appearance of dissolved sulfide and the other products of bacterial metabolism mentioned above. This ability to trace the geochemical consequences of sulfate reduction with depth, and therefore with time, makes possible the calculation of the rates at which such processes occur. Qualitatively it is easily shown that such rates are highly variable. This point is illustrated with data from two contrasting sedimentary environments. The pore water sulfate depth profile for site 34 (200 mi. west of Cape Mendicino, California) from leg 5 of the JOIDES Deep Sea Drilling Program (Manheim et al., 1970; Presley et al., 1970) shows sulfate reduction continuing over periods of several million years and depths of several hundred meters. This may be compared to data from Saanich Inlet, British Columbia (Nissenbaum et al., 1972) which indicates complete sulfate removal can occur at depths of burial less than 1 m and times on the order of hundreds of years. In this paper we focus attention on the controls of sulfate reduction rates. Laboratory results are first reviewed, followed by a brief summary of techniques for estimation of the in situ rates of reduction in sediments. We then propose that the sulfate reduction rate is positively correlated with sedimentation rate, and, finally, discuss some geochemical consequences of this correlation, including the relative abundances of dissolved sulfide and intermediate iron sulfides as well as the magnitude of the bacterial sulfur isotope effect.

Relevant Laboratory Studies

From experiments with both pure and mixed cultures under laboratory conditions, some generalizations have emerged concerning controls on the rate of bacterial sulfate reduction. This rate is known to be independent of sulfate concentration (i.e., 0th order with respect to sulfate) at concentrations $\geq$ 10 millimolar (mM) (Postgate, 1951; Harrison and Thode, 1958; Kaplan and Rittenberg, 1964; Ramm and Bella, 1974). Nakai and Jenson (1964) interpreted results of a laboratory experiment involving sulfate reduction in marine sediments as indicating a first order mechanism. However, these data are also susceptible to interpretation as 0th order (H. Sakai, quoted by Goldhaber and Kaplan, 1974). More recent laboratory studies of sulfate reduction in sediments (Martens and Berner, 1974) are consistent with a 0th order dependence.

There is clear evidence that sulfate reduction rate is dependent upon both the nature and amount of organic carbon substrate (Kaplan and Rittenberg, 1964; Sorokin, 1966; Sweeney, 1972; Ramm and Bella, 1974). Ramm and Bella have shown that at sulfate concentrations in the sea water range, the production of sulfide, S, in mg liter^{-1} day^{-1} is given by the expression $dS/dt = 77 [SOC/(650+SOC)]$, where SOC is soluble organic carbon (g/liter). This equation assumes a stable bacterial population and applies to relatively fresh organic matter consisting predominantly of only partially degraded cellular components. Organic matter in recent sediments rapidly complexes to a form humates and kerogens which are less susceptible to bacterial attack (see below). Therefore the equation is not universally applicable to all marine environments unless the kinds of organic materials are taken into account. The strong dependence of rate upon amount of carbon substrate is nonetheless emphasized by these results. The importance of the exact natures of organic matter oxidized is reflected in a number of laboratory studies (e.g., Sorokin, 1966; Kaplan and Rittenberg, 1964). For example, Kaplan and Rittenberg have shown that at constant temperature the rate of sulfate reduction normalized to a constant bacterial population may be varied by a factor of at least 50 by changing the nature of the electron donor. The range of organic molecules utilized by sulfate reducing bacteria in pure culture is rather small, and is for the most part largely limited to a few short chained carboxylic acids (Postgate, 1965). If this restriction applies in the sedimentary milieu, then sulfate reducers must depend upon a complex community of fermentative bacteria to supply these small molecules by degradation of polymerized organic matter (Sorokin, 1962; Goldhaber and Kaplan, 1974).

Additional effects on rates of sulfate reduction are temperature and pressure. Kaplan and Rittenberg (1964) found that for constant initial sulfate and carbon substrate, the rate of reduction in pure culture normalized to a constant bacterial population varies by approximately a factor of 5 between 5 and 25 C. The effect of pressure on sulfate reducing bacteria is not well documented, although in general there is some evidence for a significant decrease in bacterial metabolic rates with increasing pressure (Jannasch et al., 1971).

Laboratory investigations of the metabolic activities of sulfate reducing bacteria have also focused attention on the magnitude of the kinetic isotope effect. It has been shown (Harrison and Thode, 1958; Kaplan and Rittenberg, 1964) that the magnitude of the isotopic separation between ^{32}S and ^{34}S is inversely proportional to the rate of sulfate reduction. The greatest isotope effect (46 °/oo instantaneous separation between sulfate and sulfide) was found at the slowest achievable rate of sulfate reduction (Kaplan and Rittenberg, 1964). However, this large effect has rarely been measured in the laboratory. A great majority of the laboratory data indicate an isotope effect between 10 and 30 °/oo.

Rees (1973) has recently reviewed the subject of sulfur isotope fractionation during sulfate reduction in the light of a steady-state mathematical formulation. Steady state in this context refers to intermediary metabolic products in the biochemical pathway which can achieve constant concentrations under appropriate conditions. He argues that the large fractionation obtained by Kaplan and Rittenberg is consistent with additive fractionation effects in a multistep process. In the majority of laboratory studies, it was argued, the rate-determining step is the reduction of adenosine-5'-phosphosulfate (APS) to sulfite producing a kinetic isotope effect of 25 °/oo. A second sulfur/oxygen bond-breaking step, the reduction of sulfite, is normally fast with respect to other reactions and does not enter into the process of isotopic fractionation. Under conditions of slow metabolism, however, the reduction of sulfite may also contribute to the overall fractionation, producing an overall effect of ≈ 50 °/oo as a limit. A second possibility is that sulfate reducing bacteria may catalyze the exchange of sulfur atoms between sulfate and sulfide (Trudinger and Chambers, 1973). They argue that the extent of attainment of this equilibrium is greater at slow metabolic rates. This could lead to an isotopic separation of about 75 °/oo between SO_4^{2-} and HS^- at 25 C if equilibrium is achieved.

Estimation of Sulfate Reduction Rates in Sediments

Sulfate reduction rates in sediments have been estimated either by measuring the change in sulfate concentration with time in mud samples maintained in the laboratory (Gunkel and Oppenheimer, 1963; Nakai and Jensen, 1964; Martens and Berner, 1974) or by analyzing pore water depth profiles in terms of appropriate mathematical models (Berner, 1964a, 1972, 1974; Tsou et al., 1973; Goldhaber and Kaplan, 1980). The first technique is straightforward and presumably yields result representative of in situ values insofar as conditions of temperature and pressure resemble in situ values. For relatively rapid sulfate reduction of the type found in highly organic-matter-rich nearshore sediments, times approaching a month may bring about extensive sulfate removal. However, in many environments where sulfate reduction is considerably slower than this, times exceeding the half life of a researcher may be required to obtain any measurable reaction. The sensitivity of these laboratory studies may be improved somewhat through the use of radiosulfur (^{35}S) as a tracer (Sorokin, 1962).

The second approach, that of mathematical modeling, is appealing and convenient as a tool to gain an estimate of sulfate consumption for slow rates of reaction. The validity of this method depends upon understanding the nature and depth distribution of competing rates of processes

Table 1. Sulfate reduction rate in recent sediments.

Area	Rate (moles liter^{-1} yr^{-1})	Technique	Source
Santa Barbara Basin	5.9×10^{-4}	Modeling	Berner (1972)
Somes Sound Maine	3.7×10^{-2}	Modeling	Berner (1972)
Long Island Sound	7.1×10^{-3}	Modeling	Berner (1972)
Carmen Basin Gulf of California	3.1×10^{-4}	Modeling	Goldhaber (1974)
Pescadero Basin Gulf of California	8.4×10^{-5}	Modeling	Goldhaber (1974)
JOIDES Site 147 Cariaco Trench	3.7×10^{-5}	Modeling	Tsou et al. (1973)
JOIDES Site 148	7.3×10^{-8}	Modeling	Tsou et al. (1973)
Black Sea	$5.5 \times 10^{-1} - 4.3 \times 10^{-2}$	S^{35}	Sorokin (1962)
Littoral of Krasnovodsk Bay	$7.4 \times 10^{-2} - 1.6 \times 10^{-1}$	S^{35}	Ivanov (1968)
Littoral of Barents Sea	$1.7 \times 10^{-1} - 4.5 \times 10^{-1}$	S^{35}	Ivanov (1968)
Lab study	$2.0 \times 10^{-1} - 3.1 \times 10^{-1}$	Measured SO_4^{2-} decrease	Gunkel and Oppenheimer (1963)
Lab study	2.8×10^{-1}	Measured SO_4^{2-} decrease	Martens and Berner (1974)
Lab study	4×10^{-2}	Measured SO_4^{2-} decrease	Nakai and Jenson (1964)

affecting pore water sulfate. One possible assumption is that of a closed system—no addition of pore water sulfate other than by burial within the accumulating sediments, and no vertical redistribution of sulfate within the sediment column. The rate may be then estimated as the ratio of the decrease in sulfate concentration for the time interval over which this decrease occurs. It may be easily shown, however, that marine sediments are not closed to addition of pore water sulfate (e.g., Goldhaber and Kaplan, 1974). The gradient in SO_4^{2-} concentration with depth gives rise to a flow of matter by diffusion. The rate of sulfate addition by the combined processes of molecular diffusion and burial is of course greater than by burial alone, and the calculated rate of sulfate removal greater by a corresponding amount. The majority of results incorporating diffusion have been calculated by a formulation similar to that originally proposed by Berner (1964a). The many assumptions and limitations of these calculations are discussed in detail elsewhere in the literature (Tzur, 1971; Berner, 1976) and will not be repeated here. We wish to note two points, however. The first is the role of burrowing organisms which are ubiquitous[4] in the upper layers of recent sediments. These organisms, through their vital activities, can act to add sulfate to marine sediments (Goldhaber and Kaplan, 1980). This additional mode of sulfate addition has not previously been included in the mathematical formulation which, in turn, can underestimate the sulfate reduction rate where such macro-benthonic organisms are active. Secondly, it has been noted that very rapid sulfate reduction rates may occur in a very thin zone near the sediment-water interface (Berner, 1972; Sweeney, 1972; Goldhaber and Kaplan, 1974a). Mathematical models of pore water sulfate distribution may not

---

[4] With the obvious exception of anaerobic environments such as the Black Sea and dysaerobic environments such as Santa Barbara basin off the southern California coast.

include this zone. Both effects lead to an underestimation of the actual sulfate reduction rate. Despite such problems, we believe that mathematical modeling gives a reasonable perspective on the variability in sulfate reduction rates between different sedimentary environments.

Table 1 contains a compilation of the available rates of sulfate reduction, in marine sediments derived by various methods. The data are seen to range over many orders of magnitude. Particularly striking is the rate of a relatively slowly depositing environment depleted in organic carbon obtained from Deep Sea Drilling Project (JOIDES site 148) data.

Correlation between Sulfate Reduction Rate and Sedimentation Rate

As stated earlier the dominant controls on sulfate reduction rate under laboratory conditions are the nature and the abundance of electron donor (i.e., organic matter) available to the bacterial community. The control of organic carbon abundance on pyrite in sediments is suggested by Fig. 2, (Sweeney, 1972) for a variety of sedimentary environments and depths within the sediment column. Although there is a great deal of scatter, this plot demonstrates that there is some tendency for high organic carbon contents to be associated with high pyrite sulfur contents. A perfect correlation between sulfur and carbon on such a plot would imply that a constant proportion of organic matter deposited is bacterially utilized, resulting in release of dissolved sulfide capable of being converted to iron sulfide (Sweeney, 1972).

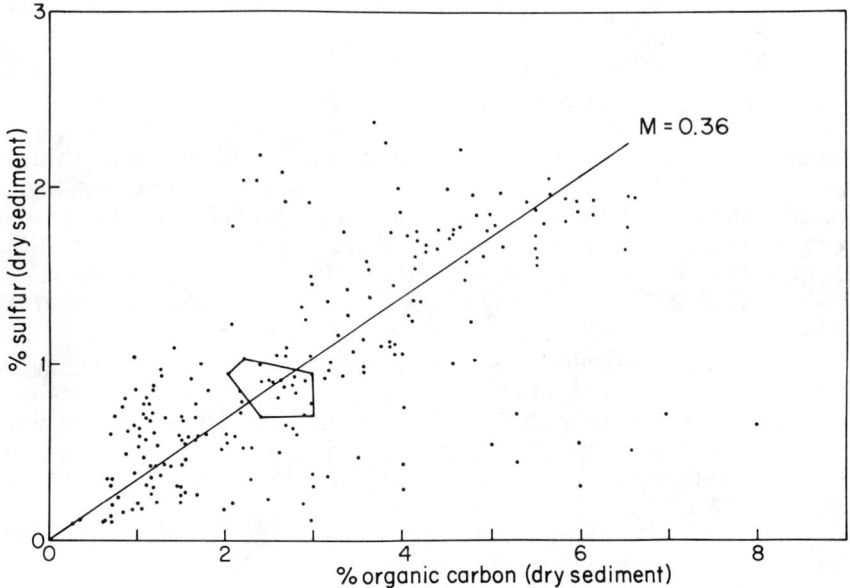

Fig. 2. Plot of percent reduced sulfur vs. organic carbon content. After Sweeney (1972).

The sensitivity of bacterial metabolic rates to the nature of organic matter (as opposed to the amount) may perhaps be inferred from the localization of high sulfate reduction rates (as determined, for example, by ^{35}S) to the vicinity of the sediment-water interface. Similarly, it is known that numbers of culturable bacteria, including sulfate reducers, fall off extremely rapidly with depth in sediments (Zobell and Feltham, 1942; Oppenheimer, 1960; Sorokin, 1962). This localization is in contrast to the abundance of organic carbon, which does not typically show such a dramatic exponential decrease with depth of sediment. The localization of bacterial activity to upper layers occurs in spite of the demonstrated ability of many bacteria including sulfate reducers to migrate through sediments at rates as high as 3 cm per day (see e.g., Oppenheimer, 1960).

Organic matter in marine sediments is derived predominantly from marine plankton although terrestrial inputs are locally important. During transport and burial, the organic matter undergoes transformations into a series of (in part operationally defined) fractions. These transformations may be schematically represented as: degraded cellular material → water soluble complexes containing amino acids, fatty acids and sugars → fulvic acids → humic acids → kerogen (Nissenbaum and Kaplan, 1972). This pathway represents increasing polymerization of the organic material. It seems reasonable to assume that the more complex material is less susceptible to bacterial attack through lack of suitable solubilizing and enzyme systems (Alexander, 1971). At least five variables are important in determining the extent of transformation of organic matter. These are: environment of deposition, source of material, environment of accumulation, temperature, and pressure. Initially, environment of deposition is of overriding importance. Irrespective of the quantity and nature of organic material being deposited, if the path through which it travels permits exposure to oxygen, a significant amount of the organic matter will be decomposed. Zsolnay (1971) compared the organic geochemistry of adjacent oxic and anoxic basins in the Baltic Sea and found that greater amounts of organic carbon, soluble amino acids, fatty acids, and hydrocarbons are preserved under anoxic conditions. Furthermore, a greater percentage of the total organic carbon was in a non-condensed form in the anoxic basin. Source of material is important, in that land-derived higher plant material containing lignin and waxes are probably not degraded to CO_2 as readily as plankton-derived organic materials (Aizenshtat et al., 1973). Environment of accumulation affects organic geochemistry because in slowly accumulating pelagic sediments organic matter near the sediment surface can be exposed for long periods of time to an oxygenated environment. This facilitates attack by aerobic bacteria and reworking by benthonic organisms. Conversely, rapidly accumulating sediments may afford protection from such exposure and allow the preservation of those compounds easily degraded by bacteria and capable of supporting rapid bacterial rates (Borodovskiy, 1965a, 1965b; Price and Calvert, 1970; Berner, 1972).

The following considerations based upon the preceding discussion have led us to suggest that total sediment accumulation rate may represent an important summary variable in determining rates of bacterial metabolism: (1) total organic carbon preserved in sediments tends to be

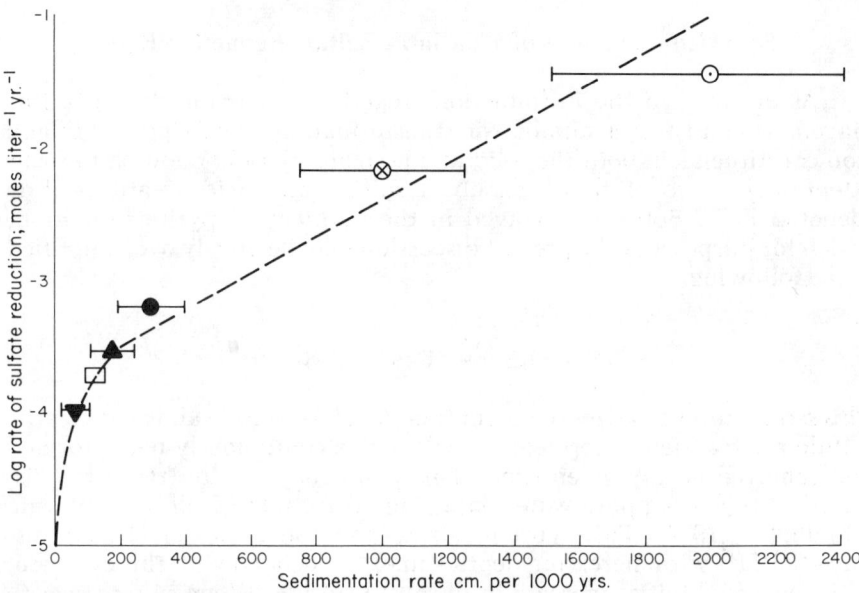

Fig. 3. Plot of log of sulfate reduction rate vs. sedimentation rate. Reduction rates taken from Table 1.
- ○ Somes Sound Maine; Sedimentation rate, Berner (1972).
- ⊗ Long Island Sound; Sedimentation rate, Berner (1972).
- ● Santa Barbara basin; Sedimentation rate, Koide et al., 1972, and Emery (1960).
- △ Carmen Basin; Sedimentation rate, Van Andel (1964).
- ▽ Pescadero Basin; Sedimentation rate, Van Andel (1964).
- □ Cariaco Trench.

positively correlated with high sedimentation rate (Berner, 1972), although grain size is an important moderating influence; (2) less complexing and rearrangement of organic material occurs during rapid burial; and (3) rapidly accumulating sediments tend to occur at relatively shallow depths (with the exception of turbidites), leading to lower pressure, shorter time of exposure to oxygenated water during transport, and generally, higher bottom water temperature.

In every case, rapid deposition would seem to act in the direction of enhancing metabolic rates. In order to verify this hypothesis, we have prepared a plot (Fig. 3) of sediment accumulation rate vs. sulfate reduction rate shown in Table 1. Only those rates calculated by modeling of pore water sulfate profiles are shown. This is done in order to present a self-consistent set of rates. It was suggested above that actual initial rates of in situ reduction are likely to be greater than those predicted by modeling. However, the ranges of both rate of reduction and sedimentation are large, and it is probable that the trend indicated by the broken line in Fig. 3 would be preserved if the actual rates were known more precisely. Although more data are needed, we accept as a working hypothesis the correlation between bacterial rate and sedimentation rate. It is obvious, however, that small-scale perturbations within a single depositional environment are less likely to be predicted than large-scale trends between different environments.

Some Consequences of a Variable Sulfate Reduction Rate

As outlined in the introduction, reactions occurring during sulfur diagenesis consist of a number of transformations involving sulfur and iron constituents in both the solid and aqueous phases. We wish to focus attention on two of these: dissolved sulfide and acid volatile sulfide (denoted FeS). Both are involved in the pathway of pyrite formation, which for purposes of the present discussion may be grossly oversimplified to the following:

$$SO_4^{2-} \xrightarrow{} H_2S \xrightarrow{\text{"Fe"}} FeS \xrightarrow{S^\circ} FeS_2$$

This sequence emphasizes the point that dissolved sulfide and acid volatile sulfide are transient intermediates which are continuously being formed and removed at any given time. This point may be illustrated for dissolved sulfide using pore water data (Fig. 4) from the Gulf of California (Goldhaber, 1974). This figure illustrates the systematic depletion of dissolved sulfate with increasing depth (time). Associated with this decrease, is the expected initial increase in dissolved sulfide. When sulfate goes to

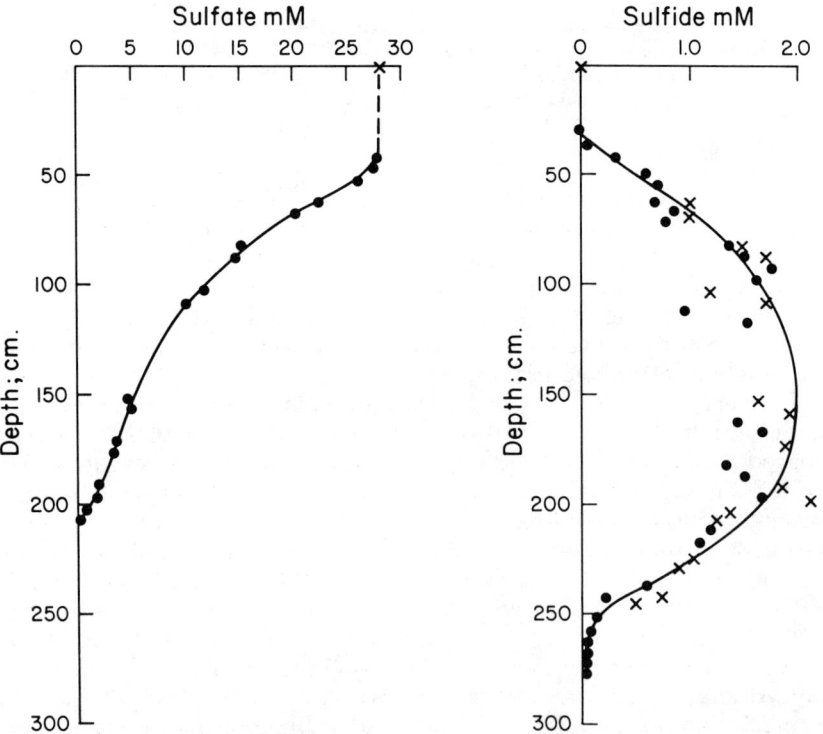

Fig. 4. Plot of dissolved sulfate and sulfide in pore waters of a core from Carmen Basin, Gulf of California. Sulfide was measured by a colorimetric (methylene blue) procedure (°) or gravimetric precipitation of Ag$_2$S (X).

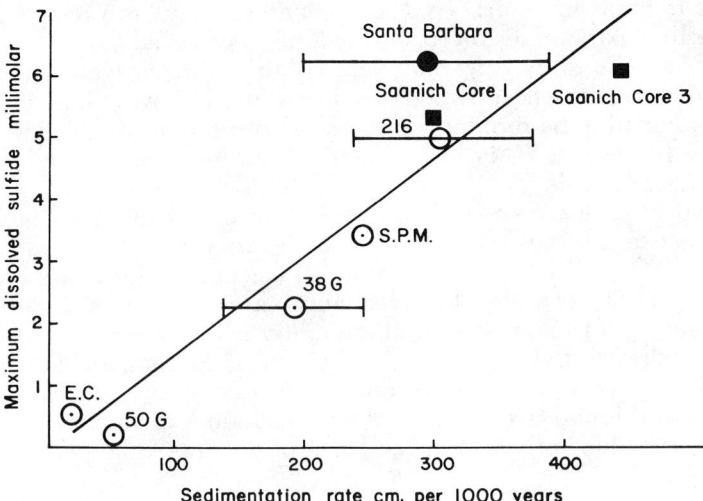

Fig. 5. Plot of maximum dissolved sulfide (for cores in which sulfate is essentially completely removed) vs. sedimentation rate.
■ Data from Nissenbaum et al. (1972). Saanich inlet British Columbia. The sedimentation rate for core 1 has been recalculated to take into account a diminished biogenic component.
● Sulfide data from Kaplan et al. (1963); sedimentation rate from Emery 1960 and Koide et al. (1972).
○ Sulfide data from Goldhaber (1974).
50G; Pescadero basin, Golf of California.
38G; Carmen basin, Gulf of California.
21G; Sal Si Puedes basin, Gulf of California.
SPM; San Pedro Martir basin, Gulf of California
Sedimentation rates from Van Andel, 1964.
E. C. East Cortez Basin Sedimentation rate from Emery, 1960.

zero, however, sulfide production from this source must cease. Below the depth of zero sulfate, dissolved sulfide decreases, reflecting progressive consumption by reaction with solid components (and diffusion of dissolved sulfide). This consumption is also occurring at depths shallower than that at which sulfate disappears, but in the upper 150 cm sulfide production rate is greater than sulfide removal rate; hence the increase in this zone (Goldhaber and Kaplan, 1974). Note that the maximum concentration of dissolved sulfide is far less than would be predicted for simple closed system sulfate reduction (i.e., ~28 mM, equal to the starting sulfate concentration) due to reaction of sulfide with the sediment to form FeS and loss of sulfide by diffusion. This maximum observed concentration is of interest in that it reflects the balance between relative rates of dissolved sulfide production and loss. The magnitude of at least one of these factors, sulfide production (i.e., sulfate reduction) rate, is highly variable and as argued above, correlates with sedimentation rate. One might therefore suppose that the maximum concentration of dissolved sulfide would correlate positively with production rate and hence sedimentation rate. This is indeed the case (Fig. 5). Data for Fig. 5 were taken only from those analyses in which sulfate had essentially gone to zero and therefore the maximum dissolved sulfide had already been reached. The variation in maximum dissolved sulfide is also portrayed in Fig. 6, which depicts the relationship with water depth of the sediment-water interface. The highest sulfide concentrations are found in shallow sediments which may be assumed to be rapidly deposited. This figure is included in order to extend the results of Fig. 5 to include cases for which sedimentation rates are not available.

We do not mean, however, to imply that the variation in maximum sulfide concentration is related solely to variation in production rate, as the nature of the dissolved sulfide removal mechanism must also be taken into account. The details of this mechanism are not well understood, but there is good evidence that it predominantly involves reaction with iron oxide or hydrated iron oxide (see Goldhaber and Kaplan, 1974a, for a review). Rickard (1974) has studied the sulfidization of the mineral goethite (FeOOH) and found that the rate of this reaction is dependent upon sulfide concentration, pH, and surface area of goethite. The percentage of

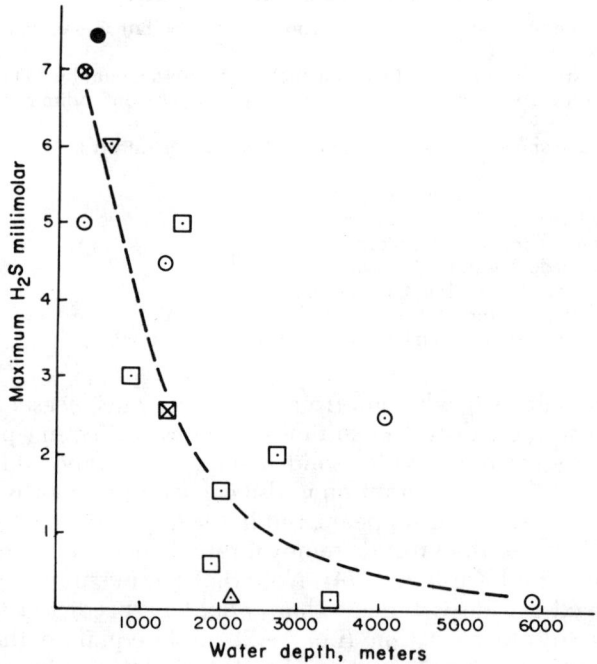

Fig. 6. Plot of maximum dissolved sulfide (for cores with essentially complete sulfate reduction) vs. water depth of the sediment water interface. The line drawn through the data is arbitrary.
  ○ Data of Rozanov et al., 1971; Northwest Pacific Ocean.
  △ Data of Hartmann et al., 1973; West Coast of Africa.
  ● Data of Berner, 1964a; Gulf of California.
  ▽ Data of Kaplan et al., 1963; Santa Barbara Basin.
  ⊙ Data of Nissenbaum et al., 1972; Saanich Inlet, British Columbia
  □ Data from Cariaco Trench; B. J. Presley, unpublished.
  ⊠ Goldhaber, 1974.

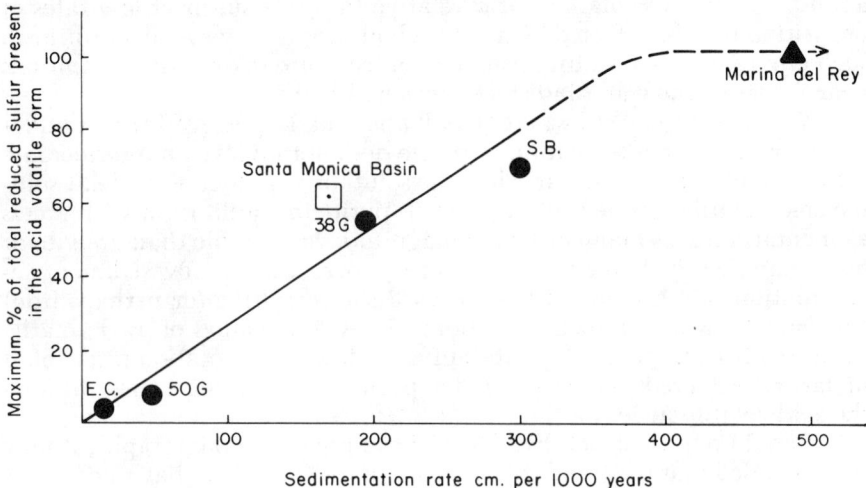

Fig. 7. Plot of maximum percent of total reduced sulfur in the acid volatile form against sedimentation rate.
▲ Data of Sweeney (unpublished) Marina Del Rey Harbor, Los Angeles; sedimentation rate is > 500 cm per 1,000 years based upon the thickness of sediment which has accumulated over the gravel bottom.
□ Data of Kaplan et al. (1963). Sedimentation rate from Emery (1960).
● Goldhaber and Kaplan (1980).

iron as iron oxide is apparently not nearly as variable for different sedimentary environments (see e.g., Gibbs, 1973) as is the sulfide production rate. Hence, variability in the consumption term from this source is small in relation to the demonstrated variability in the production rate. A clear exception is to be found in organic-carbon rich carbonate sediments (Thorstenson and Mackenzie, 1971) with very small iron oxide contents.

Even if the mechanism and rate constants for sulfide removal are constant between environments with different sedimentation rates, the time base is changing. That is, sulfide production is rapidly deposited sediments occurs in a zone representing considerably less time per unit thickness than is the case at lower rates of deposition. Therefore, there is relatively less time for sulfide to react with sediments in the rapidly accumulating regime. This trend acts in the same direction as high production rate and results in higher dissolved sulfide concentration present under these conditions.

The acid volatile sulfides are likewise transient intermediates whose abundance at any time is determined by relative rates of production and consumption. Basically, similar arguments to those made for dissolved sulfide may apply to acid volatile sulfides and a similar relationship to depositional rate is found (Fig. 7). In rapidly deposited sediments, either their rate of generation may be more rapid due to higher sulfide production rate, or the conversion to sedimentary pyrite is less complete because of shorter reaction times, or both. However, the situation is somewhat complicated in the case of acid volatile sulfur because the conversion to pyrite involves addition of elemental sulfur (Fig. 1). Therefore, the low abundance of acid volatile sulfur relative to pyrite sulfur at low rates of deposition may be affected by an abundant supply of zero-valent sulfur in such environments resulting from longer exposure to oxidizing conditions near the sediment-water interface (Berner, 1964b).

We have argued elsewhere (Goldhaber and Kaplan, 1974) that under some conditions the formation of pyrite does not involve an intermediate sulfide at all. This can occur when the solubility product of the least soluble acid volatile sulfide is not exceeded (pyrite in equilibrium with excess elemental sulfur is many orders of magnitude less soluble than greigite or mackinawite). Such undersaturation may correlate with low sediment accumulation rate because of the lower dissolved sulfide (or perhaps iron) concentrations under such conditions. Thus, low values of acid volatile sulfide sulfur compared to pyrite sulfur at slow sedimentation rates could in part reflect a reduced proportion of pyrite sulfur having passed through the acid volatile stage.

Based upon the brief discussion of the preceding paragraphs, it may be concluded that the trends displayed in Fig. 6 and 7 can have a complex origin, and further studies regarding the kinetics of reactions occurring during sulfur diagenesis are required to determine which factors in addition to variable sulfide production rate are important in explaining the distribution of sulfur in sediments. However, sedimentation rate appears to be a useful master variable against which such trends can be displayed.

An additional consequence of variable sulfate reduction rate mentioned in a preceding section, is the inverse correlation of this rate with the magnitude of the sulfur isotope effect. In order to determine whether the same inverse correlation exists in recent sediments, a plot of the $\delta^{34}S$ of initial sulfide vs. log of sulfate reduction rate was prepared in Fig. 8. By initial sulfide we mean dissolved or acid volatile sulfide generated near the sediment surface in the presence of sulfate with the sea water isotopic ratio ($+20$ $^o/oo$). Under such conditions, the $\delta^{34}S$ of the volatile sulfide in the sediment is approximately equal to that of the $\delta^{34}S$ of the sulfide being generated. At very slow rates of reduction, however, virtually all sulfide sulfur is in the pyrite form (Fig. 5 and 7). Therefore $\delta^{34}S$ of Long Basin, East Cortez Basin, and Santa Catalina Basin plotted in Fig. 8 are pyrite sulfur values. In cases where sulfate reduction rates were otherwise unavailable, they were estimated by applying the known sedimentation rate to Fig. 3. The point plotted in the upper right hand corner in Fig. 8 is a composite which illustrates the range of the vast majority of rates and isotopic values (relative to sea water sulfate) which have been achieved in the laboratory using cultures of growing sulfate reducing bacteria.[5] Based upon this plot, a reasonable case may be made for the existence of an isotopic separation with respect to sea water sulfate which increases with decreasing rate such as has been demonstrated in the laboratory. This interpretation neglects the possible existence of additional isotopic fractionating mechanisms other than bacterial sulfate reduction. For example, if the reactions $H_2S \xrightarrow{Fe} FeS$ or $FeS \xrightarrow{S^\circ} FeS_2$ are accompanied

---

[5] Resting cell suspensions, using ethanol as an electron substrate were found to produce ^{32}S enrichment of 46 $^o/oo$ (Kaplan and Rittenberg, 1964).

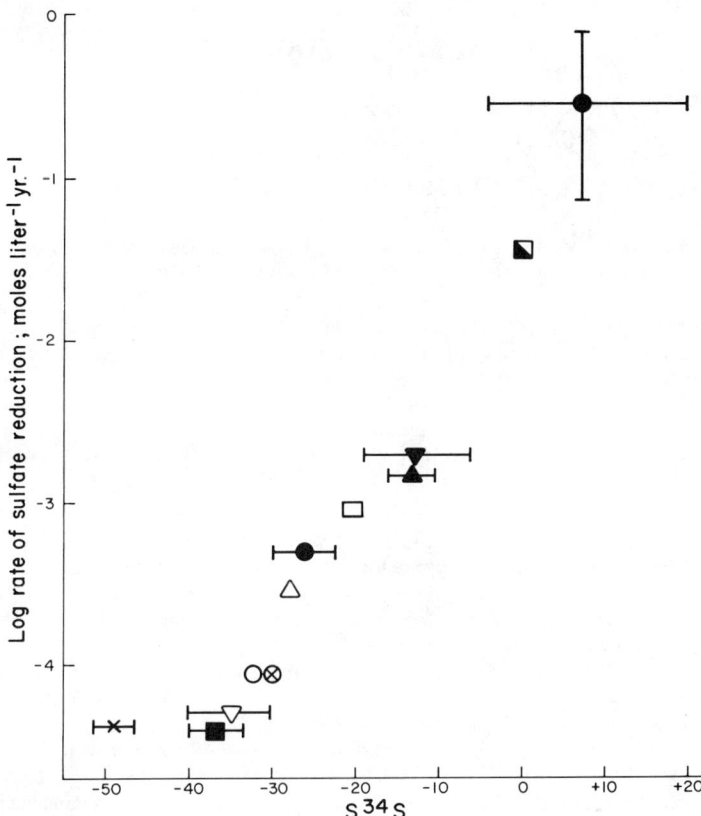

Fig. 8. Log of rate of sulfate reduction against $\delta^{34}S$ of "initial" sulfide.
Various laboratory studies using pure cultures of bacteria.
- □ Nakai and Jenson (1964).
- × Long Basin; S. California, Isotope data from Kaplan et al. (1963), Rate from Fig. 3 and sedimentation rate of Emery, 1960.
- ∇ Santa Catalina Basin; S. California; Isotope and rate data as for Long Basin.
- □ Newport Marsh, S. California; Isotope and rate data as for Long Basin.
- ■ East Cortex Basin, S. California; Isotope data Sweeney (1972), rate from Fig. 3 and sedimentation rate of Emery, 1960.
- ● Santa Barbara Basin; Isotope data Sweeney (1972), Rate from Berner (1972).
- ○ Pescadero Basin, Gulf of California; Data of Goldhaber (1974).
- △ Carmen Basin; Gulf of California; Data of Goldhaber (1974).
- ▲ Pettaquamscott River basin, Rhode Island Data of Orr and Gaines (1974).
- ▼ Marina Del Rey Harbor; Isotope data Sweeney, 1972. Rate of sulfate reduction from Fig. 3 using sedimentation rate as in Fig. 7 caption.

by either a kinetic or equilibrium isotopic effect, then the use of sulfide or pyrite sulfur to indicate the isotope effect during sulfate reduction would be invalid. We believe that such additional effects are present, though minor. For example, initial dissolved and acid volatile sulfur in a core from Carmen Basin in the Gulf of California (Goldhaber and Kaplan, 1980) were found to have virtually the same isotopic value, indicating little isotopic change in going from one to the other.

ACKNOWLEDGMENT

This study was supported by a contract from the U.S. Atomic Energy Commission AT(04-3)-34, P.A. 134.

LITERATURE CITED

1. Aizenshtat, Z., M. J. Baedecker, and I. R. Kaplan. 1973. Distribution and diagenesis of organic compounds in JOIDES sediment from Gulf of Mexico and Western Atlantic. Geochim. Cosmochim. Acta 37:1881-1898.
2. Alexander, M. 1971. Microbial ecology. John Wiley and Sons, Inc. 414 p.
3. Ben-Yaakov, S. 1973. pH buffering of pore water of recent anoxic marine sediments. Limnol. Oceanogr. 18:86-94.
4. Berner, R. A. 1964a. An idealized model of dissolved sulfate distribution in recent sediments. Geochim. Cosmochim. Acta 28:1497-1503.
5. —————. 1964b. Distribution and diagenesis of sulfur in some sediments from the Gulf of California. Marine Geol. 1:117-140.
6. —————. 1966. Chemical diagenesis of some modern carbonate sediments. Am. J. Sci. 264:1-36.
7. —————. 1970. Sedimentary pyrite formation. Am. J. Sci. 268:1-23.
8. —————. 1972. Sulfate reduction, pyrite formation and the oceanic sulfur budget. p. 347-361. In D. Dyrssen and D. Jagner (ed.) Nobel Symposium 20, The changing chemistry of the oceans. Almquist and Wiksell, Stockholm.
9. —————. 1974. Kinetic models for the early diagenesis of nitrogen, sulfur phosphorus, and silicon in anoxic marine sediments. p. 427-449. In E. D. Goldberg (ed.) The sea, Vol. 5, Marine Chemistry. John Wiley and Sons, Inc.
10. —————. 1976. Diagenetic models of dissolved species in the interstitial waters of compacting sediments. Am. J. Sci. 275:88-96.
11. Borodovskiy, O. K. 1965a. Accumulation and transformation of organic substances in marine sediments. III. Accumulation of organic matter in bottom sediments. Marine Geol. 3:33-82.
12. —————. 1965b. Accumulation and transformation of organic substances in marine sediments. IV. Transformation of organic matter in bottom sediments and its early diagenesis. Marine Geol. 3:83-114.
13. Claypool, G. E. 1974. Anoxic diagenesis and bacterial methane production in deep sea sediments. Ph.D. Thesis, University of California, Los Angeles.
14. Emery, K. O. 1960. The sea off southern California. John Wiley and Sons, Inc., New York. 366 p.
15. Gardner, L. R. 1973. Chemical models for sulfate reduction in closed anaerobic marine environments. Geochim. Cosmochim. Acta 37:53-68.
16. Gibbs, R. J. 1973. Mechanisms of trace metal transport in rivers. Science 180:71-72.
17. Goldhaber, M. B. 1974. Equilibrium and dynamic aspects of the marine geochemistry of sulfur. Ph.D. Thesis, University of California, Los Angeles.
18. —————, and I. R. Kaplan. 1974. The sulfur cycle. In E. D. Goldberg (ed.) The sea, Vol. 5, Marine Chemistry. John Wiley and Sons, Inc.
19. —————, and —————. 1980. Mechanisms of sulfur incorporation and isotope fractionation during early diagenesis in sediment of the Gulf of California. Marine Chemistry 9:95-143.
20. Gunkel, W., and C. H. Oppenheimer. 1963. Experiments regarding the sulfide formation in sediments of the Texas Gulf Coast. p. 674-683. In C. H. Oppenheimer (ed.) Symposium on Marine Microbiology. Charles C. Thomas.
21. Harrison, A. G., and H. G. Thode. 1958. Mechanism of the bacterial reduction of sulfate from isotope fractionation studies. Trans. Faraday Soc. 54:84-92.

22. Hartmann, M., and H. Nielsen. 1969. δ³⁴S-werte in Rezenten Meeressedimenten und ihre Deutung am Beispiel Einiger Sedimentprofile aus der Westluchen Ost See. Geol. Rundschau 58:621–655.
23. ———, P. Muller, E. Suess, and C. H. van der Weijden. 1973. Oxidation of organic matter in recent marine sediments. Meteor. Forsh. Ergebnisse 12:74–84.
24. Ivanov, M. W. 1968. Microbial processes in the formation of sulfur deposits. Israel Program for Scientific Trans., Jerusalem. 298 p.
25. Jannasch, H. W., K. Eimhjellen, C. O. Wirsen, and A. Farmanfarmaian. 1971. Microbial degradation of organic matter in the deep sea. Science 171:672–675.
26. Jones, G. G., and R. L. Starkey. 1957. Fractionation of stable isotopes of sulfur by microorganisms and their role in deposition of native sulfur. Appl. Microbiol. 5:111–118.
27. Kaplan, I. R., K. O. Emery, and S. D. Rittenberg. 1963. The distribution and isotopic abundance of sulfur in recent marine sediments off Southern California. Geochim. Cosmochim. Acta 27:297–331.
28. ———, and S. C. Rittenberg. 1964. Microbiological fractionation of sulfur isotopes. J. Gen. Microbiol. 34:195–212.
29. Koide, M., A. Soutar, and E. D. Goldberg. 1972. Marine geochronology with ²¹⁰Pb. Earth Plantet. Sci. Lett. 14:442–446.
30. Manheim, F. T., K. M. Chan, and F. L. Sayles. 1970. Interstitial water studies on small core samples, Deep Sea Drilling Project. Leg 5. In D. A. McManus et al. Initial Reports of the Deep Sea Drilling Project, Vol. V, U.S. Government Printing Office, Washington, DC.
31. Martens, C. S., and R. A. Berner. 1974. Methane production in the interstitial waters of sulfate depleted marine sediments. Science 186:1167–1169.
32. Nakai, N., and M. L. Jensen. 1964. The kinetic isotope effect in the bacterial reduction and oxidation of sulfur. Geochim. Cosmochim. Acta 28:1893–1912.
33. Nissenbaum, A., and I. R. Kaplan. 1972. Chemical and isotopic evidence for the in situ origin of marine humic acids. Limnol. Oceanogr. 17:570–582.
34. ———, B. J. Presley, and I. R. Kaplan. 1972. Early diagenesis in a reducing fjord, Saanich Inlet, British Columbia. I. Chemical and isotopic changes in major components of interstitial water. Geochim. Cosmochim. Acta 36:1007–1027.
35. Oppenheimer, C. H. 1960. Bacterial activity in sediments of shallow marine bays. Geochim. Cosmochim. Acta 19:244–260.
36. Orr, W. L., and H. G. Gaines, Jr. 1974. Observations on rate of sulfate reduction and organic matter oxidation in the bottom waters of an estuarine basin of the Pettaquamscutt River (Rhode Island). 6th Int. Meeting on Organic Geochemistry, Paris p. 781–812.
37. Postgate, J. R. 1951. The reduction of sulphur compounds by Desulphovibrio desulphuricans. J. Gen. Microbiol. 5:725–738.
38. ———. 1965. Recent advances in the study of the sulfate-reducing bacteria. Bacterial Revs. 29:425–441.
39. Presley, B. J., and I. R. Kaplan. 1968. Changes in dissolved sulfate, calcium and carbonate from interstitial water of near-shore sediments. Geochim. Cosmochim. Acta 32:1037–1048.
40. ———, M. B. Goldhaber, and I. R. Kaplan. 1970. Interstitial water chemistry, Deep Sea Drilling Project, Leg 5. In D. A. McManus et al. Initial Reports of the Deep Sea Drilling Project, Vol. V, U.S. Government Printing Office, Washington, DC.
41. Price, N. B., and S. E. Calvert. 1970. Compositional variation in Pacific Ocean ferromanganese nodules and its relationship to sediment accumulation rates. Mar. Geol. 9:145–171.
42. Ramm, A. E., and D. A. Bella. 1974. Sulfide production in anaerobic microcosms. Limnol. Oceanogr. 19:110–118.
43. Rees, C. E. 1973. A steady state model for sulfur isotope fractionation in bacterial reduction process. Geochim. Cosmochim. Acta 27:1141.
44. Richards, F. A. 1965. Anoxic basins and fjords. p. 611–645. In J. P. Riley and G. Skirrow (ed.) Chemical oceanography. Vol. 1. Academic Press, New York.

45. Rickard, D. T. 1974. Kinetics and mechanisms of the sulfidation of goethite. Am. J. Sci. 274:941–952.
46. Rozanov, A. G., I. I. Volkov, N. N. Zhabina, and T. A. Yagodinskiy. 1971. Hydrogen sulfide in the sediments of the continental slope, Northwest Pacific Ocean. Geochemistry International 5:333–339.
47. Sholkovitz, E. 1973. Interstitial water chemistry of the Santa Barbara basin sediments. Geochim. Cosmochim. Acta 37:2043.
48. Sorokin, Y. I. 1962. Experimental investigation of bacterial sulfate reduction in the Black Sea using S^{35}. Mikrobiologiya 31:402–410.
49. ———. 1966. Sources of energy and carbon for biosynthesis by sulfate reducing bacteria. Mikrobiologiya 35:761–766.
50. Sweeney, R. E. 1972. Pyritization during diagenesis of marine sediments. Ph.D. Thesis, University of California, Los Angeles.
51. Theede, H., A. Ponat, K. Hiroki, and C. Schlieper. 1969. Studies on the resistance of marine bottom invertebrates to oxygen-deficiency and hydrogen sulfide. Marine Biol. 2:325–337.
52. Thode, H. G., J. Monster, and H. B. Dunford. 1961. Sulfur isotope geochemistry. Geochim. Cosmochim. Acta 25:159–174.
53. Thorstenson, D. C. 1970. Equilibrium distribution of small organic molecules in natural waters. Geochim. Cosmochim. Acta 34:745–770.
54. ———, F. T. Mackenzie. 1971. Experimental decomposition of algae in seawater and early diagenesis. Nature 234:543–545.
55. Trudinger, P. A., and L. A. Chambers. 1973. Reversibility of bacterial sulfate reduction and its relevance to isotope fractionation. Geochim. Cosmochim. Acta 37:1775–1778.
56. Tsou, J. L., D. Hammond, and R. Horowitz. 1973. Interstitial water studies, Leg 15. Study of CO_2 released from stored deep sea sediments. *In* B. C. Heezen, and I. G. MacGregor et al. Initial Reports of the Deep Sea Drilling Project, Vol. XX, U.S. Government Printing Office, Washington, DC.
57. Tzur, Y. 1971. Interstitial diffusion and advection of solute in accumulating sediments. J. Geophys. Res. 76:4208–4211.
58. Van Andel, T. H. 1964. Recent marine sediments of the Gulf of California. p. 216–310. *In* T. H. Van Andel and G. G. Shor (ed.) Marine Geology of the Gulf of California. Am. Assoc. Petrol. Geol. Mem. 3.
59. Vinogradov, A. P., V. A. Gruenko, and V. I. Ustinov. 1962. Isotopic composition of sulphur compounds in the Black Sea. Geokhimiya 10:973–997.
60. Zobell, C. E., and C. B. Feltham. 1942. The bacterial ecology of a marine mud flat as an ecological factor. Ecology 23:69–78.
61. Zsolnay, V. A. 1971. Diagenesis as a function of redox conditions in nature: A comparative survey of certain organic and inorganic compounds in an oxic and anoxic Baltic basin. Kiel. Meeresforsh. XXVII:135–165.

Chapter 3

Aqueous Pyrite Oxidation and the Consequent Formation of Secondary Iron Minerals

DARRELL KIRK NORDSTROM[1]

ABSTRACT

The oxidation of pyrite in aqueous systems is a complex biogeochemical process involving several redox reactions and microbial catalysis. This paper reviews the kinetic data on pyrite oxidation, compares available data on the inorganic vs. microbial oxidative mechanisms and describes the occurrence of mineral products resulting from pyrite oxidation. Although oxygen is the overall oxidant, kinetic data suggests that ferric iron is the direct oxidant in acid systems and that temperature, pH, surface area, and the presence of iron and sulfur-oxidizing bacteria can greatly affect the rate of reaction. The vast amount of literature on the microbial and geochemical investigations on this subject have limited usefulness for understanding natural systems. Additional research is needed on the hydrologic, geologic and microbiologic characteristics of field sites where oxidation occurs. The acid water resulting from pyrite oxidation may precipitate a large suite of soluble and insoluble iron minerals depending on pH, degree of oxidation, moisture content, and solution composition.

INTRODUCTION

The oxidation of common sulfide minerals such as pyrite plays a key role in (a) the supergene alteration of ore deposits, (b) the formation of acid mine waters, (c) the formation of acid sulfate soils, (d) the source and distribution of dissolved sulfate in natural waters, (e) the source and distribution of heavy metals in the aquatic environment, and (f) the in situ solution mining and dump leaching of heavy metals for economic recovery. Pyrite oxidation is a complicated process which includes several types of oxidation-reduction reactions, hydrolysis and complex ion formation, solubility controls and kinetic effects. To interpret these reactions in

---

[1] Department of Environmental Sciences, Clark Hall, Univ. of Virginia, Charlottesville, VA 22903. Current address: Water Resources Division, U.S. Geological Survey, 345 Middlefield Road, Menlo Park, CA 94025.

Copyright © 1982 Soil Science Society of America, 677 S. Segoe Rd., Madison, WI 53711. *Acid Sulfate Weathering.*

natural aquatic systems requires the additional knowledge of microbial ecology, hydrology, and geology. It is not surprising that important aspects of this process are not clearly understood and many questions remain unresolved.

The overall process describing pyrite oxidation is commonly given by the following incongruent reaction:

$$FeS_{2(s)} + (15/4)\ O_{2(aq)} + (7/2)\ H_2O_{(l)} \rightarrow Fe(OH)_{3(s)} + 2H_2SO_{4(aq)} \quad [1]$$

in which pyrite and water, in the presence of oxygen, form insoluble ferric hydroxide and sulfuric acid. Although the ultimate driving force is atmospheric oxygen, the fundamental mechanism and the major rate-determining step(s) may not involve oxygen at all. Furthermore, many other iron minerals may form in addition to or instead of ferric hydroxide. In natural environments the relative importance of physical, chemical and microbiological factors may vary widely. This paper (1) reviews the known rates and suggested mechanisms for pyrite oxidation with some discussion of the effects of geology, hydrology, and climate and (2) describes the occurrence of hydrated iron sulfate and iron oxide minerals which result from pyrite oxidation. Illustrative examples are given from acid mine drainage studies because more research has been done in this area than any other. The conclusions, however, are just as applicable to acid sulfate soils.

Investigations on pyrite oxidation usually center on the question of whether an inorganic or microbiological mechanism is more important in controlling the rate of the oxidative process. The relative importance of these two mechanisms will dictate the preferred approach for "at-source" control of acid mine drainage and acid sulfate soil formation.

INORGANIC OXIDATION MECHANISMS

The oxidation and hydrolysis shown by reaction 1 involves the loss of 1 electron by iron, 14 electrons by sulfur, and the gain of 7½ electrons by oxygen per mole of pyrite. Also, 1 mole of iron is hydrolyzed and precipitated. Clearly, all of these changes cannot take place in one step. Kinetic studies and quantum mechanical calculations indicate that one-electron-transfer reactions are most likely to occur, two-electron reactions much less likely and more than two are highly unlikely (Basolo and Pearson, 1967). Furthermore, the oxidative half-cell reaction does not have to take place at the same rate as the reductive half-cell. A maximum of 22 electron transfer reactions are possible, and there may be the same number of possible rate-determining steps. More steps are possible if other oxidizing agents are considered. Experimental measurements on this heterogeneous system indicate that two electrons are frequently transferred and that all 22 steps do not need to be individually considered. The following discussion will begin with pyrite in aqueous solution and sequentially consider the loss of electrons from pyrite as well as the rates and mechanisms of oxidation whenever data are available.

Several investigators have shown that the first step in the breakdown of a sulfide in an aqueous solution is the dissolution of the metal and the oxidation of sulfide to neutral sulfur. High temperature (~100 to 200 C) metallurgical reactions consistently have the general stoichiometry:

$$MS_{(s)} \rightarrow M^{2+}_{(aq)} + S^{\circ}_{(s)} + 2e^- \qquad [2]$$

for a divalent metal, M (Burkin, 1966; Wadsworth, 1973). Using pyrite as an electrode, Sato (1960) made potentiometric measurements at 25 C which indicated the initial step for pH ≤ 3.0 was:

$$FeS_{2(s)} \rightarrow Fe^{2+}_{(aq)} + S^{\circ}_2 + 2e^- \qquad [3]$$

and this reaction was also found as a limiting condition in the experiments of Garrels and Thompson (1960). Current-potential measurements on pyrite by Biegler and Swift (1979) at 25 C and pH ≤ 3.0 were interpreted as a combination of

$$FeS_2 + 8H_2O \rightarrow Fe^{3+} + 2SO_4^{2-} + 16H^+ + 15e^- \qquad [4]$$

and

$$FeS_2 \rightarrow Fe^{3+} + 2S^{\circ} + 3e^- \qquad [5]$$

where reaction 5 is not an intermediate leading to reaction 4. In order to speed up the reaction rate, Biegler and Swift (1979) had to make these measurements at rather high potentials (0.95 to 1.4 V) which oxidized dissolved iron to the ferric state. Lower potentials were achieved by Bailey and Peters (1976) by increasing the temperature to 110 C, and in these experiments there were significant amounts of ferrous iron. All of these measurements were done in the absence of oxygen. In neutral to basic solutions Sato (1960) proposed the reaction:

$$FeS_{2(s)} + 2H_2O_{(l)} \rightarrow Fe(OH)_{3(s)} + S^{\circ} + 3H^+_{(aq)} + 3e^- \qquad [6]$$

based on potentiometric measurements. The neutral sulfur produced in reactions 5 and 6 would, at first, simply remain at the pyrite surface as part of the structure while ferrous ions leached out from the surrounding lattice sites. This sulfur-rich surface would eventually become unstable and do one of two things: (1) disrupt and reorganize into elemental sulfur (such as S_8 rings) or (2) disrupt into solution as dimers which would be highly unstable and would form sulfate rapidly in the presence of a strong oxidizing agent. Electrochemical measurements support these hypotheses (Peters, 1977; Biegler and Swift, 1979). Metal ions leach readily from sulfide minerals and for some the sulfur will remain behind in the original structure. This residual sulfur behaves like elemental sulfur in that it dissolves in carbon disulfide (Sullivan, 1930). There is evidence for both the presence and absence of elemental sulfur resulting from pyrite oxidation. Clark (1966) has cited several instances of pyrite-derived sulfur from coal seams. Presumably the organic matter retards the further oxidation of sulfur. Smith et al. (1968), and Stokes (1901) found that elemental sulfur was produced in the initial stages of oxidation using freshly-prepared pyrite, whereas samples conditioned by prior oxidation did not produce sulfur in the oxidation experiments. In highly oxidizing sulfide mineral deposits little or no sulfur is found (Kinkel et al., 1956; Nordstrom, unpublished data). However, Bergholm (1955) has documented the production of elemental sulfur from pyrite at low pH and in the presence of ferric iron. In Bergholm's experiments most of the sulfur coated the pyrite as elemental sulfur and retarded the oxidation. Brock et al. (1976) found that elemental sulfur did not reduce ferric iron at a pH of 1.6 and temperatures of 60 to 90 C. Decreasing temperature and decreasing pH both led to increased sulfur production at 100 to 130 C in the experiments of McKay and Halpern (1958). Also, McKay and Halpern (1958) have hypothesized two independent reactions of the same stoichiometry as Biegler and Swift (1979) which gives additional credance to the dual reaction paths, one leading to the production of sulfur and the other to sulfate. These observations suggest that the oxidation rate of free sulfur to sulfate relative to the oxidation rate of sulfide-sulfur to free sulfur determines whether free sulfur will occur or not. In turn, these rates will depend upon the transport rate of the oxidizing agents to the pyrite surface, the pH, and the temperature. The evidence cited indicates that these two rates are somewhat comparable but that sulfide to sulfur is faster.

The oxidizing agent in waters of neutral pH is oxygen since the only other likely oxidant, Fe^{3+}, is highly insoluble. The reduction of oxygen on pyrite proceeds through the following steps:

$$O_2 + e^- \rightarrow O_2^- \quad [7]$$

$$O_2^- + 2H^+ + e^- \rightarrow H_2O_2 \quad [8]$$

At pH values above 4, reaction 8 appears to be the rate-limiting step, whereas at lower pH values the rate is independent of pH and the limiting step is given by reaction 7 (Biegler et al., 1977). This conclusion is also verified by the rate data of Smith et al. (1968). The initial oxidation-reduction reaction of pyrite in water and oxygen results from combining reactions 3, 7, and 8:

$$FeS_{2(s)} + O_2 + 2H^+ \rightarrow Fe^{2+} + S_2^\circ + H_2O_2 \quad [9]$$

Elemental sulfur formation is probably faster than peroxide formation for a system containing pure pyrite in solutions of neutral pH where diffusional processes are not rate-limiting. Therefore pyrite oxidation kinetics and sulfur formation depends upon the availability and rate of oxygen reduction. In soils and bedrock, the rate is probably related to the rate of oxygen infiltration.

The three products of reaction 9 are unstable and rapidly oxidized or decomposed. At neutral pH values ferrous iron oxidizes very rapidly according to the expression:

$$d[Fe(II)]/dt = k[Fe(II)][O_2]/[H^+]^2 \qquad [10]$$

(Ghosh, 1974; Singer and Stumm, 1968; Stumm and Lee, 1961). Brackets denote concentrations, k is the rate constant and t is the time. Equation [10] holds for pH values above 4.5 and for poorly buffered solutions. Ghosh (1974) pointed out that the oxidation rate increases with increasing carbonate alkalinity and that highly buffered waters may differ by more than an order of magnitude in their ferrous oxidation rates. In poorly buffered perchlorate solutions and at a pH of 7.0, the rate constant is about 10^2 day^{-1} by extrapolating the data of Singer and Stumm (1968). When sulfate is used instead of perchlorate the rate is different; Singer and Stumm (1968) found that the rate decreased, whereas Huffman and Davidson (1956) found the rate increased with increasing sulfate concentration. The discrepancy must be resolved by additional measurements over a wider range of concentrations.

A review of pyrite oxidation kinetics by Shumate et al. (1971) pointed out that the rate dependence on oxygen concentration differs somewhat between investigators. They reported a rate expression based on an adsorption equilibrium hypothesis:

$$r = (k[O_2])/(1 + K_1[O_2] + K_2[I]) \qquad [11]$$

where r is the rate of pyrite oxidation in micromoles pyrite per hour per gram of sample, k is the rate constant, K_1 is the adsorption equilibrium constant for oxygen on pyrite, K_2 is the adsorption constant for an inert gas, I, on pyrite. Equation [11] fits the data well but does not provide any insight into the electrochemical nature of the reaction mechanism.

Once the iron is oxidized it will hydrolyze and precipitate as ferric hydroxide:

$$Fe^{2+} + (5/2)\ H_2O + \tfrac{1}{4} O_2 \rightarrow Fe(OH)_{3(s)} + 2H^+ \qquad [12]$$

or, if the hydrogen peroxide produced by reaction 7 is the oxidizing agent:

$$Fe^{2+} + (1/2)\ H_2O_2 + 2H_2O \rightarrow Fe(OH)_{3(s)} + 2H^+ \qquad [13]$$

The presence of iron hydroxide on the surface of oxidizing pyrite has been observed by Baker (1972) using Mossbauer spectroscopy.

It is interesting to note that if the sulfur produced in reaction 3 did not further oxidize there would be no change in the pH of the solution since the protons consumed in reaction 9 exactly balance the protons produced in reaction 12 or 13:

$$FeS_{2(s)} + (3/4)\ O_2 + (3/2)\ H_2O \rightarrow Fe(OH)_{3(s)} + S_2^o \qquad [14]$$

When the sulfur is oxidized to sulfate, the pH drops:

$$S_2^o + 3O_2 + 2H_2O \rightarrow 2SO_4^{2-} + 4H^+. \qquad [15]$$

Relatively little has been studied on the rates and mechanisms of reaction 15 with regard to pyrite oxidation, and this reaction could be an important rate-determining step in the initial weathering of pyrite to produce acid conditions from neutral solutions. Six electrons are transferred per sulfur atom, and the formation of several intermediate sulfur species such as thiosulfate, sulfite and polythionates are possible complicating factors. Sulfite, SO_3^{2-}, is not stable and rapidly oxidizes to sulfate in the presence of oxygen or any other oxidizing agent. Thiosulfate, $S_2O_3^{2-}$, is more stable at high pH than low pH and readily decomposes to elemental sulfur and sulfite (and hence to sulfate) in acid solutions. Polythionates are more stable than thiosulfate or sulfite, especially in acid media, but they too tend to breakdown by disproportionation to simpler species such as sulfur and sulfate. Thiosulfate can be oxidized by ferric iron to tetrathionate and the indication of thiosulfate formation on air-oxidized sulfides by Steger and Desjardins (1978) leads to the hypothesis that polythionates are important intermediates during aqueous pyrite oxidation in neutral to acidic solutions. This hypothesis is verified by the study of Nor and Tabatabai (1977) who reported that nearly all the dissolved sulfur present during the oxidation of elemental sulfur in soils was sulfate at a pH of 7.8, whereas tetrathionate was dominant at pH values of 5 to 6. Goldhaber (1977, personal communication) found tetrathionate and sulfate to be present in equal proportions from pyrite oxidation at a pH of 6 whereas thiosulfate was dominant at pH values of 8 and 9.

Reaction 15 is the initial acid-producing reaction which can reduce the pH to about 4.5 where the rate of ferrous iron oxidation slows down significantly (Singer and Stumm, 1968) and ferric hydroxide is more soluble. Ferrous iron oxidation becomes independent of pH below a pH of 3 with a rate constant of $10^{-3.5}$ day^{-1}. For example, in a 9×10^{-4} M ferrous iron solution it takes about 5½ days to oxidize 1×10^{-5} moles at a pH of 2, a P_{O_2} of 0.2 atm and a temperature of 25 C.

As the ferric concentrations increase with the increased acidity, the role of ferric iron becomes more important as an oxidizing agent. Measurements by Garrels and Thompson (1960) and by Smith et al. (1968) have shown that pyrite is rapidly oxidized by ferric iron in the absence of oxygen and at low pH values according to the stoichiometry:

$$FeS_{2(s)} + 14Fe^{3+} + 8H_2O \rightarrow 15Fe^{2+} + 2SO_4^{2-} + 16H^+ \qquad [16]$$

No residual sulfur was noted in these experiments which runs contrary to the work of earlier investigators such as Bergholm (1955). From the data of Garrels and Thompson (1960), 50% of a 2×10^{-3} m solution of ferric sulfate was reduced by pyrite in 5 hours. Singer and Stumm (1969) carred out similar experiments over a broader range of conditions and showed the first order dependence on both Fe^{3+} and FeS_2:

$$-d[Fe^{3+}]/dt = k[Fe^{3+}][FeS_2] \qquad [17]$$

where [FeS_2] represents the molar concentration of pyrite. Singer and

Stumm (1969) calculated rate constants which ranged from 0.389 day^{-1} to 17.4 day^{-1} depending on the proportions of total Fe^{3+} and FeS$_2$ to the solution volume.

Figure 1 compares the oxidation rates of ferrous ion to ferric ion by oxygen (III), pyrite to acid ferrous sulfate solution by oxygen (II), and pyrite by ferric ion (I) as a function of pH. At low pH values (≤ 3.0) ferric iron oxidizes pyrite much more rapidly than oxygen and more rapidly than dissolved ferrous can be oxidized by oxygen. At neutral to alkaline pH values the rate of ferrous oxidation rises rapidly, but the dissolved ferric concentration also decreases greatly due to the precipitation of ferric hydroxide. The results of these studies support the contention that pyrite is initially oxidized by oxygen, and the pH consequently decreases depending on the rate of oxidation of sulfur to sulfate. When the pH de-

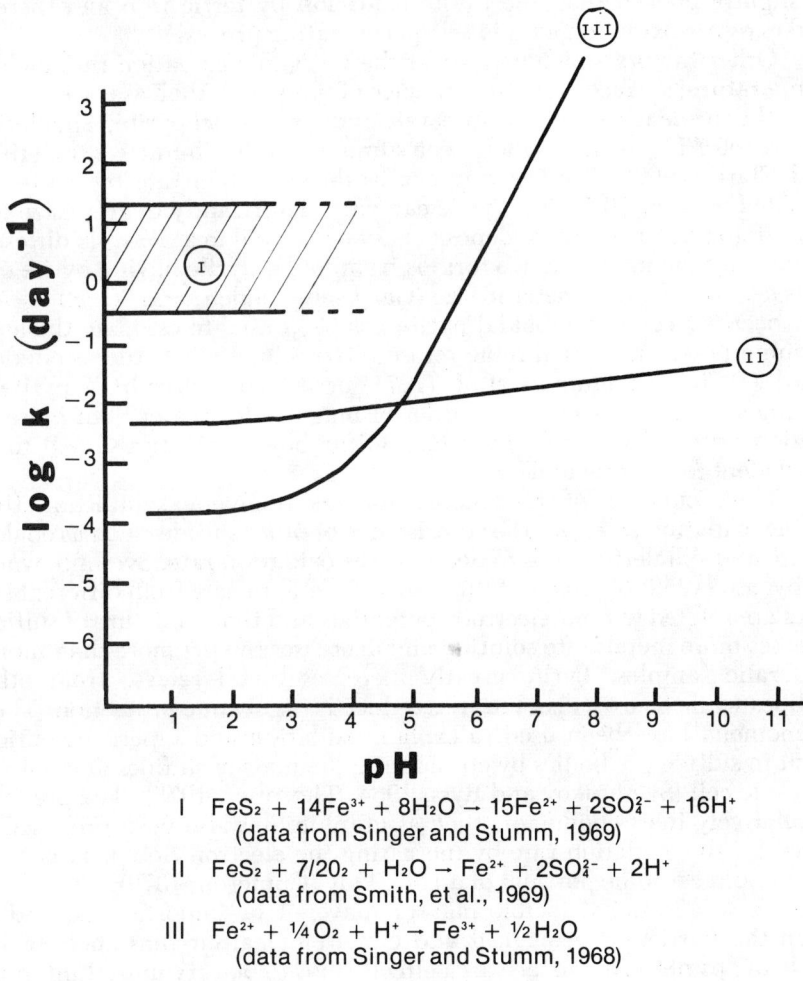

I   FeS$_2$ + 14Fe^{3+} + 8H$_2$O → 15Fe^{2+} + 2SO$_4^{2-}$ + 16H$^+$
    (data from Singer and Stumm, 1969)

II FeS$_2$ + 7/2O$_2$ + H$_2$O → Fe^{2+} + 2SO$_4^{2-}$ + 2H$^+$
    (data from Smith, et al., 1969)

III Fe^{2+} + ¼O$_2$ + H$^+$ → Fe^{3+} + ½H$_2$O
    (data from Singer and Stumm, 1968)

Fig. 1. Comparison of rate constants as a function of pH for I. The oxidation of pyrite by ferric iron, II. The oxidation of pyrite by oxygen and III. The oxidation of ferrous iron by oxygen.

creases to 4.5, ferric iron becomes more soluble and begins to act as an oxidizing agent, and below a pH of 3.0 it is the only important oxidizer of pyrite. The presence or absence of oxygen makes no difference to the oxidation rate by ferric iron (Singer and Stumm, 1969). Since pyrite can reduce ferric ion to ferrous faster than ferrous can be regenerated into ferric by oxygen, the pyrite will simply reduce all the ferric ions and then the reaction will stop. Thus, the ferrous to ferric oxidation has been called the rate-determining step in the production of acid mine drainage (Singer and Stumm, 1970). One of the strongest catalysts of this reaction is the iron-oxidizing bacterium, *T. ferrooxidans*, which is known to increase the ferrous oxidation rate by five to six orders of magnitude (Lacey and Lawson, 1970; Singer and Stumm, 1970; Nordstrom, 1976). This increase makes the ferrous oxidation rate, in the presence of the bacterium, comparable or slightly greater than the pyrite oxidation by ferric iron and thereby makes pyrite oxidation a rapid self-perpetuating process.

Other factors which can affect the inorganic oxidation rate include temperature, surface area, the presence of impurities such as trace metals, and the presence of other minerals such as chalcopyrite, sphalerite, calcite, etc. These factors have been summarized by Shumate et al. (1971) and Clark (1966). Temperature causes the oxidation rate by oxygen to double for every 10 C rise. Pyrite can vary significantly in grain size and morphology. Pyrite in coal deposits has been found in at least six different forms, and the most reactive form is framboidal pyrite having pyrite crystals less than a micrometer in size (Caruccio, Geidell, and Sewell, 1976). The occurrence of framboidal pyrite has been used to estimate the acid-forming potential of coal mine refuse (Caruccio, 1975). Studies cited by Clark (1966) and Shumate et al. (1971) described "sulfur ball" pyrite as having surface areas about an order of magnitude greater than museum grade pyrite of the same diameter. Sulfur ball pyrite is assumed to be equivalent to "framboidal" pyrite.

The occurrence of trace metals appears to have no significant effect on the oxidation rate, but the co-existence of other sulfides such as chalcopyrite and sphalerite tends to decrease the oxidation rate. Sveshnikov and Dobychin (1956) pointed out that rates of metal release from different sulfides are related to their electrode potentials and that a mixture of sulfides releases more metals into solution and decreases the pH more than monomineralic samples. Pyrite greatly increases metal release from other sulfides while its own dissolution is reduced by galvanic protection. These phenomena have been used to explain oxidation and supergene enrichment in sulfide ore bodies by considering the mass of sulfides as a natural galvanic cell (Sveshnikov and Ryss, 1964; Thornber, 1975). The presence of relatively inert conductors such as graphite has also been proposed to speed up the oxidation rate by increasing the electron flow between the anodic and cathodic portions of an ore body (Cameron, 1979).

These additional factors may or may not be important depending upon the particular geological and climatological circumstances of the oxidizing pyrite. The surface area effect seems to be very important in the initial stages of acid sulfate production, but once the pH has decreased to

3 or less it would seem to be less important. On the other hand, bacterial oxidation could be the single most important factor over the whole range of pH of natural waters.

MICROBIAL OXIDATION MECHANISMS

There are at least two well-established facts of microbial ecology: (1) microbes are ubiquitous on this planet, existing under a wide range of natural conditions (Kushner, 1978) and (2) microbes catalyze many reactions including those related to the formation and oxidation of sulfide ore deposits (Kuznetsov et al., 1963; Zajic, 1969). Very few microorganisms have been studied as much as the sulfur-oxidizing genus, *Thiobacillus*. This bacterium utilizes some form of sulfur as an energy source and reduces CO_2 for a carbon source and is therefore termed chemoautotrophic. Three species of *Thiobacillus* have been isolated from acid mine wastes: *T. ferrooxidans* which oxidizes ferrous iron and pyrite as well as sulfur, *T. thiooxidans* which oxidizes only sulfur and pyrite, and *T. acidophilus* which is a facultative autotroph (grows on either inorganic or organic substrates, Guay and Silver, 1975) oxidizing sulfur but not ferrous iron. *Thiobacillus acidophilus* cannot oxidize pyrite unless it is in a mixed culture. These bacteria are acidophilic with optimal growth conditions around a pH of 2 to 3, although they can survive up to pH values as high as 6 and 7.

The importance of *Thiobacillis* in pyrite oxidation and the mechanism of microbial attack has long been a controversial issue. *Thiobacillus ferrooxidans* has been implicated as an essential catalyst in the production of acid mine waters because (1) it can usually be isolated from acid mine waters; (2) it actively accelerates pyrite oxidation in lab experiments; and (3) it can speed up the oxidation of dissolved ferrous ions by five to six orders of magnitude over the inorganic rate, thereby providing a way of rapidly regenerating ferric ions in acid solutions. Direct evidence for microbial growth in the deeper portions of the unsaturated zone, however, has not been demonstrated. Growth may occur in the surface layer of a soil but some skepticism has been raised about the possibility of growth deeper in the subsurface such as in underground mines or in coal refuse piles. Kleinmann (1979) and Kleinman and Crerar (1979) have refuted these criticisms in an oxidation study of pyritic coals and overburden where they simulated the well-oxygenated surface layer, the unsaturated zone above the water table and the saturated zone below the water table. They found that *T. ferrooxidans* catalyzed pyrite oxidation in the unsaturated zone for a few days following rainfall infiltrations as well as in the surface layer.

There has been considerable discussion about the mechanism of microbial pyrite oxidation. *Thiobacillus ferrooxidans* can rapidly generate ferric ions from ferrous ions in acidic media; pyrite is then directly and inorganically oxidized by ferric iron. This process has been named the "indirect contact" mechanism. Some investigators (Bryner and Jameson,

1958; Beck and Brown, 1968) have questioned this mechanism and have proposed that the bacterium makes direct contact with the pyrite crystals and oxidizes them through enzymatic pathways. The bacterial population on the pyrite surface is much greater than in solution (Tuovinen and Kelly, 1972), and it seems that *T. thiooxidans* oxidizes elemental sulfur by direct contact. Thus, a direct contact mechanism has been suggested for the microbial oxidation of pyrite. This mechanism is also indicated by studies on the oxidation of non-ferrous sulfides such as sphalerite, chalcocite, and covellite in the presence of *T. ferrooxidans* (Silver and Torma, 1974; Duncan and Walden, 1972; Nielson and Beck, 1972) where iron is not present in the system. Silverman (1967) attempted to determine which mechanism was more important, and he concluded that both mechanisms were operating concurrently. However, in every experiment the bacteria were allowed to make direct contact with the pyrite grains, and in the experiments without bacteria he failed to measure and compare the reaction rates. A series of critical experiments were carried by Arkesteyn (1979) in which he not only demonstrated the direct contact mechanism but also showed that the rate of pyrite and elemental sulfur oxidation at a pH of 5.0 was faster than the oxidation rate of ferrous iron, all in the presence of *T. ferrooxidans*. In addition, when Arkesteyn (1979) separated the bacteria from pyrite by a dialysis bag the oxidation rate decreased. These results confirmed other sets of experiments in which he individually inhibited the oxidation of ferrous iron or sulfur by the organism without stopping the oxidation process.

A further point that tends to confuse the whole microbial picture is the ability of *T. thiooxidans* and *Sulfolobus acidocaldarius* (a thermophilic sulfur and iron-oxidizing bacterium) to reduce ferric iron during the aerobic oxidation of elemental sulfur (Brock and Gustafson, 1976). *Thiobaccillus ferrooxidans* also reduces ferric iron during sulfur oxidation under anaerobic conditions while utilizing the oxidized iron as an energy source. These data have very important implications concerning our understanding of sulfide mineral oxidation in subsurface environments such as in tailings piles, water-logged soils, and sulfide mineralization located below the water table. First, it must be recognized that these iron and sulfur-oxidizing bacteria are facultative with respect to their oxygen requirements. Clearly the presence of a suitable electron acceptor such as ferric ion will do just as well as oxygen. Thus, we can expect pyrite oxidation to continue by microbial activity under low or even undetectable oxygen concentrations. No one, however, has determined if the microorganisms catalyze the anaerobic oxidation process. If they do, then clearly mine sealing and flooding is an inappropriate method of preventing acid mine drainage from abandoned mines.

Another source of controversy is the possible role of microorganisms in the initial stages of pyrite oxidation. *Thiobaccillus ferrooxidans* can play an active role once the pH has decreased to 4 to 4.5. How does a neutral soil or ground water initially become acid from pyrite oxidation? One possibility is the inorganic oxidation by molecular oxygen, but this mechanism is quite slow. It is more likely that the initial step is microbially catalyzed by some of the heterotrophs and/or autotrophs which occur in the soil horizon and are known to metabolize inorganic sulfur compounds (Roy and Trudinger, 1970). Walsh and Mitchell (1972a, 1972b) have proposed that *Metallogenium*, an acid-tolerant iron-oxidizing bacterium, initiates a pH-dependent succession of bacteria. Metallogenium optimally oxidizes iron between pH values of 3.5 and 5. Once *Metallogenium* brings the pH down to about 4 then *T. ferrooxidans* takes over and reduces the pH to below 4. In his studies on simulated coal refuse environments, Kleinman (1979) found that (1) inoculation with and without *Metallogenium* made no difference in the rate of initial acidification and (2) *T. ferrooxidans* not only survives at pH values up to 7, but it also initiates pyrite oxidation to provide the acidity and ferrous ions. It should be kept in mind that although the bulk solution phase may have a neutral pH, the pH right at the surface of a pyrite grain may be considerably less and may be the site location of viable *Thiobacilli*. He suggested that in the initial stage of oxidation the direct contact mechanism was probably more important since ferric ions are too low in concentration. Several other species of *Thiobacillus* are also capable of performing the same function as initiator (Sokolova and Karavaiko, 1968; Zajic, 1969) provided that free sulfur is available from the pyrite. Arkesteyn (1980) found that the presence of *T. thiooxidans, thioparus, intermedius,* and *perametabolis* does not increase the pyrite oxidation rate over sterile blanks when starting at pH values of 6.0. There seems to be a slight enhancement of the rate when Arkesteyn (1980) used *T. ferrooxidans* but not by an amount greater than the experimental uncertainty. These data seem contrary to his experiments carried out at pH 5.0 in which microbial catalysis by direct contact was evident. In addition, new evidence summarized by Kelly et al. (1979) indicates that several other bacteria are important in pyrite oxidation and that mixed cultures can be more effective in oxidizing pyrite than single species cultures. Arkesteyn (1980) was also unable to isolate *Metallogenium* from some acid sulfate soils. The conclusion from all these investigations is that the initiation of pyrite oxidation in soils or sulfide refuse piles probably occurs more rapidly by microbial catalysis than by a purely inorganic mechanism and that *T. ferrooxidans* and possibly other microorganisms which normally inhabitat the soil environment may catalyze these reactions. The potential role of *Thiobacilli* species in initiating pyrite oxidation is another area of research worthy of further investigation. The importance of the microbial vs. the inorganic mechanisms for the initiation step is still subject to controversy.

If the rate-determining steps are controlled primarily by *T. ferrooxidans*, then the oxidation rate depends upon the factors controlling the bacterial growth kinetics such as the availability of oxidizing agents, carbon dioxide, and nutrients. Periodic rainwater infiltration provides the oxygen and carbon dioxide requirements in the unsaturated subsurface (Tuovinen and Kelly, 1972; Kleinmann, 1979). Regular flushing not only provides the needed aeration, but it transports oxidation products away from the reaction zone so that fresh pyrite surfaces are exposed. Nutrient requirements (such as nitrate and phosphate) for *T. ferrooxidans* are not limiting to growth because extremely small concentrations are quite sufficient for growth. *Thiobaccillus ferrooxidans* grows in the absence of any added N compound, and it has been suggested that it may fix atmospheric N_2 (Tuovinen and Kelly, 1972). Nitrogen fixation has recently been demonstrated by Mackintosh (1978). Trace quantities of phosphate and Mg are sufficient for growth. Sulfate is a requirement for iron oxidation (Lazaroff, 1963), but it does not limit growth because it is always available as an oxidation product. Since the only important limiting factors are oxygen and carbon dioxide, it follows that no growth occurs below the oxygenated water table. Kleinman (1979) confirmed that there is no significant growth in saturated environments, and field observations have shown that when deep coal mines are flooded by mine sealing techniques there is usually a significant, albeit incomplete, reduction of acidity (Foreman, 1972). Complete inhibition of the oxidation process may never be possible because of the availability of ferric iron as an alternate oxidizing agent.

Microbial catalysis of pyrite oxidation is an established fact, and the reaction rate may be primarily determined by the growth kinetics of *T. ferrooxidans*. However, recent data from Silver (1978) demonstrates that the DNA base composition of *T. ferrooxidans* can vary depending on the substrate used for growth. *Thiobaccillus ferrooxidans* may be a heterogeneous culture and other bacteria, as yet unidentified, may be more crucial to the oxidation process. Alternatively, mixed cultures may have symbiotic relationships which could produce different oxidation rates than those obtained from pure cultures adapted to specific media. In addition to oxygen or ferric iron concentrations or pH, rates should also be expressed in terms of environmental factors such as hydrologic flow path, residence time, pyrite surface area, temperature, climatic patterns, and soil properties if they are to be applied to actual field situations.

ENVIRONMENTAL INFLUENCES

Pyrite oxidation investigators have emphasized the rate-determining step in a very complex heterogeneous system. The results indicate that the governing factor is the growth kinetics of iron and sulfur-oxidizing bacteria, chiefly *T. ferrooxidans*. Laboratory studies on batch or continuous cultures (chemostat cultures), however, are not appropriate simulations of natural systems. They are useful in obtaining rate data, comparing rates and deriving mechanisms, but our understanding of natural systems also requires knowledge of cyclic processes such as seasonal, diurnal, and rainstorm influences. These processes strongly influence reactions by regulating temperature, pH, oxygen, water and nutrient availability. Additional information on microbial ecology and the environmental factors which influence growth are needed to determine the rate at which pyrite oxidizes under natural conditions. Unfortunately, very little work has been done in this area, and few field studies contain quantitative data on the microbiology, mineralogy, geology, hydrology, and climate.

Reviews on the microbiology of acid mine waters (Lundgren et al., 1972; Nordstrom, 1977) cite many papers containing species identification but none describing the relative proportions of species. Dugan et al. (1970a, 1970b) and Nordstrom (1977) have described acid slime streamers which contain an abundance of both motile and non-motile bacilli (see color Plate 1A). The slime-excreting bacterium has not been isolated and its function in acid mine waters is not known. Natural acid sulfate waters contain mixed cultures of many species and several families of microorganisms whose growth rates may be quite different from lab growth rates with single species. An example of the interdependency of mixed cultures is a study by Arkesteyn (1980b)[2] on the association of *T. ferrooxidans* and *T. acidophilus*. Apparently *T. ferrooxidans* cannot grow in the presence of many common organic compounds without the presence of *T. acidophilus* which can utilize those compounds as an energy source. It is also interesting to note that *T. acidophilus* was present in approximately equal numbers as *T. ferrooxidans* in all the cultures which Arkesteyn (1980b)[2] examined.

A study has been made on the dissolved ferrous ion oxidation rate in a mountainous stream containing acid mine wastes (Nordstrom, 1976, 1977). The rate was found to be nearly identical to that determined for optimal growth of *T. ferrooxidans* in 9K culture media. The data indicates that the bacteria are growing at optimal rates in this dynamic aqueous environment because iron oxidation can be directly related to cell population growth (Silverman and Lundgren, 1959). After a rainstorm, the oxidation rates in the stream decreases significantly due to dilution and to the flushing effect of high flow. This study is at least one example of the role which climate and hydrology can play.

The geologic structure and composition of the rock strata are very important to the development of acid mine waters. Hollyday and McKenzie (1973) have shown how the positioning of carbonate strata with respect to oxidizing pyritic coals and shales can reduce the production of acid waters. Parizek and Tarr (1972) give an excellent account of several hydrogeologic techniques which can be used to reduce or prevent acid mine drainage. These techniques include diverting water flow away from the pyritic layers by pumping and diverting water through carbonate strata so that they develop high alkalinity before infiltrating pyritic strata. Geidel and Caruccio (1977) have further investigated the rate of alkalinity accumulation compared to acid accumulation by water infiltration through limestones, sandstones, and shales. Their results demonstrate that more frequent infiltration tends to reduce acid development. When longer dry periods are permitted there is a continual buildup of pyrite oxidation products (soluble sulfate salts) which cause higher acidity in the leachates when the next rainfall occurs. Alkalinity production, however, tends to level off with time. These conclusions are contrary to the work of Kleinman (1979) and Kleinman and Crerar (1979) who showed that the activity of *T. ferrooxidans* and the acidity production decrease if the dry intervals between infiltrations are increased. The processes occurring during infiltration are extremely important, and more research on short-term and seasonal variability is needed.

---

[2] Arkesteyn, G. J. M. W. 1980b. *Thiobacillus acidophilus*: a study of its presence in *Thiobacillus ferrooxidans* cultures. Unpublished manuscript.

OXIDATION PRODUCTS: THE FORMATION OF HYDRATED IRON MINERALS

A large number of iron minerals may be found in acid, unsaturated pyrite-rich soils, on sulfide mine dumps, and on the surfaces of exposed sulfide ores and pyritic shales. These minerals have a wide range of properties from very insoluble iron hydroxides to very soluble iron sulfate hydrates. The more soluble sulfates are most commonly formed during dry periods as evaporation promotes the rise of subsurface waters to the uppermost soil surfaces by capillary action. As these waters reach the upper portions of the soil, they become progressively more concentrated and finally precipitate various salts in an efflorescence. This phenomenon is quite comparable to the buildup of efflorescent crusts associated with closed basin lakes in arid environments (Eugster and Jones, 1979). The formation of these efflorescent iron sulfates is an important intermediate step preceding the precipitation of the more common insoluble iron minerals such as goethite and jarosite.

Hydrated ferrous sulfate minerals frequently occur on the surface of weathering pyrite where moisture is present. When conditions become sufficiently dry, the dissolved ferrous and sulfate ions produced by reaction 14 first reach saturation with respect to melanterite, $FeSO_4 \cdot 7H_2O$. Under continued dryness, melanterite dehydrates to either rozenite, $FeSO_4 \cdot 4H_2O$, or szomolnokite, $FeSO_4 \cdot H_2O$. The mineral ferrohexahydrite, $FeSO_4 \cdot 6H_2O$, is much less common and has a very limited stability. The pentahydrate, siderotil, requires copper in its structure to be stable. If these minerals are still in contact with soil water or humid air and warm temperatures, they oxidize to copiapite, $Fe^{2+}Fe_4^{3+}(SO_4)_6(OH)_2 \cdot 20H_2O$ (see color Plate 4C) which can be stable for long periods of time when protected from rainfall or flowing water. Copiapite may contain several other divalent and trivalent metals substituted into the ferrous and ferric sites. It also has a very yellow color making it difficult to distinguish from sulfur. From the author's experience this mineral is one of the most abundant minerals found on sulfide mine tailings and on oxidizing sulfide ore minerals.

The transformation to copiapite may proceed according to the reaction:

$$5FeSO_4 \cdot 7H_2O + O_2 + H_2SO_4 \rightarrow Fe^{2+}Fe_4^{3+}(SO_4)_6(OH)_2 \cdot 20H_2O + 15H_2O \qquad [18]$$

which indicates that lower pH values as well as partial oxidation is required. Copiapite may precipitate directly from acid sulfate waters, but no field evidence for this has yet been found. Copiapite overgrowths on melanterite, however, have been observed (Nordstrom and Dagenhart, 1978) and a specimen from Shasta County, California (Nordstrom, 1977) has been reproduced in color Plate 1B. The bulk of the specimen is cuprian melanterite (blue). Small amounts of rozenite occur as a white

Overall Stoichiometry

$$FeS_2 + 15/4 O_2 + 7/2 H_2O \longrightarrow Fe(OH)_3 + 2 H_2SO_4$$

ACID MEDIA (pH≤3)

SLIGHTLY ACID TO BASIC MEDIA (pH≥4)

initiation phase
$$\begin{cases} FeS_2 \longrightarrow Fe^{2+} + S_2 + 2e^- \\ O_2 + e^- \longrightarrow O_2^- \end{cases} \quad \begin{cases} FeS_2 + 3H_2O \longrightarrow Fe(OH)_3 + S_2 + 3H^+ + 3e^- \\ O_2 + 2e^- + 2H^+ \longrightarrow H_2O_2 \end{cases}$$

acid-generating phase: $S_2 + 3O_2 + 2H_2O \longrightarrow 2SO_4^{2-} + 4H^+$

catalytic phase: $FeS_2 + 14Fe^{3+} + 8H_2O \longrightarrow 15Fe^{2+} + 2SO_4^{2-} + 16H^+$

Fig. 2. The major steps during the course of pyrite oxidation.

frosting and the small yellow crystals are copiapite. The sulfate hydrate coquimbite, $Fe_2(SO_4)_3 \cdot 9H_2O$, is intimately associated with copiapite in oxidizing sulfide deposits. Melanterite, rozenite, szomolnokite, and copiapite are probably the most abundant efflorescences associated with oxidizing coal deposits (Nuhfer, 1967).[3] All of these sulfates are highly soluble and may be at least partially responsible for the increased acidity and dissolved solids load in receiving streams during rainstorm events (Nordstrom and Dagenhart, 1978).

As the iron becomes fully oxidized in acid mine waters it eventually reaches saturation with respect to either ferrihydrite[4] or jarosite, $KFe_3(SO_4)_2(OH)_6$. Jarosite is stable at lower pH values than ferric hydroxide (Brown, 1971). Both of these minerals have been observed as precipitates in acid mine drainage (Nordstrom et al., 1979) but are not stable for more than a season. Color Plate 2A shows a bright yellow deposit of jarosite precipitation from acid mine water. Fresh precipitates of iron in acid mine drainage which produce the so-called "yellow boy" may in fact be jarosite. As jarosite weathers or is exposed to dilute waters with higher pH, it will gradually decompose to ferrihydrite or goethite. Mine tailings may often be in equilibrium with both jarosite and ferrihydrite thereby providing a buffer system (Miller, 1979). Miller (1979) has estimated the pH of this buffer as 3.19 (± 0.17). Studies on acid mine drainage environments suggest that the soluble hydrated sulfates form during periods of dry weather near the vicinity of oxidizing pyrite in unsaturated soil horizons. Ferrihydrite, goethite, and jarosite are spatially distributed further away from the pyrite and commonly form by precipitation from aqueous systems such as the saturated zone or in receiving streams as in Plate 2A.

---

[3] Nuhfer, E. B. 1967. Efflorescent minerals associated with coal. M.S. Thesis. West Virginia University, Morgantown. 74 p.

[4] I have adopted the mineral name "ferrihydrite" here in place of ferric hydroxide, whether amorphous or otherwise, following the recommendation of Schwertman and Taylor (1977) and Schwertman (1979).

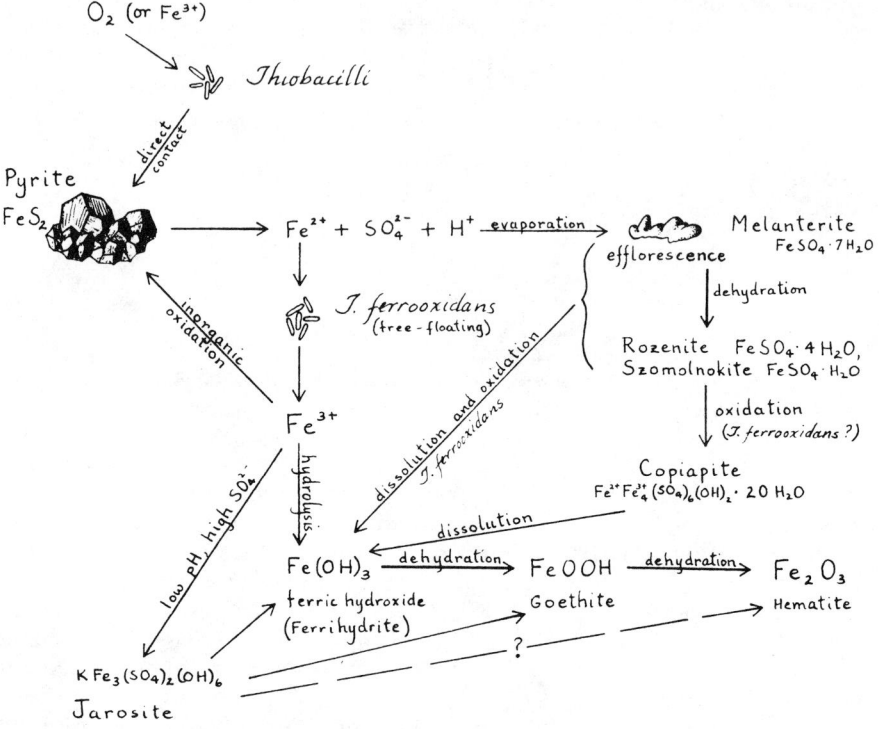

Fig. 3. The overall sequence of mineral reactions for pyrite oxidation showing the relationships between oxidizing agents, catalysts, and mineral products.

SUMMARY

The oxidation of pyrite involves many electron transfer reactions with the consequent reduction of oxygen in an aqueous environment and ultimately produces ferric iron, sulfate, and high acidity. The key reactions are summarized in Fig. 2 where the inorganic rate-limiting steps are pH dependent for values greater than 4.0. Below a pH of 3.0 the oxidation rates are independent of pH. The transition zone of pH = 3 to 4 includes both steps. The first step is the initiation phase in which elemental sulfur and ferrous iron are produced. At the higher pH values ferrous iron is oxidized to ferric hydroxide. The acid-generating phase in which sulfuric acid is produced from elemental sulfur follows next. Actual accumulation of sulfur may depend upon the concentration and availability of oxidizing agents such as sulfur-oxidizing bacteria. Once sufficient acid has been generated (pH < 3.0), ferric iron rapidly continues the oxidation of pyrite (catalytic phase). The most important catalyst is the iron-oxidizing bacterium, *T. ferrooxidans*, which greatly speeds up the oxidation of ferrous to ferric thereby regenerating the supply of ferric irons in solution. The scheme presented here is similar to the reaction sequence suggested by

Kleinmann (1979). The oxidation products include several possible iron sulfate, iron oxide, and iron hydroxide minerals whose genesis depends upon water content, degree of oxidation, and time. The grand sequence of reactants, products, and catalysts for pyrite oxidation is shown in the schematic illustration of Fig. 3. This picture attempts to tie together the roles played by (1) the oxidizing agents ferric iron and oxygen, (2) the catalyzing agent *T. ferrooxidans* which utilizes oxygen for respiration, and (3) the mineral products.

ACKNOWLEDGMENTS

I am grateful to Bob Kleinman and Dave Crerar for their helpful discussions on the chemistry of acid mine waters; to Don Thorstenson and Martin Goldhaber for their reviews of the manuscript; and to Katherine Baker, Carl Moses, and Aaron Mills for their stimulating ideas and suggestions. However, the interpretations presented are solely my responsibility. This manuscript was completed with the support of the U.S. Geological Survey, Reston, the Virginia State Water Control Board and NSF Grant EAR-7911144. I particularly thank Nico van Breemen for bringing Arkesteyn's work to my attention and to both Delvin Fanning and Nico for inviting me to rethink this whole problem through again.

LITERATURE CITED

1. Arkesteyn, G. J. M. W. 1979. Pyrite oxidation by *Thiobacillus ferrooxidans* with special reference to the sulfur moiety of the mineral. Antonie van Leeuwenhoek J. Microbiol. Serol. 45:423–435.
2. ———. 1980. Microbial processes in the formation of acid sulfate soils. Plant and Soil (In press).
3. Bailey, L. K., and E. Peters. 1976. Decomposition of pyrite in acids by pressure leaching and anodization: the case for an electrochemical mechanism. Can. Metall. Q. 15:333–334.
4. Baker, R. A. 1972. Evaluation of pyritic oxidation by Mossbauer spectrometry. Water Res. 6:9–17.
5. Basolo, F., and R. G. Pearson. 1967. Mechanisms of inorganic reactions: A study of metal complexes in solution. 2nd ed. John Wiley and Sons, Inc., N.Y. 701 p.
6. Beck, J. V., and G. D. Brown. 1968. Direct sulfide oxidation in the solubilization of sulfide ores by *Thiobacillus ferrooxidans*. J. Bacteriol. 96:1433–1434.
7. Bergholm, A. 1955. Oxidation of pyrite. Jernkontorets Ann. 139:531–549.
8. Biegler, T., D. A. J. Rand, and R. Woods. 1977. Oxygen reduction on sulphide minerals. p. 291–302. In J. O'M. Bockris, D. A. J. Rand, and B. J. Welch (ed.) Trends in electrochemistry. Plenum Press, N.Y.
9. ———, and D. A. Swift. 1979. Anodic behavior of pyrite in acid solutions. Electrochim. Acta 24:415–420.
10. Brock, T. D., S. Cook, S. Petersen, and J. L. Mosser. 1976. Biogeochemistry and bacteriology of ferrous iron oxidation in geothermal habitats. Geochim. Cosmochim. Acta 40:493–500.
11. ———, and J. Gustafson. 1976. Ferric iron reduction by sulfur- and iron-oxidizing bacteria. App. Environ. Microbial 32:567–571.

12. Brown, J. B. 1971. Jarosite-goethite stabilities at 25°C, 1 atm, Mineral. Deposita, 6: 245-252.
13. Bryner, L. C., and A. K. Jameson. 1958. Microorganisms in leaching sulfide minerals. Appl. Microbiol. 6:281-287.
14. Burkin, A. R. 1966. The chemistry of hydrometallurgical processes. Van Nostrand, Inc., N.J. 157 p.
15. Cameron, E. M. 1979. Effect of graphite on the enhancement of surficial geochemical anomalies originating from the oxidation of sulphides. J. Geochem. Expl. 12:35-43.
16. Caruccio, F. T. 1975. Estimating the acid potential of coal mine refuse. p. 197-205. *In* M. J. Chadwick and G. T. Goodman (ed.) The ecology of resource degradation and renewal. Blackwell Sci. Pub., London.
17. ―――, G. Geidell, and J. M. Sewell. 1976. The character of drainage as a function of the occurrence of framboidal pyrite and ground water quality in eastern Kentucky. p. 1-16. *In* 6th Symp. Coal Mine Drainage Res., Louisville, Ky.
18. Clark, C. S. 1966. Oxidation of coal mine pyrite. J. Sanit. Eng. Div. Am. Soc. Civ. Eng., 92 Proc. Paper 4802. p. 127-145.
19. Duncan, D. W., and C. C. Walden. 1972. Microbiological leaching in the presence of ferric iron. Dev. Ind. Microbiol. 13:66-75.
20. Dugan, P. R., C. B. MacMillan, and R. M. Pfister. 1970a. Aerobic heterotrophic bacteria indigenous to pH 2.8 acid mine water: Microscopic examination of acid streamers. J. Bacteriol. 101:972-981.
21. ―――, ―――, and ―――. 1970b. Aerobic heterotrophic bacteria indigenous to pH 2.8 acid mine water: Predominant slime-producing bacteria in acid streamers. J. Bacteriol. 101:982-988.
22. Eugster, H. P., and B. F. Jones. 1979. Behavior of major solutes during closed-basin brine evolution. Am. J. Sci. 279:609-631.
23. Foreman, J. W. 1972. Evaluation of mine sealing in Butler County, Pennsylvania. 4th Symp. Coal Mine Drainage Res., Louisville, Ky. p. 83-95.
24. Garrels, R. M., and M. E. Thompson. 1960. Oxidation of pyrite in ferric sulfate solution. Am. J. Sci. 258:57-67.
25. Ghosh, M. M. 1974. Oxygenation of ferrous iron (II) in highly buffered waters. p. 193-217. *In* A. J. Rubin (ed.) Aqueous-environmental chemistry of metals. Ann Arbor Science Pub. Inc., Mich.
26. Geidel, G., and F. T. Caruccio. 1977. Time as a factor in acid mine drainage pollution. 7th Symp. Coal Mine Drainage Res., Louisville, Ky. p. 41-50.
27. Guay, R., and M. Silver. 1975. *Thiobacillus acidophilus* sp. nov.; isolation and some physiological characteristics. Can. J. Microbiol. 21:281-288.
28. Hollyday, E. F., and S. W. MacKenzie. 1973. Hydrogeology of the formation and neutralization of acid waters draining from underground coal mines of western Maryland. Dep. Nat. Res., Maryland Geol. Survey Rep. Invest. 20. 50 p.
29. Huffman, R. E., and N. Davidson. 1956. Kinetics of the ferrous iron-oxygen reaction in sulfuric acid solution. J. Am. Chem. Soc. 78:4836.
30. Kelly, D. P., P. R. Norris, and C. L. Brierley. 1979. Microbiological methods for the extraction and recovery of metals. p. 263-308. *In* Bull, Ellwood and Ratledge (ed.) Microbial technology. Soc. Gen. Microbiol.
31. Kinkel, A. R., Jr., W. E. Hall, and J. P. Albers. 1956. Geology and base-metal deposits of West Shasta copper-zinc district Shasta County, California. U.S. Geol. Survey Prof. Paper 285. 156 p.
32. Kleinmann, R. L. P. 1979. The biogeochemistry of acid mine drainage and a method to control acid formation. Ph.D. Thesis. Princeton Univ., Princeton. 104 p. (Diss. Abstr. 7919776).
33. ―――, and D. A. Crerar. 1979. *Thiobacillus ferrooxidans* and the formation of acidity in simulated coal mine environments. Geomicrobial. J. 1:373-388.
34. Kushner, D. J. (ed.). 1978. Microbial life in extreme environments. Academic Press, N.Y. 465 p.

35. Kuznetsov, S. I., M. V. Ivanov, and N. N. Lyalikova. 1963. C. H. Oppenheimer (ed.) Introduction to geological microbiology. McGraw-Hill Book Co., N.Y. 252 p.
36. Lacey, D. T., and F. Lawson. 1970. Kinetics of the liquid-phase oxidation of acid ferrous sulfate by the bacterium *Thiobacillus ferrooxidans*. Biotech. Bioeng. 12:29–50.
37. Lazaroff, N. 1963. Sulfate requirement for iron oxidation by *Thiobacillus ferrooxidans*. J. Bacteriol. 85:78–83.
38. Lundgren, D. G., J. R. Vestal, and F. R. Tabita. 1972. The microbiology of mine drainage pollution. In R. Mitchell (ed.) Water pollution microbiology. Wiley-Interscience, N.Y. 147 p.
39. Mackintosh, M. E. 1978. Nitrogen fixation by *Thiobacillus ferrooxidans*. J. Gen. Microbiol. 105:215–218.
40. McKay, D. R., and J. Halpern. 1958. A kinetic study of the oxidation of pyrite in aqueous suspension. Trans. Metall. Soc. AIME 212:301–309.
41. Miller, S. D. 1979. Chemistry of a pyritic strip-mine spoil. Ph.D. Dissertation. Yale University, New Haven. 189 p. (Diss. Abstr. 7926660).
42. Nielson, A. M., and J. V. Beck. 1972. Chalcocite oxidation and coupled carbon dioxide fixation by *Thiobacillus ferrooxidans*. Science 175:1124–1126.
43. Nor, Y. M., and M. A. Tabatabai. 1977. Oxidation of elemental sulfur in soils. Soil Sci. Soc. Am. J. 41:736–741.
44. Nordstrom, D. K. 1976. Kinetic and equilibrium aspects of ferrous iron oxidation in acid mine waters. Abstract. Geol. Soc. Am. Ann. Mtg., Denver, Colo.
45. ———. 1977. Hydrogeochemical and microbiological factors affecting the heavy metal chemistry of an acid mine drainage system. Ph.D. Dissertation. Stanford Univversity, Stanford. 210 p. (Diss. Abstr. 7718232).
46. ———, and T. V. Dagenhart. 1978. Hydrate iron sulfate minerals associated with pyrite oxidation: Field relations and thermodynamic properties. Abstract. Geol. Soc. Am. Ann. Mtg., Toronto, Canada.
47. ———, E. A. Jenne, and J. W. Ball. 1979. Redox equilibria of iron in acid mine waters. In E. A. Jenne (ed.) Chemical modeling in aqueous systems: Speciation, sorption, solubility and kinetics. Am. Chem. Soc. Symp. Series 93:51–79.
48. Parizek, R. R., and E. G. Tarr. 1972. Mine drainage pollution prevention and abatement using hydrogeological and geochemical systems. p. 56–82. In 4th Symp. Coal Mine Drainage Res., Pittsburgh, Pa.
49. Peters, E. 1977. The electrochemistry of sulphide minerals. p. 267–290. In J. O'M. Bockris, D. A. J. Rand, and B. J. Welch (ed.) Trends in electrochemistry. Plenum Press, N.Y.
50. Roy, A. B., and P. A. Trudinger. 1970. The biochemistry of inorganic compounds of sulphur. Cambridge University Press. 400 p.
51. Sato, M. 1960. Oxidation of sulfide ore bodies, II. Oxidation mechanisms of sulfide minerals at 25°C. Econ. Geol. 55:1202–1231.
52. Schwertman, U. 1979. Is there amorphous iron oxide in soils? Abstract. Ann. Mtg. Soil Sci. Soc. Am., Ft. Collins, Colo.
53. ———, and R. M. Taylor. 1977. Iron oxides. p. 145–180. In J. B. Dixon and S. B. Weed (ed.) Minerals in soil environments. Soil Sci. Soc. Am., Madison, Wis.
54. Shumate, K. S., E. E. Smith, P. R. Dugan, R. A. Brant, and C. I. Randles. 1971. Acid mine drainage formation and abatement. E.P.A. Rep. 14010 FPR 04/71.
55. Silver, M. 1978. Metabolic mechanisms of iron-oxidizing *Thiobacilli*. p. 3–17. In Matal-Murr, A. E. Torma, and J. A. Brierley (ed.) Metallurgical applications of bacterial leaching and related microbiological phenomena. Academic Press, N.Y.
56. Silver, M., and A. E. Torma. 1974. Oxidation of metal sulfides by *Thiobacillus ferroxidans* grown on different substrates. Can. J. Microbiol. 20:141–147.
57. Silverman, M. P. 1967. Mechanism of bacterial pyrite oxidation. J. Bacteriol. 94:1046–1051.

58. Silverman, M. P., and D. G. Lundgren. 1959. Studies on the chemoautotrophic iron bacterium *Ferrobacillus ferrooxidans*. J. Bacteriol. 77:642–647.
59. Singer, P. C., and W. Stumm. 1968. Kinetics of the oxidation of ferrous iron. Second Symp. Coal Mine Drainage Res., Pittsburgh, Pa. 12–34.
60. ———, and ———. 1969. Oxygenation of ferrous iron. FWQA Rep. 14010-06/69.
61. ———, and ———. 1970. Acid mine drainage: the rate determining step. Science 167:1121–1123.
62. Smith, E. E., K. Svanks, and K. S. Shumate. 1968. Sulfide to sulfate reaction studies. 2nd Symp. Coal Mine Drainage Res., Pittsburgh, Pa. p. 1–11.
63. Sokolova, G. A., and G. I. Karavaiko. 1968. Physiology and geochemical activity of *Thiobacilli* (Russian trans.), Israel Prog. for Scientific Transl. 283 p.
64. Steger, H. F., and L. E. Desjardins. 1978. Oxidation of sulfide minerals, 4. Pyrite, chalcopyrite and pyrrhotite. Chem. Geol. 23:225–237.
65. Stokes, H. N. 1901. On pyrite and marcasite. U.S. Geol. Survey Bull. 186.
66. Stumm, W., and G. F. Lee. 1961. Oxygenation of ferrous iron. Ind. Eng. Chem. 53:143.
67. Sullivan, J. D. 1930. Chemistry of leaching chalcocite. U.S. Bur. Mines. Tech. Paper 473. 24 p.
68. Sveshnikov, G. B., and S. L. Dobychin. 1956. Electrochemical solution of sulfides and dispersion aureoles of heavy metals. Geokhimiya 1956. No. 4. p. 413–419.
69. ———, and Yu. S. Ryss. 1964. Electrochemical processes in sulfide deposits and their geochemical significance. Geochem. Intl. p. 198–204.
70. Thornber, M. R. 1975. Supergene alteration of sulphides. I. A chemical model based on massive nickel sulphide deposits at Kambalda, Western Australia. Chem. Geol. 15:1–14.
71. Tuovinen, O. H., and D. P. Kelly. 1972. Biology of *Thiobacillus ferrooxidans* in relation to the microbiological leaching of sulphide ores. Z. Allg. Mikrobiol. 12:311–346.
72. Wadsworth, M. E. 1973. Kinetics of heterogeneous systems. Ann. Rev. Phys. Chem. 23:355–384.
73. Walsh, F., and R. Mitchell. 1972a. An acid-tolerant iron-oxidizing *Metallogenium*. J. Gen. Microbiol. 72:369–376.
74. ———, and ———. 1972b. A pH-dependent succession of iron bacteria. Environ. Sci. Technol. 6:809–812.
75. Zajic, J. E. 1969. Microbial biogeochemistry. Academic Press, N.Y. 345 p.

Plate 1. A. Slime-excreting bacteria growing along the stream bottom of a creek containing acid mine drainage in Shasta Co., California. These "acid slime streamers" contain very dense populations of bacilli enmeshed in a fibrillar polymer network similar to those found by Dugan, MacMillan and Pfister (1970a). Distance lengthwise across area shown by photo is about 0.8 m (See Chapter 3). Plate 1. B. Specimen of efflorescent growth containing copiapite (yellow) overgrowths on rozenite (white) which has dehydrated from cuprian melanterite (blue). This specimen measures about 1 cm across. (See Chapter 3).

Plate 2. A. Jarosite (yellow) precipitation from an acid mine water (pH 2) in Shasta Co., California. Grey color on stream bed is a coating of bacterial slime. Scale is provided by a ring of keys in center foreground. (See Chapter 3). Plate 2. B. Face of lignite surface mine showing oxidized (about upper 1/5 of exposed soil-geologic column) and reduced zones of the pre-mining column. (See Chapter 10).

Plate 3. A. Soil profile of a Hapludult in Maryland with jarosite (pale yellow) and iron "oxide" mottles in its C horizons, which commence at a depth of about 60 cm. Depth increments on tape are in decimeters. The soil was on a slope of about 30% where the profile occurred. The area is presently forested, but it apparently was once a severely eroded pasture. (See Chapter 7). Plate 3. B. Closer view of jarosite and iron "oxide" mottles at depth of from 70 to 120 cm in C horizons of profile shown in Plate 3A. Matrix materials are glauconitic "greensands". (See Chapter 7). Plate 3. C. Soil profile (Profile 1 of Wagner et al.) with a sulfuric horizon (0–30 cm) at its surface, formed on a scalped (by man) land surface in a highway cloverleaf at Central Avenue and the Washington Beltway, Prince Georges County, MD. Depth increments on tape are in decimeters. Jarosite and iron "oxide" mottles are evident in the sulfuric horizon. Dark, sulfide bearing glauconitic material below 30 cm contains (white) calcareous fossil shell fragments. (See Chapter 7). Plate 3. D. Close up of barite as nodular, white crystalline masses in soil matrix at a depth of about 1 m in an Albaqualf near College Station, TX. Finger gives scale. (See Chapter 8).

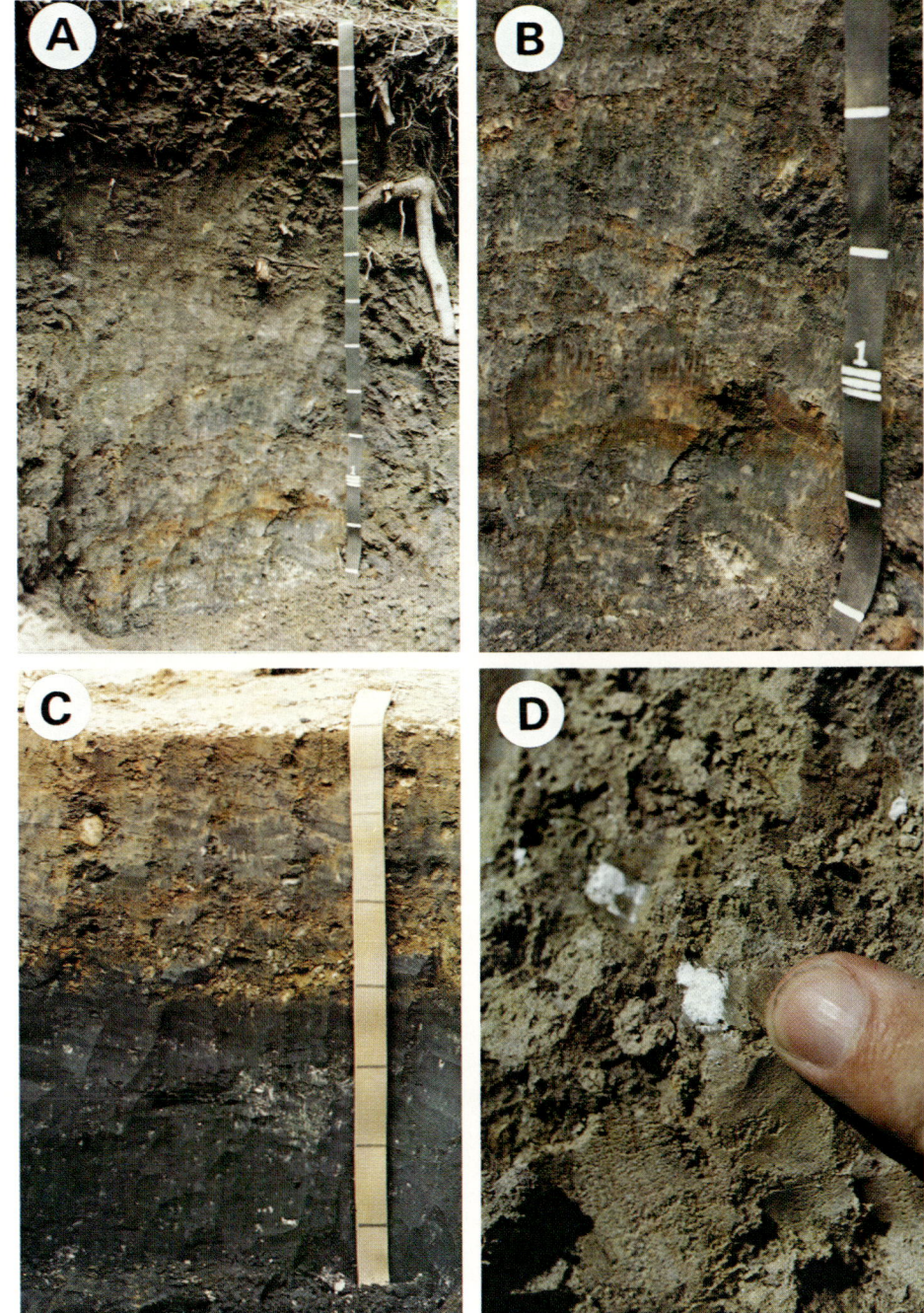

Plate 4. A. Jarosite (yellow) and "free iron oxides" (yellowish brown) on plate faces of hand specimen of oxidized zone of Eocene shale collected from scalped land surface (cut) at College Station, TX. (See Chapters 8 and 10). Plate 4. B. Hand specimen gypsum collected from surface of spoil at text pit for San Miquel lignite mine in Yequa formation in Atascosa County, TX. (See Chapters 8 and 10). Plate 4. C. Copiapite (lemon yellow) efflorescence on hand specimen of underclay of coal seam. Specimen collected beneath overhanging rock in roadcut on Interstate 64 near Cannonsburg, KY. (See Chapters 3, 7, and 10). Plate 4. D. Rozenite (white) occurring on coal hand specimen. Specimen collected a few cm above specimen shown in C. (See Chapter 10). Plate 4. E. Hand specimen of siderite (pale yellow zone shown in broken face) weathered to a crust of goethite on surfaces. Specimen collected from a larger mass exposed in a scalped land surface (cut) in Eocene shale at College Station, TX. (See Chapter 10). Plate 4. F. Rozenite (white efflorescence) at surface of lignite mine spoil in Texas. (See Chapter 10).

Chapter 4

Microbiological Transformations of Iron and Sulfur and Their Applications to Acid Sulfate Soils and Tidal Marshes[1]

K. C. IVARSON, G. J. ROSS, AND N. M. MILES[2]

ABSTRACT

Laboratory experiments show that the acidophilic iron-oxidizing bacterium *Thiobacillus ferrooxidans* is invariably isolated from acidic environments (pH 1.9 to 3.4) containing pyrite and basic ferric sulfates. When solutions of $FeSO_4$ (pH 2.9) containing either K^+, NH_4^+, or Na^+ are inoculated with the bacterium, Fe^{2+} oxidation, and formation of basic ferric sulfates begins within a few days. Their rates of formation are in accord with analyses of acid sulfate soils. Thus it is likely that the iron-oxidizing bacterium takes part in the formation of basic ferric sulfates in situ and plays a major role in the genesis of acid sulfate soils. In tidal marsh areas where some acid sulfate soils are subjected to prolonged submergence, *Desulfovibrio desulfuricans* (a sulfate-reducing bacterium) aids in the pyritization of the basic ferric sulfates. Hence in such areas there appears to be a generic relationship between pyrite and basic ferric sulfates and the above two microbes help to maintain this relationship by cycling sulfur and iron between the two minerals.

INTRODUCTION

In the mid-1960's the senior author while working with metallurgical engineers (17), employed the iron-oxidizing bacterium *Thiobacillus ferrooxidans* to promote leaching of uranium from uranium-bearing ores containing pyrite. Besides using ferrous iron as an energy source (33), the organism can also use elemental sulfur, reduced inorganic sulfur compounds, and metal sulfides (29). To culture the bacterium in the laboratory, air was blown through inoculated Fe^{2+} medium (9K) of Silverman and Lundgren (30). As observed by other microbiologists, a voluminous reddish-yellow precipitate accompanied the microbiological oxidation of

---

[1] Contribution No. 1117 Chemistry and Biology Research Institute, Agriculture Canada, Ottawa, K1A 0C6.
[2] Soil microbiologist and soil mineralogists, respectively.

Copyright © 1982 Soil Science Society of America, 677 S. Segoe Rd., Madison, WI 53711.
Acid Sulfate Weathering.

Fe^{2+}. This precipitate has been referred to as hydrated oxides of iron (10), ferric salts (8), and oxidized iron (30). A similar deposit as seen in the bacterial culture was observed to be very abundant on the stope floors of the uranium mine. In the uranium mine and in acid bituminous coal mines these deposits were locally called yellow boy (23). On the basis of suggestions by Leathen et al. (23), Ehrlich (14), and Duncan and Bruynesteyn (12) that the yellow deposits associated with acid-pyritic environments were the result of microbiological activity and were perhaps composed of basic ferric sulfates, microbiological studies (18) were undertaken to investigate the theory and its possible significance in the formation of acid sulfate soils. In later studies (21) we investigated the rates at which different basic ferric sulfates would crystallize during microbial oxidation of Fe^{2+} and (19, 20) whether or not, in areas where acid sulfate soils become flooded (13), the basic ferric sulfates could be microbiologically reduced back to a sulfide.

MATERIALS AND METHODS

Microbiological Formation of Basic Ferric Sulfates

In an attempt to determine if iron-oxidizing organisms were associated with acid-pyritic deposits containing yellow sediments of basic ferric sulfates the following four natural deposits were examined.

Deposit 1. A yellowish-brown sediment (pH 2.2) was collected from the stope floor of a uranium mine at Elliot Lake, Ontario. The main sulfide minerals in the ore were pyrite and pyrrhotite.

Deposit 2. Yellow mottles (pH 2.7 to 2.9) in the form of tubes (Fig. 1) were obtained from poorly drained soils (cat clays) of British Columbia. Similar yellow mottles from this area (9) contained 8% pyrite and 59% basic ferric sulfates.

Deposit 3. A known deposit (pH 1.9) of jarosite associated with pyrite veins in a road cut (Fig. 2), near Sharbot Lake, Ontario was sampled.

Deposit 4. Samples of a black pyritiferous and fissile shale, belonging to the Ordovician system were collected in Ottawa from a hand-dug pit (Fig. 3) beneath a heaving basement slab. Many of the shale surfaces were coated with a yellowish sediment (pH 3.4) of jarosite (25).

Isolation and Cultivation of Iron-oxidizing Organism

Enrichment cultures of *T. ferrooxidans* were obtained by inoculating 9K medium with moist intact samples of the deposits. Pure cultures were obtained by streaking enrichment cultures on solidified agar (2.5%) containing the salts of medium 9K, and the tan to brown colonies that developed on the agar were transferred to sterile solutions of 9K medium. Morphological studies were made using Gram stain, wet mounts, and phase contrast microscopy.

Fig. 1. Yellow tubes of jarosite (2 to 8 cm in length) removed from the profile of a cat clay.

Fig. 2. Profile in a roadcut, showing 3 vertical yellow deposits (arrows) of jarosite adjacent to pyrite veins imbedded in feldspathic metamorphic rock.

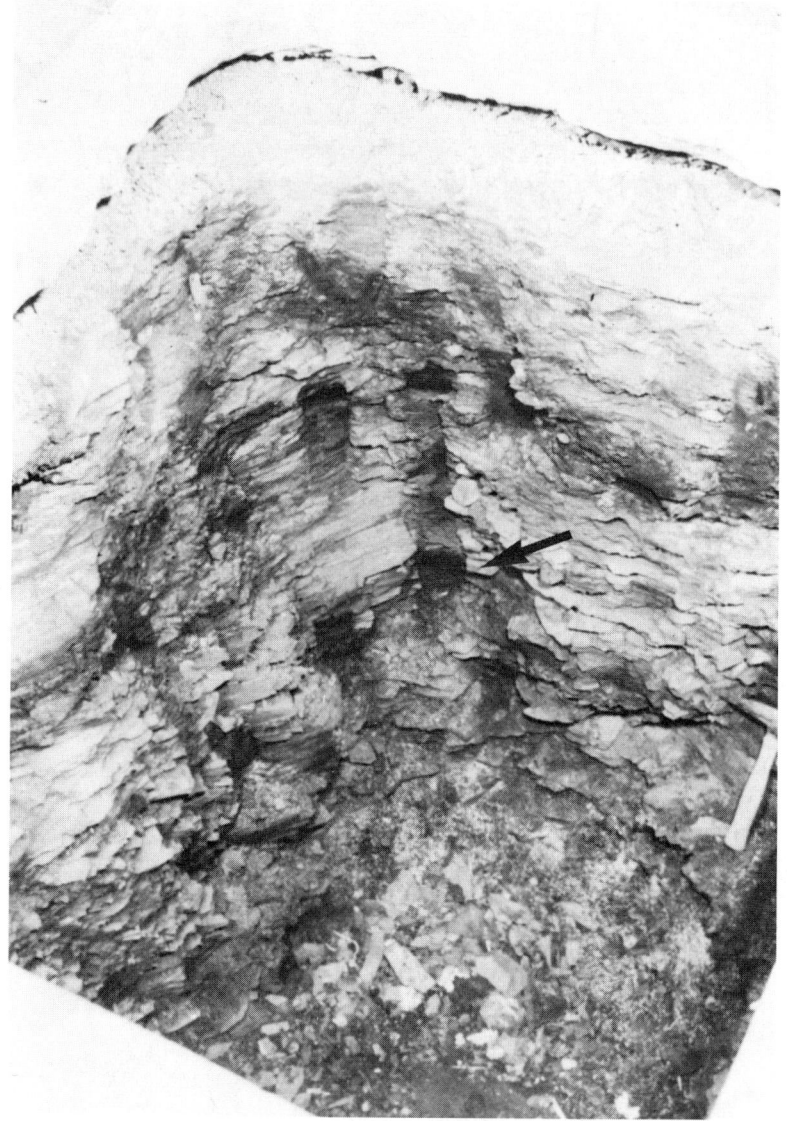

Fig. 3. A hand-dug pit beneath a heaving concrete slab. Arrow shows coatings of yellow jarosite covering a black shale deposit.

Manometric Studies

Convential manometric techniques (34) were used to study oxygen uptake by the organism when grown on a variety of substrates and different levels of ferrous iron.

Preparation and Analysis of Samples

Separation of the yellow deposits from most of their surrounding material was performed as previously described (18) and their X-ray diffraction patterns were made with a Philips X-ray diffraction unit using Fe-filtered Co radiation. Infrared spectra were recorded using a Beckman IR 12 spectrophotometer.

Rates of Formation of Basic Ferric Sulfates

Culture Systems. One and one-half liters of modified medium 9K (pH 2.95) were added to each of three 2-liter reagent bottles. The basal medium (21) contained $MgSO_4 \cdot 7H_2O$, 0.5%; $Ca(NO_3)_2 \cdot 4H_2O$, 0.3%; and $FeSO_4 \cdot 7H_2O$, 4.4%. For jarosite $[KFe_3(SO_4)_2(OH)_6]$ formation, 0.02% of K_2HPO_4 and 0.01% KCl were added to the basal medium. For natrojarosite $[NaFe_3(SO_4)_2(OH)_6]$, Na_2HPO_4 and NaCl were added and for ammoniojarosite $[NH_4Fe_3(SO_4)_2(OH)_6]$, $(NH_4)_2HPO_4$ and NH_4Cl were used. In the latter two systems the Na^+ and NH_4^+ were added on an equivalent K^+ basis. After inoculating with *T. ferrooxidans* the cultures were incubated for 42 weeks. During this period no tests on the bacterium's viability were made, for previous unpublished studies showed that under similar conditions the bacterium was still viable even after 5 years of incubation.

Analyses for Mineral Formation

At various times, during the incubation period, pH of solution was determined and samples of the liquid and precipitated basic ferric sulfates were removed, analyzed for K^+, NH_4^+, Na^+, and Fe^{3+} and the precipitate X-rayed. In presenting the X-ray diffractograms only the essential patterns, up to and including the period where maximum peak heights were obtained, are shown.

Microbiological Reduction of Basic Ferric Sulfates

Culture Systems. Sterile Starkey's medium (32), pH 7.8, and free of sulfate salts but supplemented with 1.2% NaCl and 0.15% $MgCl_2 \cdot 6H_2O$ was used to fill 1,000-ml serum bottles containing either 10 or 50 g of jarosite (the sulfate source) and 1 ml of a culture of a halo-tolerant strain of a sulfate-reducing bacterium *Desulfovibrio desulfuricans*. Uninoculated bottles served as controls. The incubation temperature was 30 C. At weekly intervals a portion of the bottom sediment in the bottles was removed and X-rayed, the pH of the cultures adjusted to 7.8 and more bacterial energy source (3.3 ml of a 60% sodium lactate sol.) added. After 33 weeks, 5 g of elemental sulfur was added to the bottle originally containing 10 g of jarosite and contents incubated another 4 weeks and sediments X-rayed. At this stage, sediments were removed and aged at 120 C in an evacuated sealed quartz tube. After 3 days the tube was opened and contents X-rayed.

RESULTS AND DISCUSSION

Microbiological Formation of Basic Ferric Sulfates

The infrared spectra of the yellow deposits collected from the shale, uranium mine, cat clay, road cut, and the reddish-yellow deposit from the laboratory bacterial growth medium showed that the five substances (Fig. 4) had patterns similar to those reported by Omori and Kerr (24) for sulfates containing hydroxyl groups. The X-ray diffractograms (Fig. 5) confirmed the data in Fig. 4. According to ASTM card no. 22-827, the bacterial precipitate and the road cut deposit were samples of pure

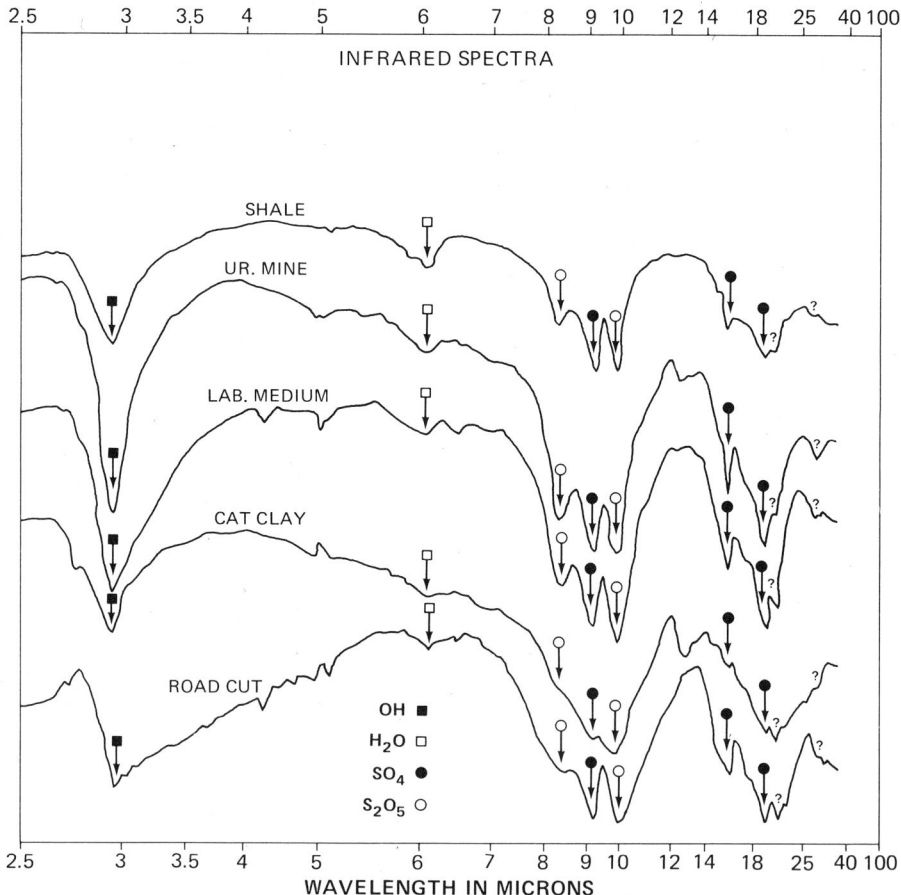

Fig. 4. Infrared spectra of yellow deposits studied.

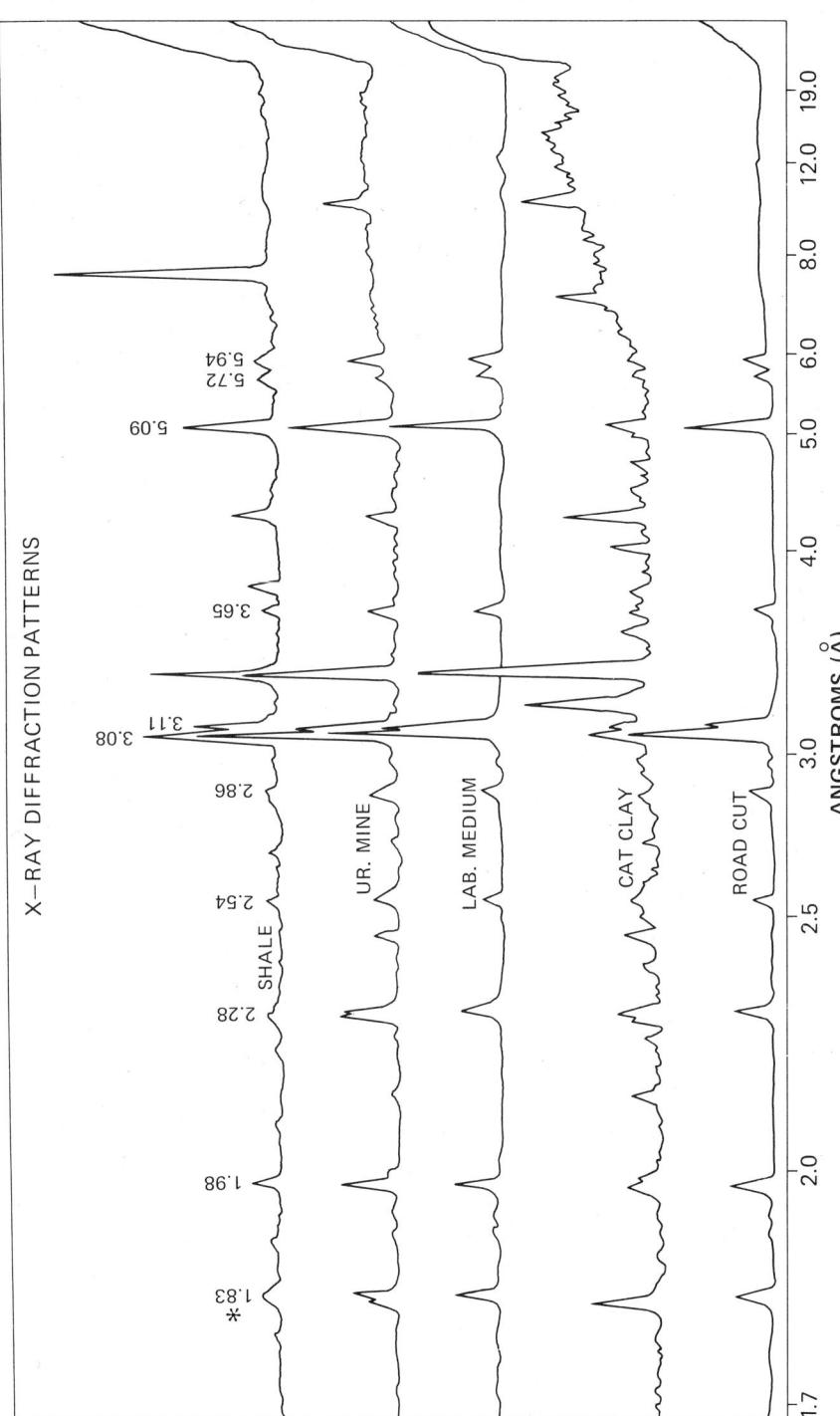

Fig. 5. X-ray diffraction patterns of yellow deposits studied. * Angstrom units for jarosite.

jarosite. Beside jarosite the remaining three yellow deposits had lines that were attributable to quartz, mica, feldspar, pyrite, and in one instance gypsum.

In 4 to 10 days the four yellowish-green solutions of medium 9K, inoculated with the yellow deposits, turned red, indicating that the Fe^{2+} was being oxidized. The pH of the solutions decreased to about 2.2. The color and the pH of the uninoculated solutions (controls) remained unchanged. Microscopic examination of wet mounts and stained preparations showed that the inoculated solutions contained many motile, rod-shaped (0.4 × 1.0 μm), gram-negative bacteria resembling *T. ferrooxidans* (11, 30). No bacteria were observed in uninoculated controls.

Manometric studies (Fig. 6) showed that at an initial pH of 3.9, a 0.5 ml suspension of *T. ferrooxidans* would oxidize 28, 14, 7, and 3.5 mg of Fe^{2+} in about 4, 2, 1, and 0.5 hours, respectively. In contrast, without the

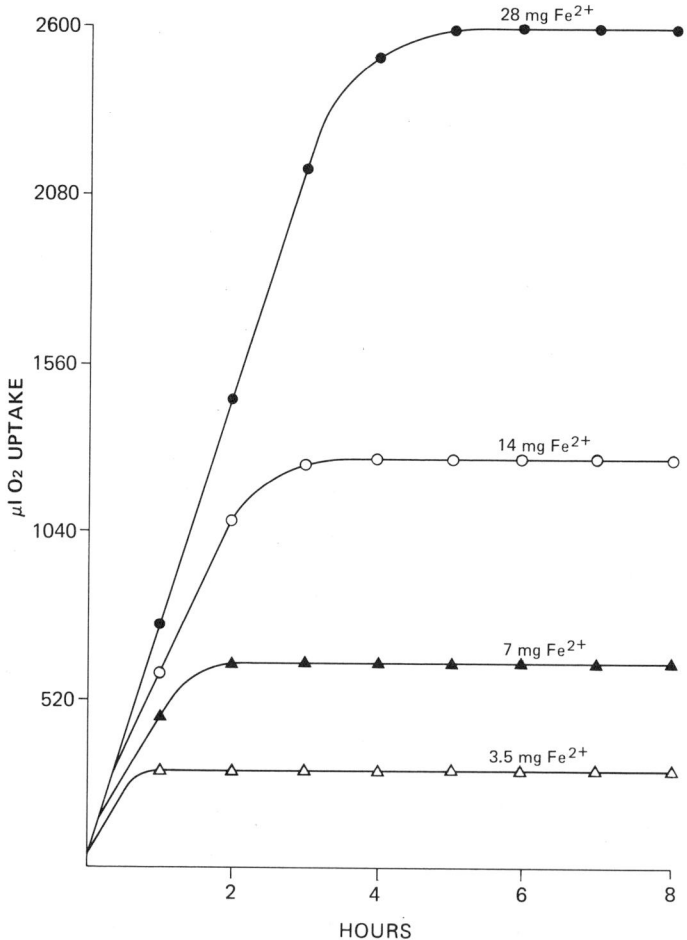

Fig. 6. Rate of iron oxidation (initial pH 3.9) in presence of *Thiobacillus ferrooxidans*.

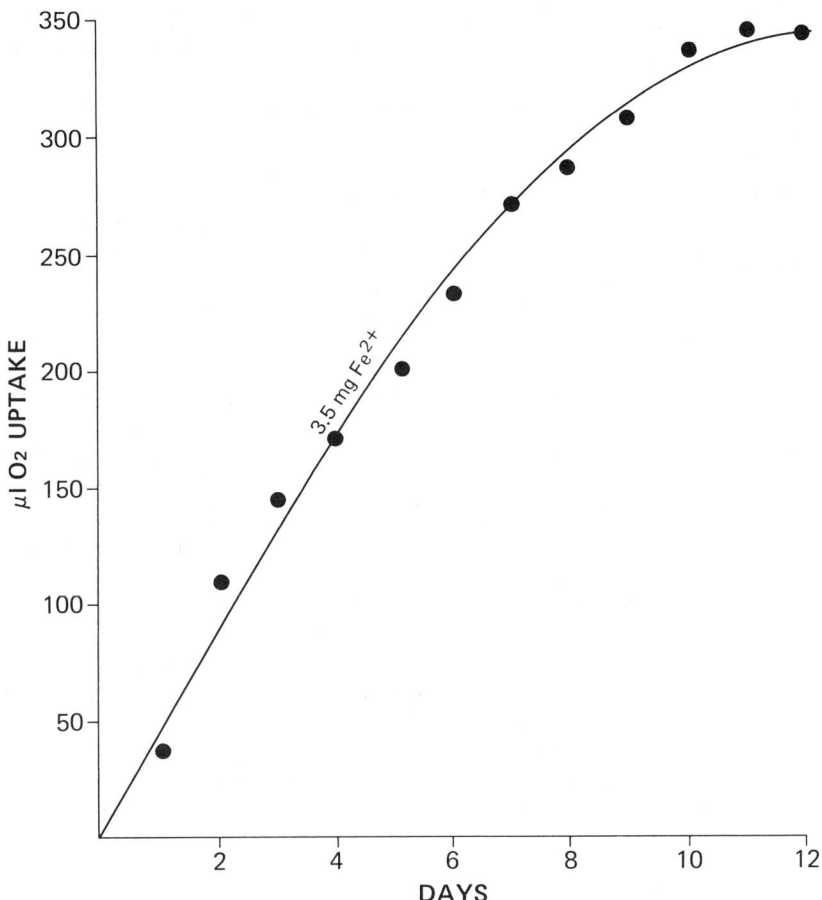

Fig. 7. Rate of iron oxidation (initial pH 4.0) in absence of *Thiobacillus ferrooxidans*.

bacterium and at pH 4.0 (Fig. 7) it required 12 days to oxidize 3.5 mg of Fe^{2+}, which is about 600 times as slow as in its presence. Singer and Stumm (31) have provided stronger support for the microbial acceleration of Fe^{2+} oxidation. While studying this process in a mine-water environment, they showed that abiotic oxidation was a function of pH and at pH values close to 3, microbial oxidation accelerated the reaction by a factor of 10^6. This shows that under the conditions encountered in most acid sulfate soils and in the presence of the bacterium the amount of Fe^{2+} oxidized in 1 day, would in its absence require a period of about 30,000 years.

From the above data one can conclude that (1) in the laboratory the bacterial oxidation of Fe^{2+}, at pH values approximating those of most acid sulfate soils, is accompanied by the formation of basic ferric sulfates; (2) the bacterium *T. ferrooxidans* is invariably isolated from natural acidic deposits of basic ferric sulfates associated with pyrite; (3) in such surroundings, abiotic oxidation of Fe^{2+} is too slow to be of any importance;

and (4) thus in nature the bacterium undoubtedly takes part in the formation of basic ferric sulfates in situ and plays a major role in the genesis of acid sulfate soils.

Other workers in this field, notably Bloomfield and Coulter (4), have arrived at similar conclusions. The processes involved may briefly be described accordingly. Following the chemical oxidation of pyrite to produce $FeSO_4$, further action by the bacterium yields $Fe_2(SO_4)_3$. In presence of suitable cations; $Fe_2(SO_4)_3$ is chemically oxidized to produce the yellow mottles and high acidity of these soils, by the reaction:

$$3Fe_2(SO_4)_3 + 1/2\ O_2 + 11H_2O + 2K^+ \rightarrow 2KFe_3(SO_4)_2(OH)_6 + 5H_2SO_4.$$

Rates of Basic Ferric Sulfate Formation

Within a few days microbial oxidation of Fe^{2+} began in the modified 9K medium containing either K^+, NH_4^+, or Na^+-salts. During the first 14 weeks of oxidation (Table 1) the pH in the three systems decreased at different rates. Their relative rates of acid production were $K^+ > NH_4^+ > Na^+$. After 14 weeks no further decrease in pH occurred. The X-ray diffractograms of each precipitate that formed in these three single alkali cation systems indicated a different basic ferric sulfate pattern. Although the patterns differed only slightly, the position of the (003) reflection at about 5.6 to 5.8 Å and the (021) and (113), doublet reflections in the 3.1 Å area, permitted identification. The d(Å) spacings for jarosite and natrojarosite were essentially the same as given in ASTM card nos. 22-827 and 11-302, respectively, while those for ammoniojarosite agreed very well with the refined diffractogram pattern of Ivarson et al. (20). The increase in the d(Å) spacings for the (113) and (003) reflections of the Na, K, and NH_4 jarosites reflects the increase in ion sizes of the respective alkali cations (5).

The comparative rates at which the three alkali cations affected crystallization of basic ferric sulfates paralleled that for acid production. Thus, the incubation times at which the diffractogram patterns of jarosite (Fig. 8), ammoniojarosite (Fig. 9) and natrojarosite (Fig. 10) reached their maximum peak heights were 3, 5, and 10 weeks, respectively. This parallelism between rates of basic ferric sulfate crystallization and acid production is in accord with the earlier mentioned equation, i.e., the greater the amount of basic ferric sulfates crystallized, the greater the amount of acid produced. The results of the calculated amounts of each

Table 1. Effect of cations in solution on pH change.[†]

Cation	Weeks of incubation			
	2	3	5	14
K^+	1.88	1.57	1.48	1.48
NH_4^+	2.07	1.88	1.75	1.65
Na^+	2.18	1.99	1.93	1.81

[†] Initial pH 2.95.

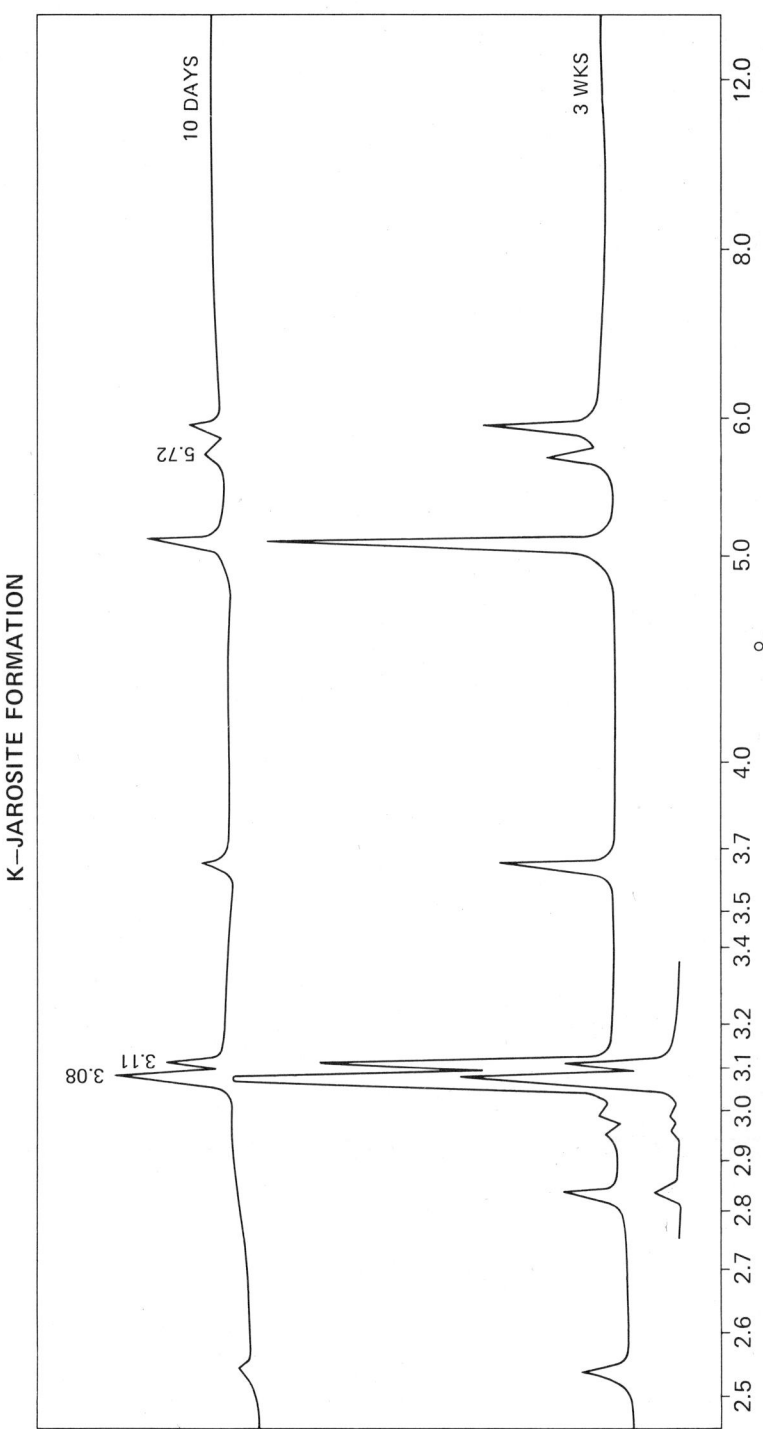

Fig. 8. X-ray diffractograms of the jarosite formed in an incubated solution containing FeSO$_4$, K$^+$, and an iron-oxidizing bacterium.

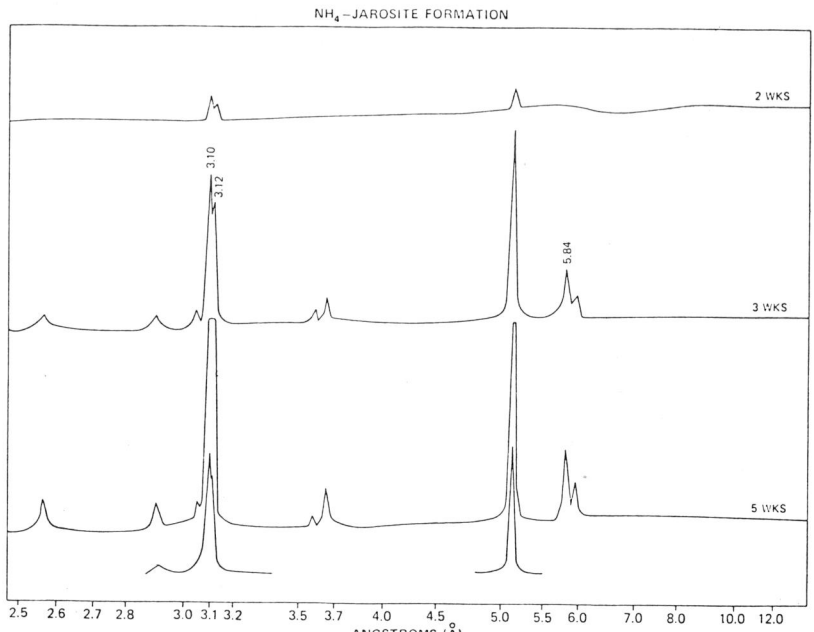

Fig. 9. X-ray diffractograms of the ammoniojarosite formed in an incubated solution containing $FeSO_4$, NH_4^+, and an iron-oxidizing bacterium.

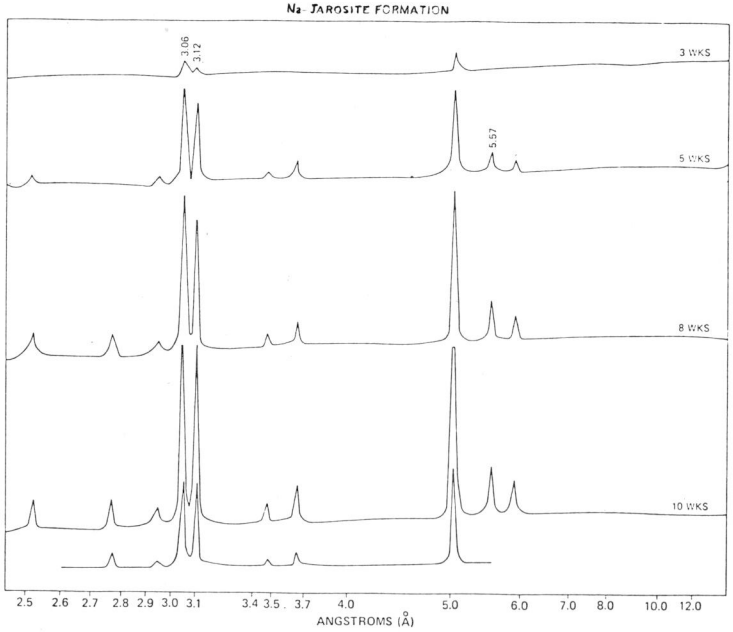

Fig. 10. X-ray diffractograms of the natrojarosite formed in an incubated solution containing $FeSO_4$, Na^+, and an iron-oxidizing bacterium.

basic ferric sulfate that crystallized during incubation (Fig. 11), in turn agreed very well with X-ray data in Fig. 8, 9, and 10. Thus after 3 to 4 weeks of incubation, virtually all of the K^+ had reacted to form jarosite. In contrast, about 55% of the NH_4^+ and only 1 to 3% of the Na^+ had reacted to form ammoniojarosite and natrojarosite, respectively. Thereafter, the amounts of NH_4^+ and Na^+ that continued to form their resultant jarosites was fairly slow. This is easily seen by extrapolating the NH_4^+ curve and the results then show that about 20% of the NH_4^+ and 40% of the Na^+ remain in solution, after 42 weeks of incubation.

The above results are in good agreement with synthetic and naturally occurring basic ferric sulfates. For example, in the formation of synthetic basic ferric sulfates (5), there was a strong preference for K^+ over Na^+ and H_3O^+ in the structure and a survey of acid sulfate soils (>8,000 km²) in Thailand (35) showed that K-jarosite was far more abundant than the Na or H_3O-form. Our results show also that there is a strong preference for NH_4^+ over Na^+ in the jarosite structure.

Microbiological Reduction of Basic Ferric Sulfates

After 3 weeks of incubation, the inoculated bottle containing 50 g of jarosite showed zones of blackening in the precipitated yellow material. The liquid above the jarosite was turbid and the odor of H_2S was very noticeable when the weekly pH determinations were made. The pH was generally approaching 6 and was adjusted to 7.8 with NaOH. The pH decrease was likely due to the microbiological production of acetic acid and CO_2 and the production of sulfate and bisulfite ions upon dissolution of

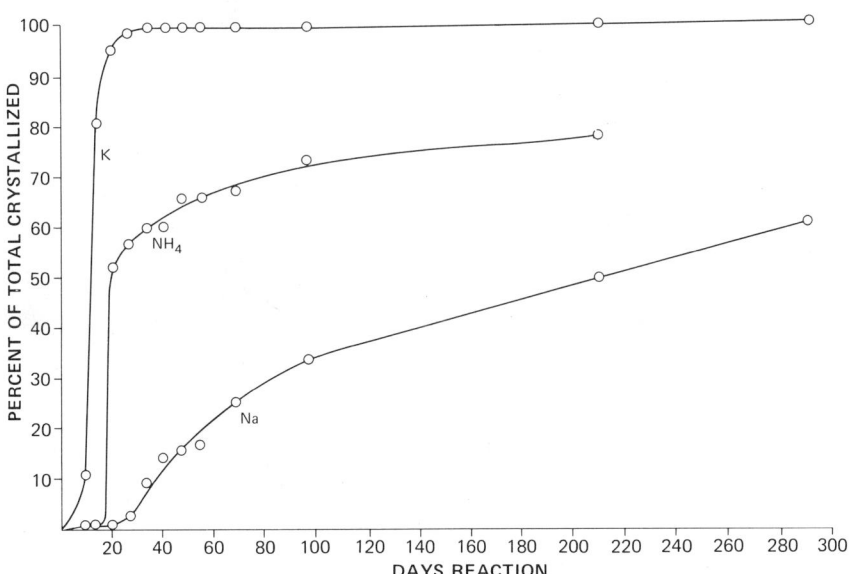

Fig. 11. Cumulative formation of K, NH_4, and Na jarosite in incubated systems, each containing a single alkali cation. Calculated from amount of cation remaining in solution.

jarosite (6). After 6 weeks all the material in the inoculated bottle was black while that of the control (Fig. 12) appeared unchanged.

Except for the absence of two strong peaks at 3.08 and 3.11 Å, the pattern for the unreduced jarosite (Fig. 13, trace 1) agreed closely with ASTM card no. 22-827 for jarosite. These two strong peaks were visible on the film but were not resolved on the tracing and appeared as a single strong peak at 3.09 Å.

The sample in which bacterial sulfate reduction occurred for 8 weeks (trace 2) still showed most of the peaks for the unreduced jarosite. However, the higher background, broader and less well-defined peaks indicated a decrease in its crystallinity. The pattern also showed peaks for vivianite ASTM card no. 3-700 or possibly its polymorph metavivianite (27). Phosphate ions were present in the medium as K_2HPO_4 (32).

After 20 weeks of reduction, no jarosite diffraction peaks (trace 3) were discernible and the peaks for vivianite were stronger. Eight d-values which matched those of mackinawite (ASTM card no. 15-37) were present. Their board lines were characteristic of recently formed mackinawite (1).

For the second experiment where 10 g of jarosite were used, the reduction time decreased considerably. In this case all the X-ray peaks for jarosite disappeared after 8 weeks of incubation. After 33 weeks, except

Fig. 12. Appearance of solutions containing sodium lactate and jarosite. Left, uninoculated; right, inoculated with *Desulfovibrio desulfuricans*.

for more prominent mackinawite and vivianite peaks indicating greater crystallinity, the pattern of the reduced material remained the same as at the 8 week period. However, at this stage when elemental sulfur was added and the system incubated an additional 4 weeks, the mackinawite

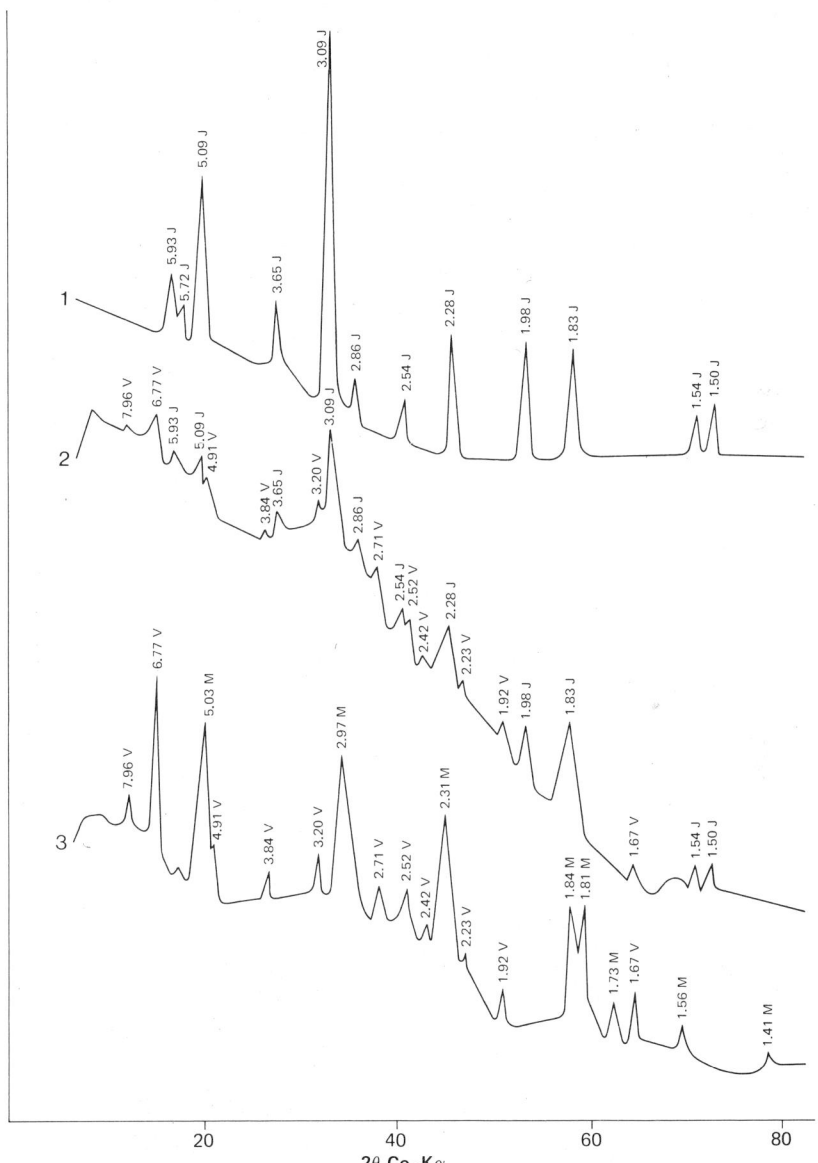

Fig. 13. Smoothed photodenistomer traces of X-ray film patterns: J = jarosite; M = mackinawite; V = vivianite.
Trace 1: fresh unreduced, bacterially formed jarosite.
Trace 2: jarosite bacterially reduced for 8 weeks.
Trace 3: jarosite bacterially reduced for 20 weeks.

and vivianite (Fig. 14, trace 1) was transformed into greigite (trace 2). Vivianite disappeared and six faint peaks of sulfur were present, indicating that all of the sulfur had not been consumed. However, when the material was aged for 3 days under increased pressure and temperature, greigite and sulfur disappeared, and appreciable amounts of well crystallized pyrite and traces of marcasite were formed (trace 3). The re-

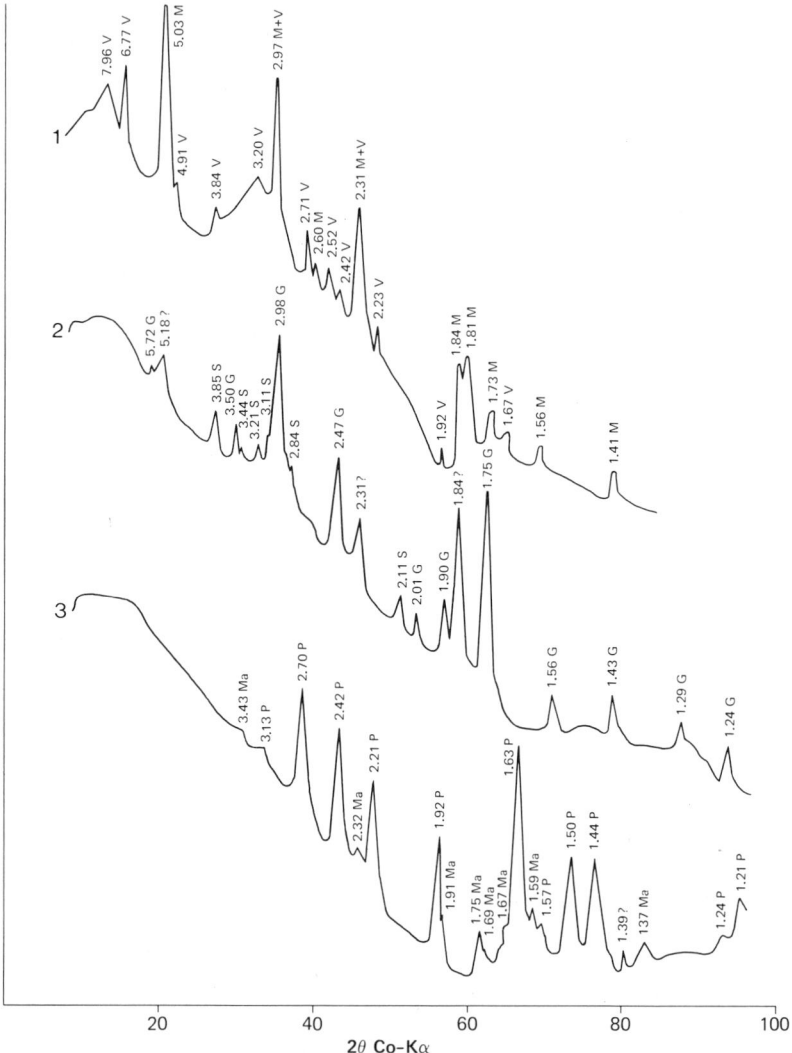

Fig. 14. Smoothed photodensitometer traces for X-ray film patterns. Trace 1, jarosite bacterially reduced for 33 weeks; trace 2, material in trace 1, incubated with elemental sulfur for 4 weeks; trace 3, material in trace 2, aged under increased temperature and pressure for 3 days. M, mackinawite; V, vivianite; G, greigite; S, sulfur; P, pyrite; Ma, marcasite; ?, unknown.

sults show that a basic ferric sulfate can be bacteriologically transformed by the stepwise sulfidation process: basic ferric sulfate → mackinawite → greigite → pyrite.

The first reaction involved is probably the transformation of jarosite to goethite. Brown (7) has provided evidence that jarosite outside its stability field (pH < 3) is slowly converted to goethite. Also Van Breemen and Harmensen (36) observed that after periodic flooding, jarosite in an acid sulfate soil hydrolyzed to an X-ray amorphous ferric oxide which transformed to goethite. The goethite in a neutral to basic reducing environment is then transformed into mackinawite (1) by the reaction:

$$2HFeO_2 + 3H_2S \rightarrow 2FeS + S° + 4H_2O.$$

Then the final reactions (2) may be written as follows:

$$3FeS \text{ mackinawite} + S° \text{ rhombic} \rightarrow Fe_3S_4 \text{ greigite}$$

$$Fe_3S_4 \text{ greigite} \rightarrow 2\ FeS \text{ mackinawite} + FeS_2 \text{ pyrite}$$

$$FeS \text{ mackinawite} + S° \text{ rhombic} \rightarrow FeS_2 \text{ pyrite}$$

$$Fe_3S_4 \text{ greigite} + 2S° \text{ rhombic} \rightarrow 3FeS_2 \text{ pyrite}$$

In the last experiment the sulfidation of mackinawite stopped after 33 weeks and elemental sulfur had to be added before the process would resume. This was perhaps due to the fact that mackinawite has a 1:1 atomic ratio of Fe:S, while jarosite has a 3:2 ratio. Therefore, not enough sulfur is present in the jarosite to satisfy all the iron. This is in agreement with the above equations and laboratory experiments which have shown that either elemental sulfur (3, 22) or an excess of H_2S (15, 28) is needed to complete the transformation of mackinawite into pyrite.

From the above results it might be asked when acid sulfate soils become inundated with sea water, is there enough H_2S or elemental sulfur present to complete the pyritization process? As demonstrated by Ivarson et al. (20) copious amounts of H_2S are produced when basic ferric sulfates, decaying algae, and sulfate-reducing bacteria are placed in sea water. Also it has been observed (16, 26) that in presence of decomposing organic matter, mats of sulfur, created by the disposition of sulfur within the cells of the bacterium *Beggiatoa*, are laid down on sea floor sediments.

Thus, on the basis of the data presented in this paper it would appear that in an environment of sea water flooding and decomposing organic matter, basic ferric sulfates can be microbiologically reduced back to pyrite. Therefore in some marshy coastal areas where acid sulfate soils are subjected to reduction-oxidation cycles (19), there exist a generic relationship between basic ferric sulfates and pyrite and the two microbes (*T. ferrooxidans* and *Desulfovibrio desulfuricans*) help to maintain this relationship by cycling sulfur and iron between the two minerals.

LITERATURE CITED

1. Berner, R. A. 1964. Iron sulfides formed from aqueous solution at low temperatures and atmospheric pressure. J. Geol. 72:293–306.
2. ———. 1967. Thermodynamic stability of sedimentary iron sulfides. Am. J. Sci. 265:773–785.
3. ———. 1970. Sedimentary pyrite formation. Am. J. Sci. 268:1–23.
4. Bloomfield, C., and J. K. Coulter. 1973. Genesis and management of acid sulfate soils. Adv. Agron. 25:256–326.
5. Brophy, G. P., and M. F. Sheridan. 1965. Sulfate studies. IV. The jarosite-natrojarosite-hydronium jarosite solid solution series. Am. Mineral. 50:1595–1607.
6. Brown, J. B. 1970. A chemical study of some synthetic potassium-hydronium jarosites. Can. Mineral. 10:696–703.
7. ———. 1971. Jarosite-geothite stabilities at 25°C, 1 atm. Mineral. Deposita 6:242–252.
8. Bryner, L. C., J. V. Beck, D. B. Davis, and D. G. Wilson. 1954. Microorganisms in leaching sulfide minerals. Ind. Eng. Chem. 46:2587–2592.
9. Clark, J. S., C. A. Gobin, and P. N. Sprout. 1961. Yellow mottles in some poorly drained soils of the lower Fraser Valley, British Columbia. Can. J. Soil Sci. 41:218–227.
10. Colmer, A. R., and M. E. Hinkle. 1947. The role of microorganisms in acid mine drainage. A preliminary report.
11. Colmer, A. R., K. L. Temple, and M. E. Hinkle. 1950. An iron-oxidizing bacterium from the acid drainage of some bituminous coal mines. J. Bacteriol. 59:317–328.
12. Duncan, D. W., and A. Bruynesteyn. 1971. Enhancing bacterial activity in a uranium mine. Can. Inst. Mining Metallurgy Trans. 64:116–120.
13. Edleman, C. H., and J. M. Van Staveren. 1958. Marsh soils in the United States and the Netherlands. J. Soil Water Conserv. 13:5–17.
14. Ehrlich, H. L. 1964. Microbial transformations of minerals. p. 43–60. *In* H. Heukelekian and N. C. Dondero (ed.) Principles and applications of aquatic microbiology. John Wiley and Sons, Inc., New York.
15. Hallberg, R. O. 1972. Iron and zinc sulfides formed in a continuous culture of sulfate-reducing bacteria. Neues Jahrb. Mineral. 11:481–500.
16. ———, L. E. Bagander, A. G. Engvall, and F. A. Schippel. 1972. Method for studying geochemistry of sediment-water interface. Ambio. 1:71–72.
17. Harrison, V. F., W. A. Gow, and K. C. Ivarson. 1966. Leaching of uranium from Elliot Lake ore in the presence of bacteria. Can. Min. J. 87:64–67.
18. Ivarson, K. C. 1973. Microbiological formation of basic ferric sulfates. Can. J. Soil Sci. 53:315–323.
19. ———, and R. O. Hallberg. 1976. Formation of mackinawite by the microbial reduction of jarosite and its application to tidal sediments. Geoderma 16:1–7.
20. ———, ———, and T. Wadsten. 1976. The pyritization of basic ferric sulfates in acid sulfate soils: A microbiological interpretation. Can. J. Soil Sci. 56:393–406.
21. ———, G. J. Ross, and N. M. Miles. 1979. The microbiological formation of basic ferric sulfates. 2. Crystallization in presence of potassium, ammonium, and sodium salts. Soil Sci. Soc. Am. Proc. 43:908–912.
22. Kaplan, I. R., K. O. Emery, and S. C. Rittenberg. 1963. The distribution and isotopic abundance of sulfur in recent marine sediments of Southern California. Geochim. Cosmochim. Acta. 27:297–331.
23. Leathen, W. W., S. A. Braley, and L. D. McIntyre. 1953. The role of bacteria in the formation of acid from certain sulfuritic constituents associated with bituminous coal. I. *Thiobacillus thiooxidans*. Appl. Microbiol. 1:61–64.
24. Omori, K., and P. F. Kerr. 1963. Infrared studies of saline sulfate minerals. Geol. Soc. Am. Bull. 74:709–734.
25. Penner, E., J. E. Gillot, and W. J. Eden. 1970. Investigation of heave in Billings shale by mineralogical and biogeochemical methods. Can. Geotech. J. 7:333–338.

26. Pfenning, N. 1975. The phototrophic bacteria and their role in the sulfur cycle. Plant and Soil 43:1-16.
27. Ritz, C., E. J. Essene, and D. R. Peacor. 1974. Metavivianite $Fe_3(PO_4)_2 \cdot 8H_2O$ a new mineral. Am. Mineral. 59:896-899.
28. Roberts, W. M. B., A. L. Walker, and A. S. Buchanan. 1969. The chemistry of pyrite formation in aqueous solution and its relation to the dispositional environment. Mineral. Deposita (Berl.) 4:18-29.
29. Silver, M. 1978. Metabolic mechanisms of iron-oxidizing *Thiobacilli*. p. 3-17. *In* L. E. Murr, A. E. Torma and J. A. Brierley (ed.) Metallurgical applications of bacterial leaching and related microbiological phenomena. Academic Press, New York.
30. Silverman, M. P., and D. G. Lundgren. 1959. Studies on the chemoautotrophic iron bacterium *Ferrobacillus ferrooxidans*. 1. An improved medium and a harvesting procedure for securing high yields. J. Bacteriol. 77:642-647.
31. Singer, P. C., and W. Stumm. 1970. Acid mine drainage: The rate-determining step. Science 167:1121-1123.
32. Starkey, R. L. 1938. A study of spore formation and other morphological characteristics of *Vibrio desulfuricans*. Arch. Mikrobiol. 9:268-304.
33. Temple, K. L., and A. R. Colmer. 1951. The autotrophic oxidation of iron by a new bacterium: *Thiobacillus ferrooxidans*. J. Bacteriol. 62:605-611.
34. Umbreit, W. W., R. H. Burris, and J. F. Stauffer. 1964. Manometric techniques. Burgess Publishing Co., Minneapolis.
35. Van Breeman, N. 1976. Genesis and solution chemistry of acid sulfate soils in Thailand. Ph.D. Thesis. Wageningen, Netherlands. Agric. Res. Rep. (Versl. Landbouw. Onderz.) 848, ISBN 90 220 0600X.
36. ————, and K. Harmensen. 1975. Translocations of iron in acid sulfate soils. I. Soil morphology and the chemistry and mineralogy of iron in a chronosequence of acid sulfate soils. Soil Sci. Soc. Am. Proc. 39:1140-1148.

Chapter 5

Microbial Formation of Basic Ferric Sulfates in Laboratory Systems and in Soils

G. J. ROSS, K. C. IVARSON, AND N. M. MILES[1]

ABSTRACT

Microbial formation of basic ferric sulfates in laboratory systems and in soils was investigated. It was shown that in laboratory systems incubated at 28 C and containing *Thiobacillus ferrooxidans*, ferrous sulfate, feldspars, micas, and montmorillonites, the alkali cations required for basic ferric sulfate formation were supplied by the minerals. During this process feldspars and illite dissolved congruently and released Na and K non-preferentially but glauconite released K preferentially producing a nontronite phase. In the presence of feldspars and micas the rate of basic ferric sulfate formation depended on the weatherability of these minerals; in the presence of K, NH_4, and Na-saturated montmorillonite the rate followed the order jarosite > ammoniojarosite > natrojarosite. Hydronium jarosite formed slowly in the presence of Li-saturated montmorillonite. In systems containing dissolved salts of $K + NH_4$, $K + Na$, and $NH_4 + Na$, solid solutions of basic ferric sulfates containing these cation pairs were formed. The rapid formation of jarosite as compared with other forms of basic ferric sulfates agrees with its reported more common occurrence in acid sulfate soils. However, in four out of six acid sulfate soils from widely separated areas of Canada amounts of natrojarosite were dominant or equal to amounts of jarosite.

INTRODUCTION

Acid sulfate soils occupy appreciable areas throughout the world of which approximately 20 million ha have been mapped (2, 15). They may be recognized by the presence of conspicuous yellow jarosite mottles (4, 7, 18, 20). Jarosite is often associated with kaolinite in acid sulfate soils and illite, montmorillonite, chlorite, and feldspars are also commonly present in these soils. Glauconite has been identified in fossil acid sulfate soils (3).

---

[1] Chemistry and Biology Research Institute, Agriculture Canada, Ottawa, Ontario K1A 0C6. C.B.R.I. Contribution No. 1116.

Copyright © 1982 Soil Science Society of America, 677 S. Segoe Rd., Madison, WI 53711. *Acid Sulfate Weathering.*

It has been suggested that the iron and sulfur of the basic ferric sulfates were derived from the chemical oxidation of pyrite and that the alkalies originated from clay minerals by acid attack (8, 21). Some support for this hypothesis was provided by the deterioration of clay minerals, particularly chlorite, in acid sulfate soils (14).

The acid attack on clay minerals in acid sulfate soils is facilitated by the relatively large amounts of acid produced during jarosite formation. The processes involved may briefly be described as follows: Following the chemical oxidation of pyrite to produce $FeSO_4$, further oxidation (strongly enhanced by the bacterium *Thiobacillus ferrooxidans*) yields $Fe_2(SO_4)_3$. This compound reacts with suitable cations to produce the soil's yellow mottles and its strong acidity by the reaction $6Fe_2(SO_4)_3 + O_2 + 22 H_2O + 4 K \rightarrow 4 K Fe_3(SO_4)_2(OH)_6 + 10 H_2SO_4$.

The available evidence indicates that the K form of basic ferric sulfate, jarosite, is most abundant in acid sulfate soils (18) but the occurrence in sediments and soils of the Na form, natrojarosite, and of mixtures or solid solutions of K, Na, and H_3O basic ferric sulfates have also been reported (13, 19). Because the X-ray diffraction patterns of the various basic ferric sulfates show relatively small differences, they may be difficult to identify, particularly in the presence of other soil minerals. In some cases, therefore, analyses of solutions used for removal of iron oxides (5) may be helpful.

The main objective of the work reported here was to determine whether or not and in what way feldspars, micas, and clay minerals are altered during microbial oxidation of $FeSO_4$ and, if so, whether or not this alteration supplies the alkali cations required for basic ferric sulfate formation. Another objective was to study solid solution formation of basic ferric sulfates that crystallize in solutions containing $FeSO_4$, *T. ferrooxidans*, and salts of K, NH_4, and Na. The third objective was to analyze for the occurrence of basic ferric sulfates in some Canadian acid sulfate soils. Particular attention was given to establishing the nature and identity of the basic ferric sulfates found in these soils.

MATERIALS AND METHODS

Experiments on the Microbial Formation of Basic Ferric Sulfates in the Presence of Minerals. The minerals used in these experiments were < 2 µm glauconite, illite, muscovite, montmorillonite, and 2 to 5 µm albite and microcline (9). Aliquots of the montmorillonite (from Clay Spur, Wyoming) were treated with 1N Cl salt solutions of the respective cations to prepare samples saturated with K, NH_4, Na, Li, Mg, and Al cations.

The minerals were inoculated separately in reagent bottles containing a modified medium 9K (17). The solutions were adjusted to pH 3.6 with H_3PO_4 and each bottle was inoculated with *T. ferrooxidans*. The bottles were aerated at 28 C.

Aliquots of the suspensions were taken periodically for X-ray and chemical analyses. X-ray diffraction analysis was done using mainly a Philips diffractometer with Fe-filtered, Co radiation on samples sedimented on glass slides. Amounts of jarosite in some of the dried sediments

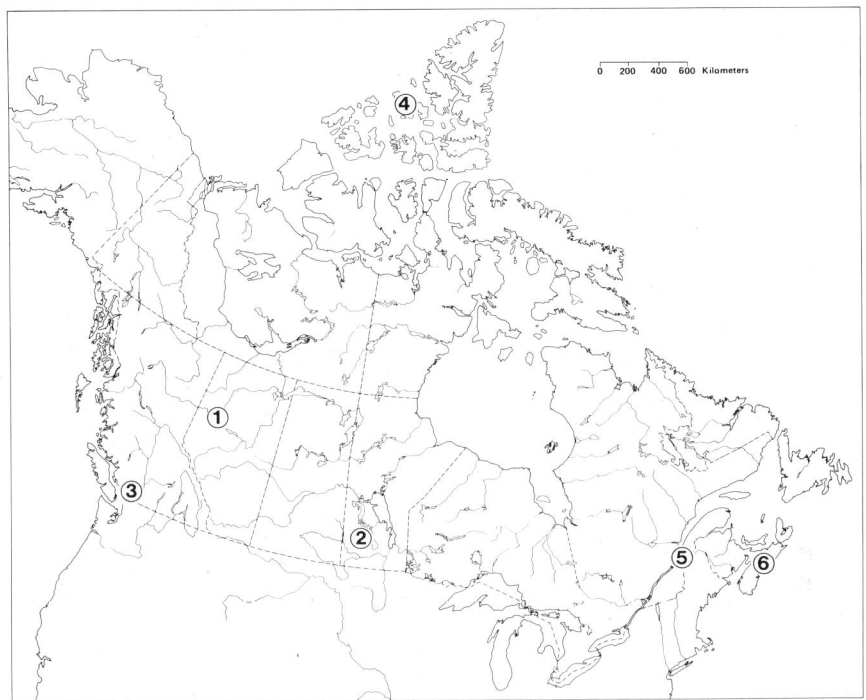

Fig. 1. Locations of Canadian soils sampled for analysis of basic ferric sulfates. 1—Peace River area, Alberta; 2—Grandview area, Manitoba; 3—Lower Fraser Valley, British Columbia; 4—Ellef Ringnes Island, North Canada; 5—St-Lawrence River Valley, Quebec; 6—near Harvey Marsh, Nova Scotia.

were estimated from K contents in the dithonite extracts of these sediments. Elemental analysis of the HF-HClO$_4$ decomposed original and reacted samples was done to compare their chemical composition before and after incubation (9).

Experiments on the Microbial Formation of Basic Ferric Sulfates in Solutions of K, NH$_4$, and Na Salts. For these experiments three bottles were used, each containing 1.5 liters of the modified medium 9K, equivalent concentrations of K + Na, K + NH$_4$, and Na + NH$_4$, the bacterium *T. ferrooxidans*, and H$_2$SO$_4$ to lower the pH to 2.95. The alkali cations were added as HPO$_4$ and Cl salts. The bottle with K + Na contained 0.0125% of K$_2$HPO$_4$ and 0.005% KCl and equivalent amounts (in terms of Na concentration) of Na$_2$PO$_4$ and NaCl. The same equivalents of K + NH$_4$ and of Na + NH$_4$ were present in the other two bottles. The suspensions were aerated at 28 C.

Chemical analysis was done on the solids and supernatants of 25 ml suspensions that were periodically withdrawn. The solids were dissolved in 1.5 N HCl and these solutions and the supernatants were analyzed for K and Na by atomic absorption and for NH$_4$ by the method of Keay and Menage (12).

X-ray diffraction analysis of the solids was done on 40 mg samples dispersed in 1 ml H_2O and sedimented on 25 × 37 mm glass slides using a Philips diffractometer with Fe-filtered, Co radiation. The (003) and the (021) and (113) reflections were resolved by using a 0.5° divergent and antiscatter slit, 0.024" receiving slit, and a scanning rate of 0.5° per min (10).

Analyses on the Occurrence and Identity of Basic Ferric Sulfates in Canadian Acid Sulfate Soils. Several soils, known or suspected to be acid sulfate soils, were sampled in widely separate areas of Canada (Fig. 1). The Boundary soil (No. 1) is situated near Grimshaw in the Peace River region of Alberta. It is a Podzol and has developed on extremely acid shale material (16). The Keld soil (No. 2) is a Gleysol from the Grandview area of Manitoba and has developed on till that originated from strongly acid shale and granitic rock (6). The Nicomekl soil (No. 3) is a Peaty Gleysol developed on the poorly drained deltaic deposits in the Lower Fraser Valley of British Columbia (4). The Arctic soil (No. 4) has developed on shale material on Ellef Ringnes Island in northern Canada. The "sols de l'Anse" (No. 5) are Peaty-Gleysols situated in Quebec along the St. Lawrence River slightly above or below the high-tide water level on parent materials that contained high concentrations of salts and sulfur prior to drainage (1). The Acadia soil (No. 6) near Harvey Marsh in Nova Scotia is a "dykeland" soil developed on marine (intertidal) parent material.[2]

To obtain a higher concentration of basic ferric sulfates in the soil samples, material from the yellow mottles and specks were picked out with a spatula. This material was Mg saturated by centrifuge-washing five times with 1 N $MgCl_2$ and then with H_2O until free of excess Cl.

The Mg-saturated, air-dried material was analyzed by X-ray diffraction using the same instrument, procedure, and conditions as were used for the basic ferric sulfates that crystallized from solutions containing K, NH_4, and Na salts, as described in the previous section.

Because X-ray diffraction analysis indicated the presence of both jarosite and natrojarosite in the samples, the photolytic ammonium oxalate method for removal of free iron oxides from soils (5) was used to estimate the total and proportional amounts of jarosite and natrojarosite in the samples. X-ray diffraction results showed that this method also dissolves basic ferric sulfates. Prior to this treatment the samples were thoroughly Mg saturated by centrifuge-washing five times with 1 N $MgCl_2$ and then with distilled H_2O until free of excess Cl. A 100 mg Mg-saturated, air-dry sample was added to a 250 ml beaker with 100 ml ammonium oxalate (pH 4.4, consisting of 0.02 M oxalic acid and 0.255 M ammonium oxalate) stirred with a magnetic stirrer and irradiated from above by a 125 W mercury arc lamp. After the extraction was completed (approximately 1 hour), the suspension was centrifuged, the K and Na in the supernatants determined by atomic absorption and allocated to jarosite and natrojarosite by taking 7.81% as the K content of jarosite and 4.73% as the Na content of natrojarosite.

---

[2] A description of this profile may be obtained from the author.

RESULTS AND DISCUSSION

Microbial Formation of Basic Ferric Sulfates in the Presence of Minerals. The X-ray diffraction patterns for these experiments show that basic ferric sulfates were formed in the presence of all the minerals. In the feldspar systems natrojarosite crystallized in the presence of albite, and jarosite formed in the presence of microcline (9). In the mica and montmorillonite systems jarosite formed in the presence of illite (Fig. 2), muscovite (not shown), glauconite (Fig. 3), and K montmorillonite (Fig. 4). Ammoniojarosite crystallized in the presence of NH_4 montmorillonite (Fig. 5) and natrojarosite in the presence of Na montmorillonite (Fig. 6). If $(NH_4)_2SO_4$ instead of $Ca(NO_3)_2$ was used as the microbial N source, formation of only ammoniojarosite occurred and the formation of jarosite was suppressed (e.g., Fig. 2). Natrojarosite was recognized by its (003) reflection at 5.57Å and its (021) and (113) doublet at 3.12 and 3.06Å (ASTM card no. 11-302). Jarosite was identified by its (003) reflection at 5.72Å and its (021) and (113) doublet at 3.11 and 3.08Å, respectively. Ammoniojarosite was indicated by its (003) peak at 5.80Å and its (021) and (113) doublet at 3.12 and 3.10Å which appeared almost as one peak. Hydronium jarosite formed in the Li montmorillonite system and was identified by its (003) peak at 5.54Å and its (021) and (113) doublet at 3.13 and 3.09Å (Fig. 7). Small amounts of hydronium jarosite were also identified in the Mg and Al-montmorillonite systems (not shown). The presence of hydronium jarosite was supported by chemical analysis of the precipitate at the end of the experiment which showed the virtual absence of Na and K impurities from which basic ferric sulfates could have formed formed.

The rate of basic ferric sulfate crystallization may be estimated from the development of peak intensities during incubation and, for three of the minerals, from the amounts of jarosite formed at the end of the experiment (Table 1). In the feldspar systems natrojarosite formed faster in the presence of albite than did jarosite in the presence of microcline (9), and in the mica systems jarosite crystallized more rapidly in the presence of glauconite than of illite (cf., Fig. 2 and 3). In the montmorillonite systems the rate of formation was jarosite in the K montmorillonite system > ammoniojarosite in the NH_4 montmorillonite system > natrojarosite in the Na montmorillonite system > hydronium jarosite in the Li montmorillonite system > hydronium jarosite in the Mg and Al montmorillonite systems (Fig. 4, 5, 6, and 7).

Table 1. Amounts of jarosite formed in the presence of micas and feldspars after 4 months reaction.

Mineral in system	Percent jarosite formed of total sediment†
Glauconite	12.0
Illite	6.0
Microcline	2.5

† Calculated by allocating amounts of K in dithionite extracts of sediments to jarosite (taking 7.81% as K content of jarosite) after subtracting the amounts of K in dithionite extracts of the original clay minerals. Amounts of jarosite expressed as percent of air-dry sample prior to dithionite treatment.

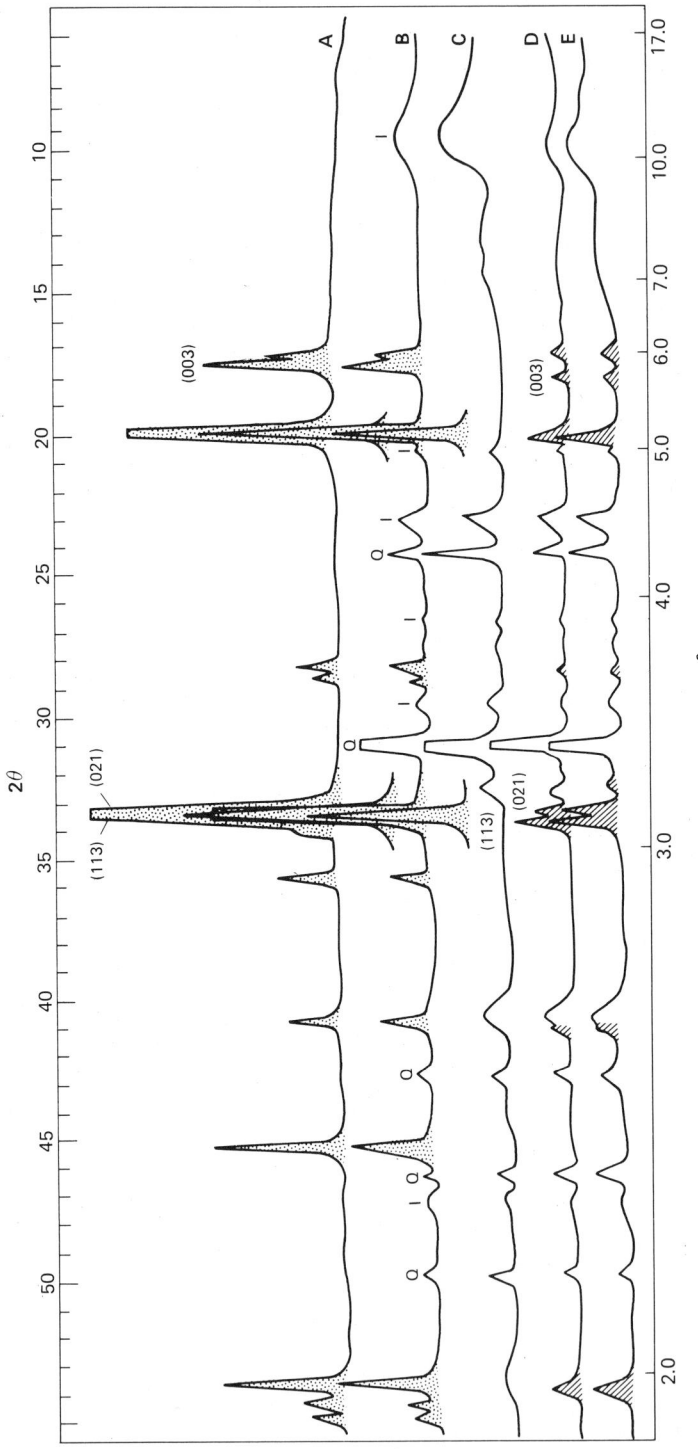

Fig. 2. X-ray diffractograms of illite after incubation with an iron-oxidizing bacterium: (A) ammoniojarosite; (B) incubated for 4 months with $(NH_4)_2SO_4$; (C) original illite; (D) incubated 2 months with $Ca(NO_3)_2$; (E) incubated 4 months with $Ca(NO_3)_2$. Q—quartz; I—illite. Samples in random orientation.

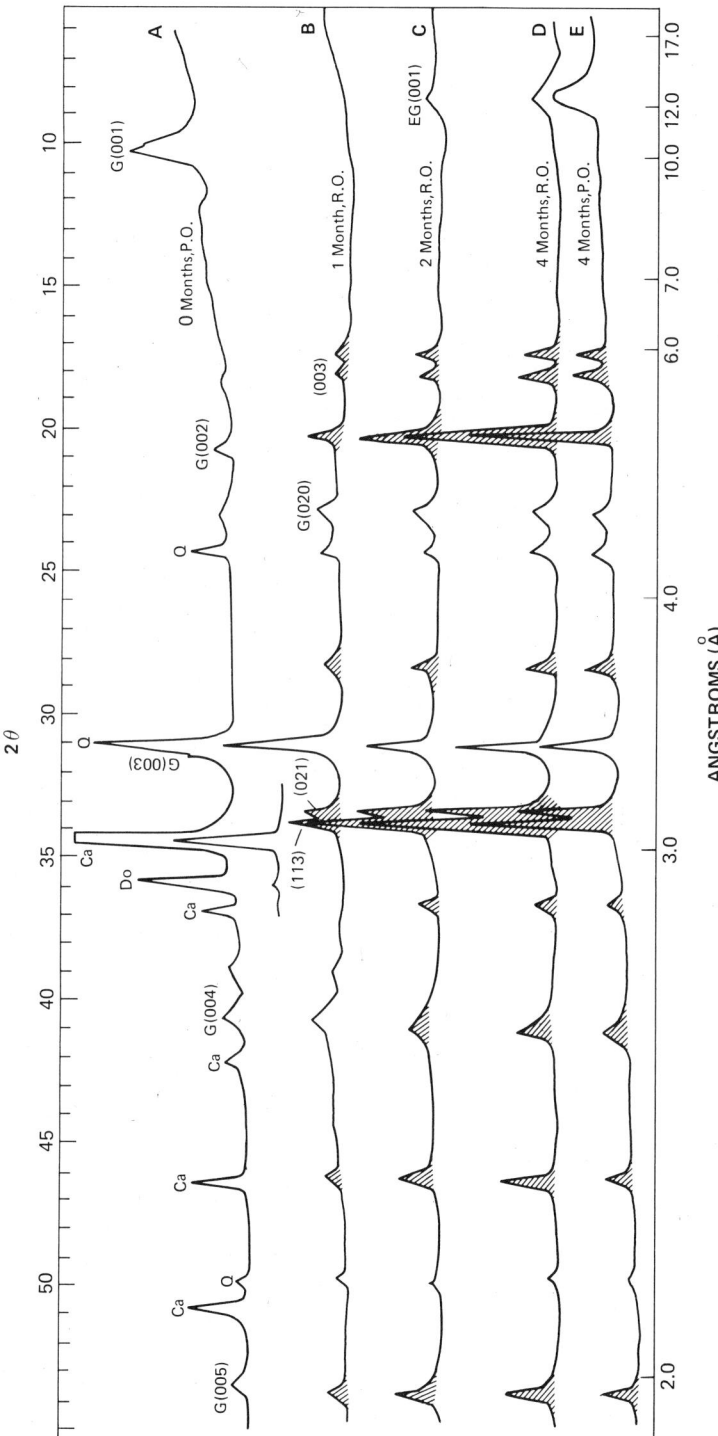

Fig. 3. X-ray diffraction patterns of glauconite after incubation with an iron-oxidizing bacterium. G—glauconite; P.O.—preferred orientation; R.O.—Random orientation. Ca—calcite; Q—Quartz; Do—dolomite; EG—expanded glauconite. The shaded peaks are for jarosite.

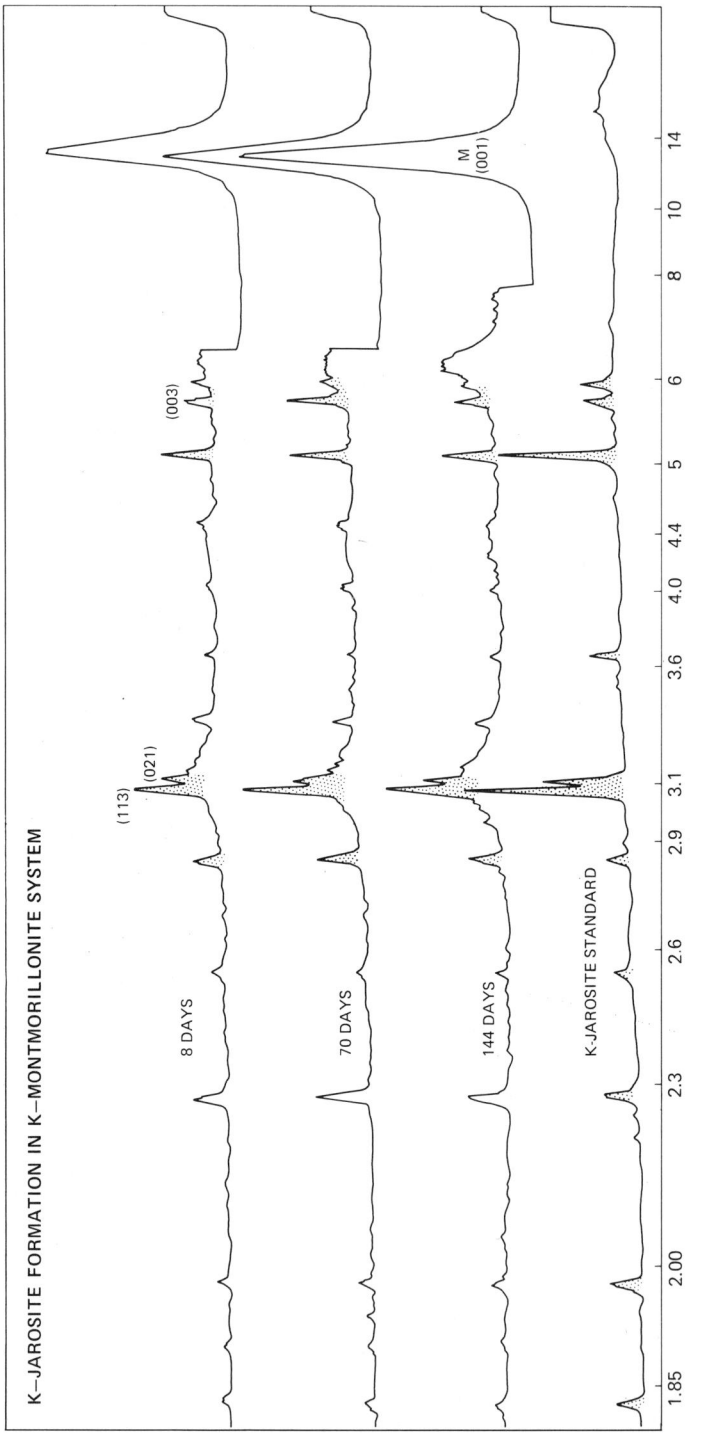

Fig. 4. X-ray diffraction patterns of K montmorillonite during incubation in a FeSO$_4$ solution containing *T. ferrooxidans*. The shaded peaks are for jarosite.

FORMATION OF BASIC FERRIC SULFATES

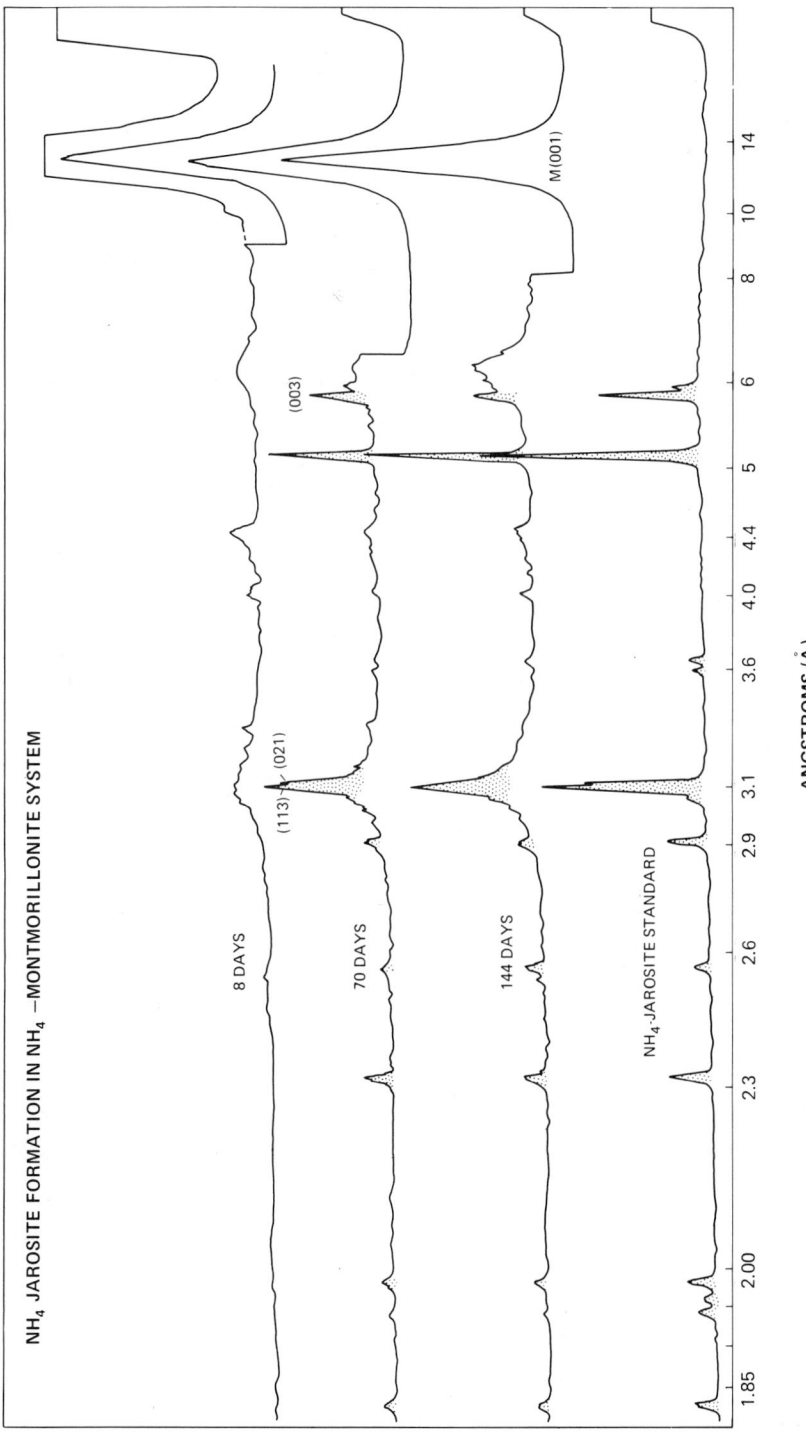

Fig. 5. X-ray diffraction patterns of NH_4 montmorillonite during incubation in a $FeSO_4$ solution containing *T. ferrooxidans*. The shaded peaks are for ammoniojarosite.

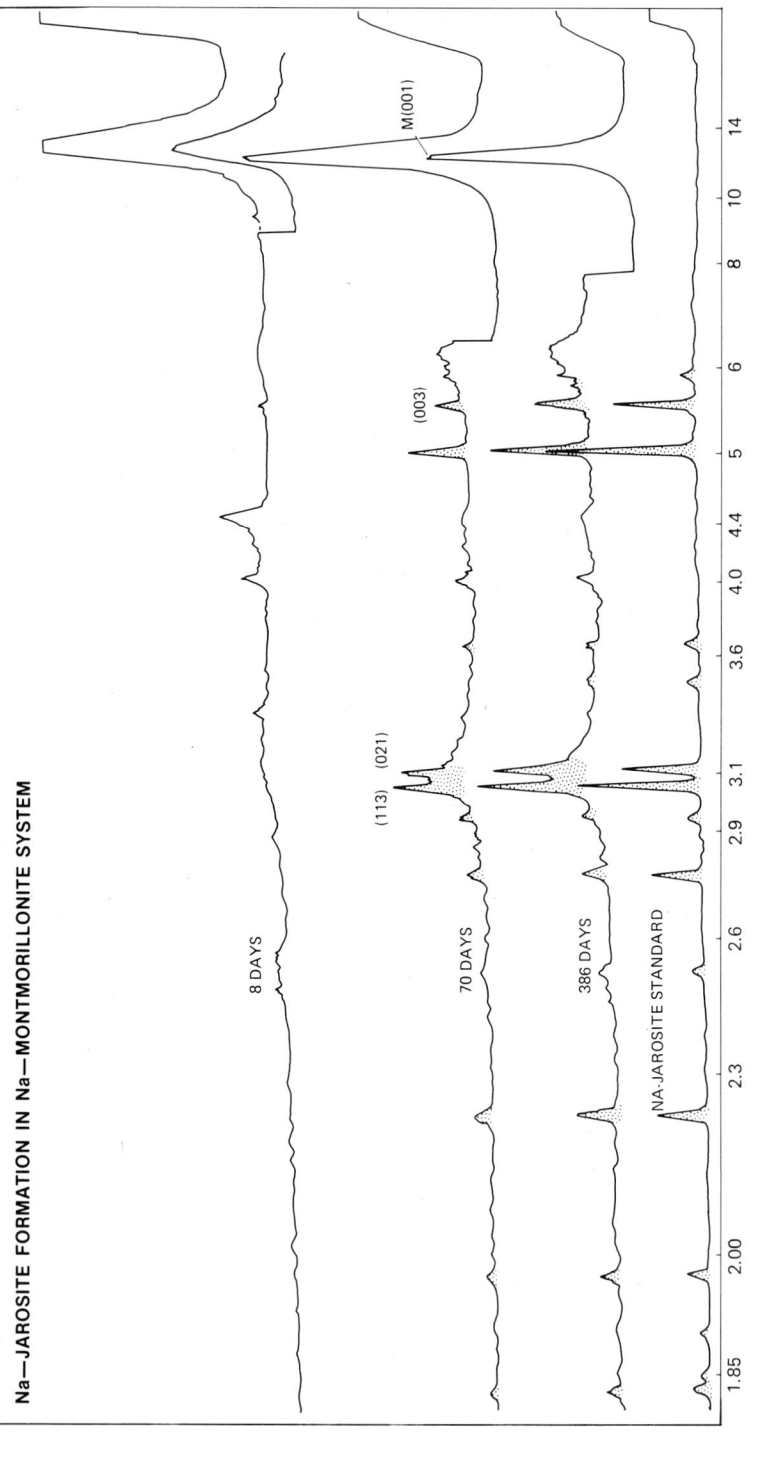

Fig. 6. X-ray diffraction patterns of Na montmorillonite during incubation in a FeSO$_4$ solution containing *T. ferrooxidans*. The shaded peaks are for natrojarosite.

FORMATION OF BASIC FERRIC SULFATES

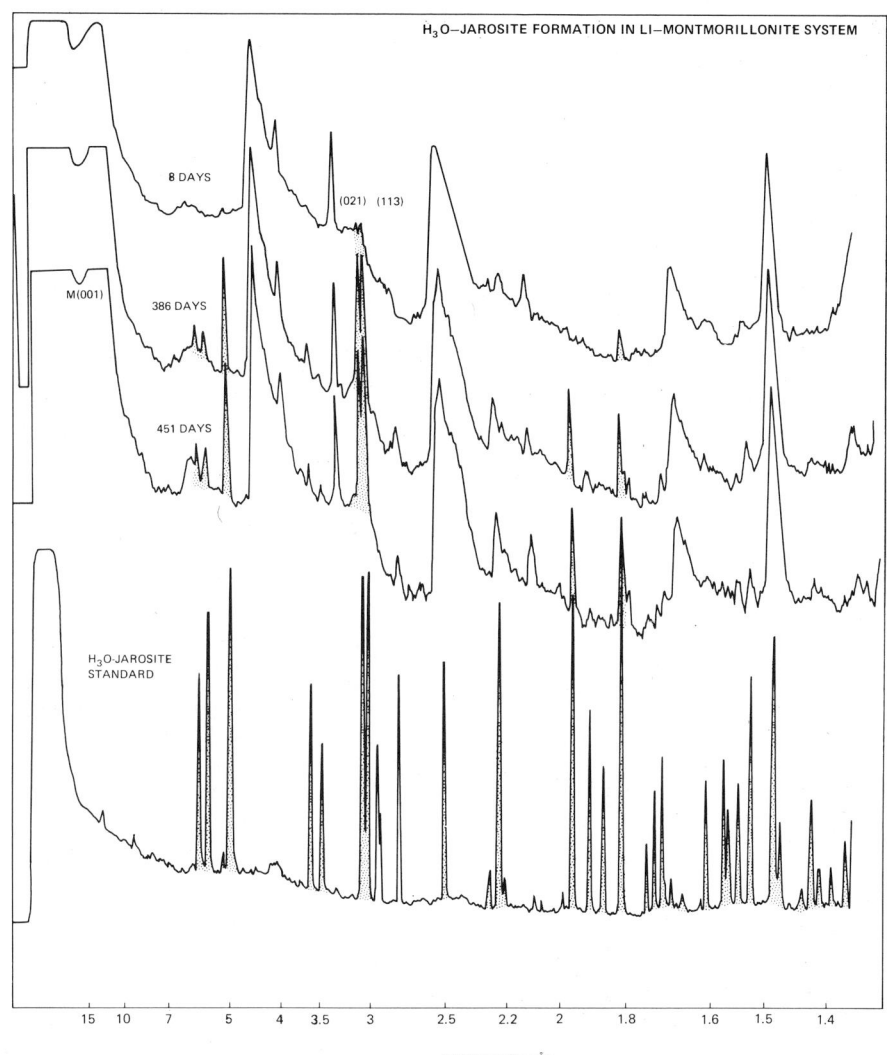

Fig. 7. Photodensitometer traces of photographs obtained with the Guinier-de Wolff camera of Li montmorillonite during incubation in $FeSO_4$ solution containing *T. ferrooxidans*. The shaded peaks are for hydronium jarosite.

The X-ray diffraction patterns for illite before and after incubation are not appreciably different (Fig. 8) but those for glauconite show that the original mineral had partially altered to nontronite (Fig. 9). This indicates that interlayer K of illite was dissolved non-preferentially along with other structural ions and that interlayer K of glauconite was dissolved preferentially. This is also evident from the chemical data, particularly the elemental ratios, in Table 2 which show that for glauconite after incubation the proportion of K lost was much greater than the proportional loss of other structural ions. For illite this difference is not appar-

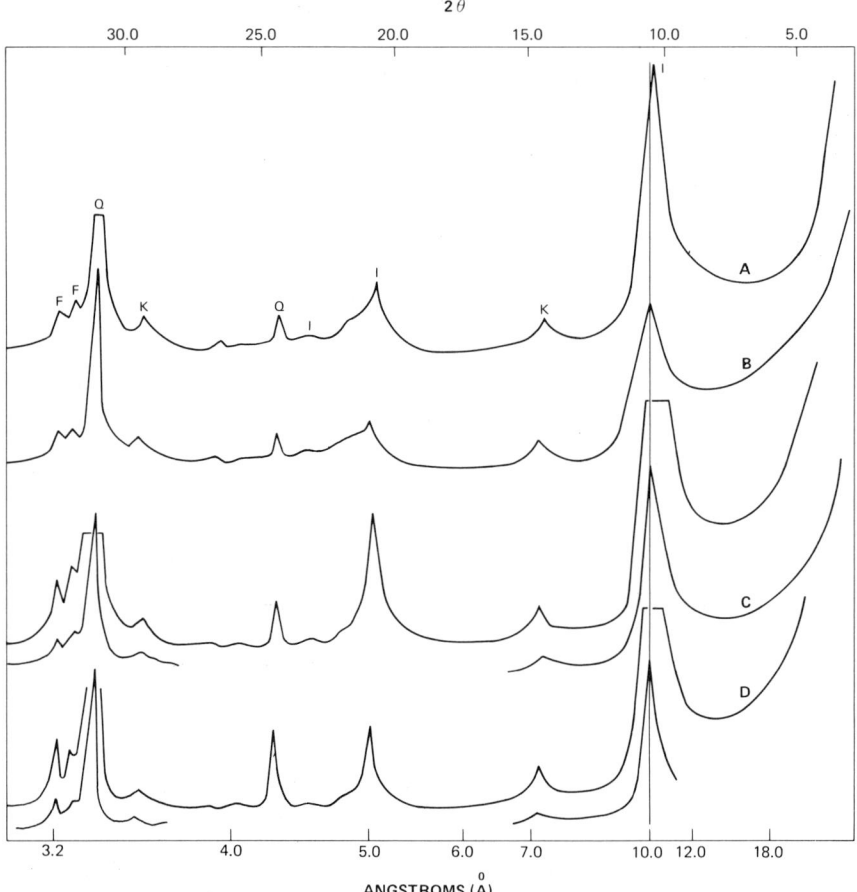

Fig. 8. X-ray diffractograms of illite. (A) glycerolated original sample; (B) glycerolated sample after incubation with an iron-oxidizing bacterium for 4 months; (C) heat-treated original sample; and (D) heat-treated sample after incubation with iron-oxidizing bacterium for 4 months. F—feldspar; Q—quartz; I—illite; K—kaolinite. Samples in preferred orientation.

ent. The apparent proportional loss of other structural ions may be attributed to dilution with amorphous silica residue from some structural decomposition at the relatively low pH of the incubated systems. The X-ray diffraction patterns for the feldspars and montmorillonite before and after incubation also remained the same although some decrease in crystallinity was evident from a reduction in peak intensities which was probably also due to the accumulation of silica residues.

The evidence shows therefore that the alkali cations required for basic ferric sulfate formation may be derived from structural and exchange positions of common soil minerals and that hydronium jarosite may be formed in the absence of alkali cations. If the alkali cations were derived from structural positions, the rate of basic ferric sulfate formation depended on the weatherability of the minerals. Thus the rate was greater

FORMATION OF BASIC FERRIC SULFATES

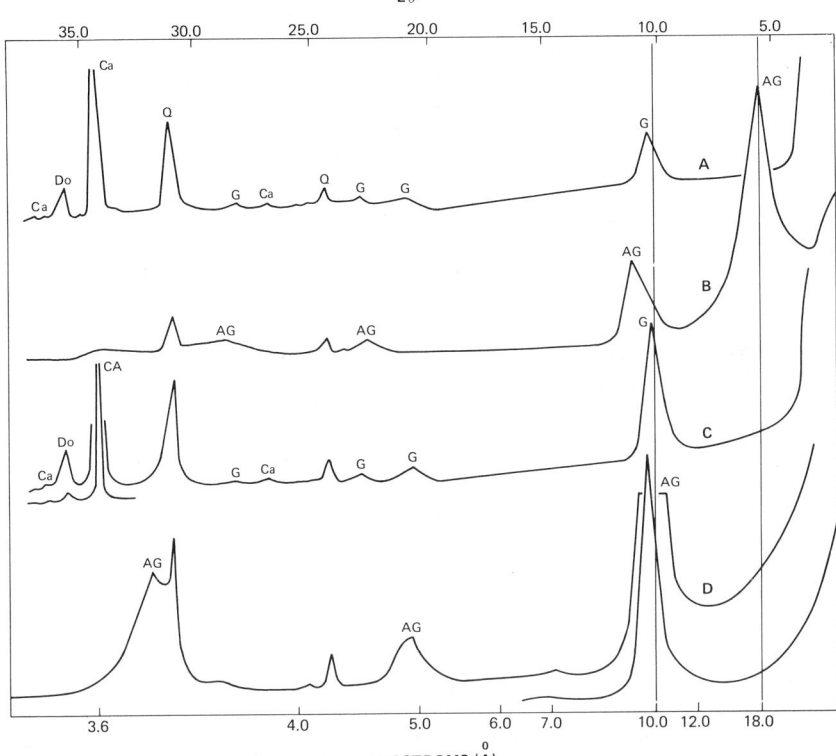

Fig. 9. X-ray diffractograms of glauconite. (A) glycerolated original sample; (B) glycerolated sample after incubation with an iron-oxidizing bacterium for 4 months; (C) heat-treated original sample; and (D) heat-treated sample after incubation with iron-oxidizing bacterium for 4 months. Ca—calcite; Do—dolomite; Q—quartz; G—glauconite; AG—altered glauconite. Samples in preferred orientation.

in the presence of albite than of microcline and it was greater in the presence of glauconite than of illite or muscovite. If the alkali cations were derived from exchange positions, the rates followed the same order as that with alkali cations in solution (10), namely, jarosite > ammoniojarosite > natrojarosite. Hydronium jarosite formed least rapidly. The alkali and other structural cations from the feldspars (9) and illite appeared to be dissolved at similar rates, but the K from glauconite was released preferentially, producing an 18 Å nontronite phase.

Microbial Formation of Basic Ferric Sulfates in Solutions of K-, NH_4- and Na Salts. In the three systems, each containing equivalent amounts of the two alkali salts (K + Na, K + NH_4, and Na + NH_4), K disappeared from the solutions first, followed by NH_4 and then by Na. These results were similar to those observed for the systems each containing one alkali salt of K, NH_4, and Na and showed that under the same conditions the rate of crystallization is jarosite > ammoniojarosite > natrojarosite (10). According to chemical analysis at the end of the experiment after 290

Table 2. Elemental composition of micas and feldspars before and after their involvement in the microbial formation of jarosites.†

Element	Glauconite (before)	Glauconite (after)	Illite (before)	Illite (after)	Microcline‡ (before)	Albite‡ (before)
			%			
Al	5.29	4.95 (0.94)§	8.80	7.84 (0.89)§	7.80	8.68
Fe	10.40	9.43 (0.91)	3.01	2.18 (0.72)	0.06	0.04
Mg	2.10	2.12 (1.01)	1.22	0.95 (0.78)	0.00	0.00
K	4.42	2.28 (0.52)	3.48	2.86 (0.82)	10.30	0.20
Na	0.12	0.11 (0.91)	0.52	0.68 (1.31)	2.02	8.19

† Free carbonates and free iron and aluminum oxides were removed prior to elemental analysis.
‡ Elemental composition was not determined after jarosite formation.
§ Ratios of elemental contents before and after incubation of glauconite and illite.

days, the solid in the $NH_4 + Na$ system contained 66% ammoniojarosite and 34% natrojarosite, the solid in the $NH_4 + K$ system contained 46% ammoniojarosite and 54% jarosite, and the solid in the $Na + K$ system contained 39% natrojarosite and 61% jarosite. Although the solids in these systems gave strong basic ferric sulfate diffraction patterns, the reflections for each of the two components in each solid were not separate and distinct suggesting the formation, at least partly, of solid solutions. This is evident from a comparison of (003) reflections (Fig. 10) and of (021) and (113) reflections (Fig. 11) of the precipitates with those of 50/50 mechanical mixtures.

The results showing the rapid crystallization of jarosite as compared with other basic ferric sulfates agree with reported analyses of acid sulfate soils showing jarosite to be the most abundant basic ferric sulfate (18). Careful X-ray and chemical analyses, however, may show other basic ferric sulfates, particularly natrojarosite, to be more abundant in acid sulfate soils than previously indicated. Some reports have cited the occurrence of natrojarosite and hydronium jarosite in soils (19). The formation and identification of basic ferric sulfate solid solutions in the conditions used in these experiments suggest that such solid solutions may not be rare in acid sulfate soils and could be recognized by appropriate analyses.

Occurrence and Identity of Basic Ferric Sulfates in Canadian Acid Sulfate Soils. Figure 1 shows the approximate locations of the six acid sulfate soils that were sampled in widely separated areas in Canada, and Fig. 12 contains X-ray diffraction patterns and data showing amounts of jarosite and natrojarosite present in the mottles of these samples. The amounts were estimated by allotting Na and K contents of ammonium oxalate extracts to natrojarosite and jarosite, respectively. The pH of the soil samples before separating the yellow mottles varied from 3.5 to 4.1 and the presence and activity of *T. ferrooxidans* was verified in all these samples. The data in Fig. 12 show a range from predominantly natrojarosite in the Boundary soil to predominantly jarosite in the Acadia soil. The chemical data showing this range are supported by the X-ray data for the (003) and (113) spacings. The (003) spacings range from 5.56Å for the dominant natrojarosite in the Boundary soil to 5.71Å for the dominant jarosite in the De l'Anse and Acadia soils. The (113) spacings follow the

FORMATION OF BASIC FERRIC SULFATES

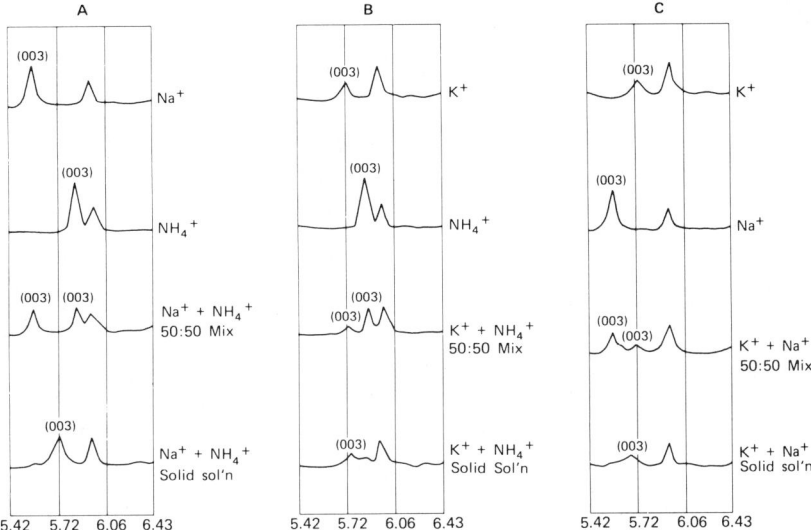

Fig. 10. X-ray diffraction patterns showing (003) peaks of basic ferric sulfates formed during 42 weeks incubation in Na + NH$_4$, K + NH$_4$, and K + Na systems. Patterns produced using a scanning speed of 0.5°/min. and compared with basic ferric sulfates formed in single Na, NH$_4$, and K systems.

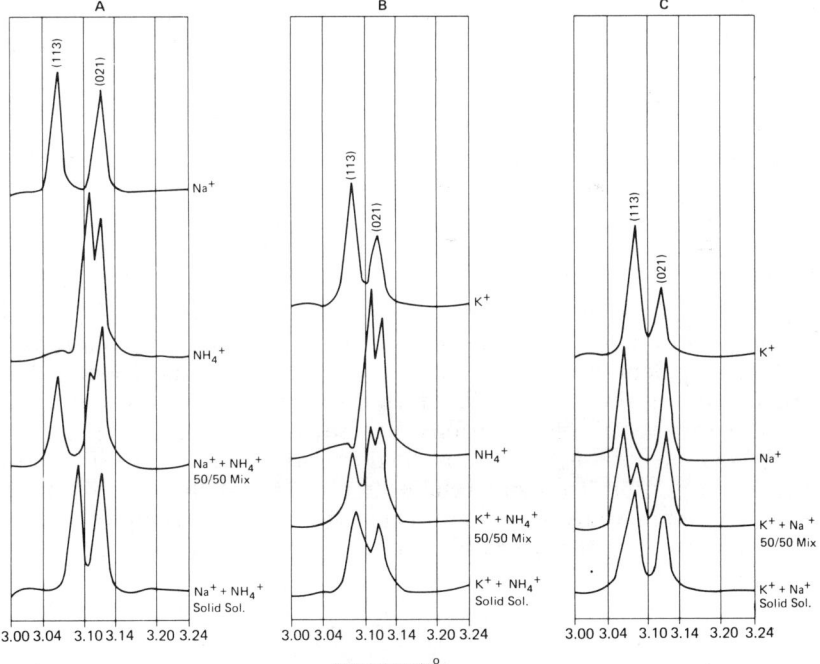

Fig. 11. X-ray diffraction patterns showing the (021) and (113) doublet peaks of basic ferric sulfates formed during 42 weeks incubation in Na + NH$_4$, K + NH$_4$, and K + Na systems. Patterns produced using a scanning speed of 0.5°/min and compared with basic ferric sulfates formed in single Na, NH$_4$, and K systems.

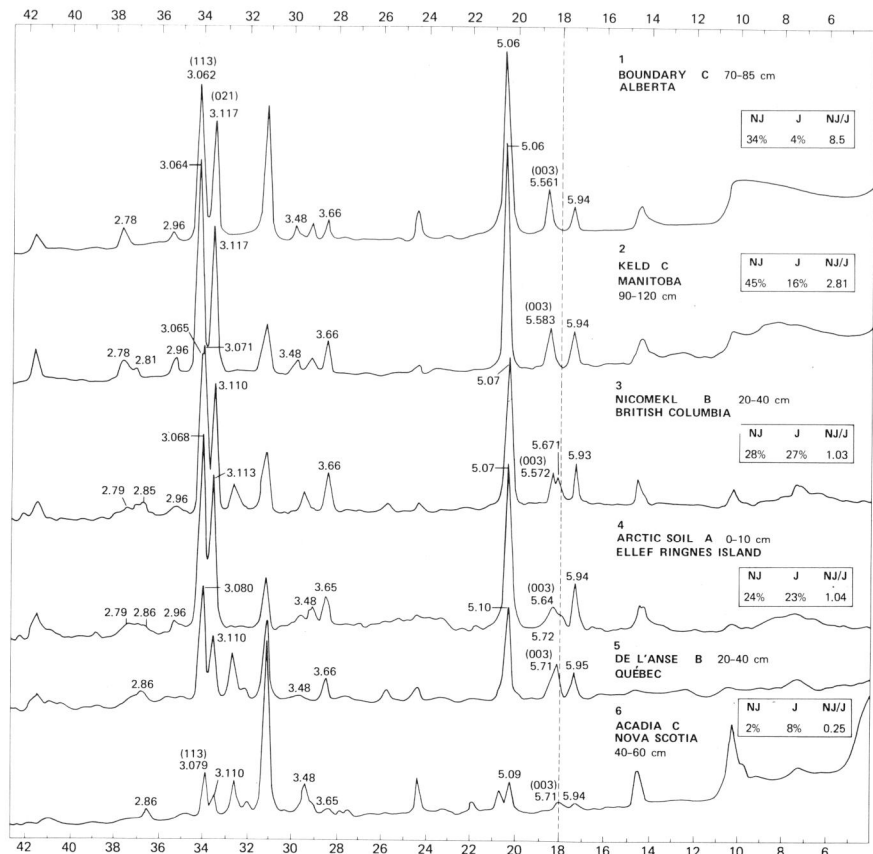

Fig. 12. X-ray diffraction patterns of material separated from yellow mottles of the soils indicated. Only spacings (in Å) of distinct jarosite or natrojarosite peaks are shown. Spacings of (003), (021), and (113) peaks were verified on additional X-ray patterns run at 0.5°/min. NJ is natrojarosite, J is jarosite.

same trend and range from 3.062 to 3.080Å. Solid solution of Na and K basic ferric sulfate may occur in the Boundary and Keld soils but separate phases are evident from the separate (003) peaks in the Nicomekl and probably also in the Arctic soil. More detailed analyses are required to verify the presence of solid solutions in these samples.

There is no obvious reason for the dominance of natrojarosite in some of these samples but it may be related to the ratio of Na to K (in ppm) in the soil solutions. Analyses of water extracts of bulk soil samples showed a Na/K ratio of 8.1 for the Keld soil and of 3.3 for the Acadia soil.

CONCLUSIONS

The main conclusions from the work presented here are that during the microbial formation of basic ferric sulfates the required alkali cations may be derived from common soil minerals such as feldspars, micas, and montmorillonites. Feldspars and most micas appear to dissolve and degrade during this process but sometimes new mineral phases are formed as in the case of nontronite from glauconite.

Solid solutions of basic ferric sulfates are readily formed from mixed K, NH_4, and Na salts during the microbial oxidation of $FeSO_4$. The different basic ferric sulfates and their solid solutions may be recognized by careful X-ray and chemical analyses.

X-ray analysis and chemical extraction procedures showed that both natrojarosite and jarosite were present in samples of six acid sulfate soils from widely separated areas in Canada. Amounts of natrojarosite were dominant or equal to amounts of jarosite in four of these samples. The predominance of natrojarosite may be related to the Na/K ratio of the soil solution.

ACKNOWLEDGMENTS

The authors thank R. G. Hill, G. C. Scott, and H. Malinowski for technical help and G. J. Beke, R. G. Eilers, A. E. Foscolos, J. D. Lindsay, and M. R. Bullen for obtaining the soil samples.

LITERATURE CITED

1. Baril, R., and B. Rochefort. 1965. Etude pedologique du comte de Kamourask (Quebec) Min. de l'Agric. de la Prov. de Quebec, P.Q., Canada. p. 47.
2. Bloomfield, C., and J. K. Coulter. 1973. Genesis and management of acid sulfate soils. Adv. Agron. 25:256–326.
3. Buurman, P., N. Van Breemen, and A. G. Jongsman. 1973. A fossil acid sulfate soil in ice-pushed Tertiary deposits near Uelsen, (Kreis Nordhorn), Germany. p. 52–80. In H. Dost (ed.) Acid sulfate soils. ILRI Publ. 18, Vol. II, Inst. for Land Reclamation and Improvement. Wageningen, Netherlands.
4. Clark, J. S., C. A. Gobin, and P. N. Sprout. 1961. Yellow mottles in some poorly drained soils of the lower Fraser Valley, British Columbia. Can. J. Soil Sci. 41:218–227.
5. De Endredy, A. S. 1963. Estimation of free iron oxides in soils and clays by a photolytic method. Clay Min. Bull. 5:209–217.
6. Ehrlich, W. A., L. E. Pratt, and F. P. Leclaire. 1959. Report of reconnaissance soil survey of Grandview map sheet area. Manitoba Dep. of Agric. and Conservation, Winnipeg, Manitoba, Canada. p. 44.
7. Fleming, J. F., and L. I. Alexander. 1961. Sulfur acidity in Southern Carolina tidal marsh soils. Soil Sci. Soc. Am. Proc. 25:94–95.
8. Furbush, W. J. 1963. Geologic implications of jarosite, pseudomorphic after pyrite. Am. Mineral. 48:703–706.
9. Ivarson, K. C., G. J. Ross, and N. M. Miles. 1978. Alterations of micas and feldspars during microbial formation of basic ferric sulfates in the laboratory. Soil Sci. Soc. Am. J. 42:518–524.
10. ————, ————, and ————. 1979. The microbial formation of basic ferric sulfates. 2. Crystallization in presence of K-, NH_4-, and Na salts. Soil Sci. Soc. Am. J. 43:908–912.
11. ————, ————, and ————. 1980. The microbiological formation of basic ferric sulfates. 3. Influence of clay minerals on crystallization. Can. J. Soil Sci. 60:137–140.

12. Keay, J., and P. M. A. Menage. 1970. Automated determination of ammonium and nitrate in soil extracts by distillation. Analyst (London) 95:379–382.
13. Kubisz, J. 1961. Natural hydronium jarosites. Bull. Acad. Polon. Sci., Series Sci. Chim. Geol. Geogr. 9:195–200.
14. Lynn, W. C., and L. D. Whittig. 1966. Alteration and formation of clay minerals during cat clay development. Clays Clay Miner. 14:241–248.
15. Ponnamperuma, F. N., T. Attanandana, and G. Beye. 1973. Amelioration of three acid sulfate soils for lowland rice. p. 391–406. In H. Dost (ed.) Acid sulfate soils, ILRI. Publ. 18, Vol. II Inst. for Land Reclamation and Improvement, Wageningen, Netherlands.
16. Scheelar, M. D., and W. M. Odynsky. 1968. Soil survey of the Grimshaw and Notikewin area. Dep. of Extension, The Univ. of Alberta, Edmonton, Alberta, Canada. p. 49.
17. Silverman, M. P., and D. G. Lundgren. 1959. Studies on the chemoautotrophic iron bacterium *Ferrobacillus ferrooxidans*. I. An improved medium and a harvesting procedure for securing high yields. J. Bacteriol. 77:642–647.
18. Van Breemen, N. 1976. Genesis and solution chemistry of acid sulfate soils in Thailand. Ph.D. Thesis. 263 p. Wageningen Netherlands. Agric. Res. Rep. (Versl. landbouwk. Onderz.) 848, ISBN 90 220 0600 X.
19. ―――, and K. Harmsen. 1975. Translocations of iron in acid sulfate soils. I. Soil morphology and the chemistry and mineralogy of iron in a chronosequence of acid sulfate soils. Soil Sci. Soc. Am. Proc. 39:1140–1148.
20. Van Der Kevie, W. 1973. Physiography, classification and mapping of acid sulfate soils. p. 204–222. In H. Dost (ed.) Acid sulfate soils. ILRI. Publication 18, Vol. 1, Inst. for Land Reclamation and Improvement, Wageningen, Netherlands.
21. Warshaw, C. M. 1958. The occurrence of jarosite in underclays. Am. Mineral. 41:288–296.

Chapter 6

Genesis, Morphology, and Classification of Acid Sulfate Soils in Coastal Plains[1]

N. VAN BREEMEN[2]

ABSTRACT

Acid sulfate soils form when potentially acid pyritic marshes are drained and tidal influence decreases, either naturally or by man. During oxidation of pyrite to ferrous sulfate, ferric oxide, or jarosite and sulfuric acid, the supply of O_2 is rate-limiting. Upon slow oxidation of pyritic soil in situ, buffering by clay minerals, jarosite, and ferric oxide, keeps pace with acid formation, and the pH remains between 3 and 4. Lower pH values may develop with rapid oxidation as in excavated soil and aerated pyritic soil samples. Compared to non-acid marine soils, acid sulfate soils exhibit retarded physical development due to lower evapotranspiration by a less luxuriant vegetation. A well-developed acid sulfate soil shows, from bottom to top, an unoxidized pyritic substratum, a jarositic horizon due to oxidation of ferrous sulfate diffusing upward from oxidizing pyrite, and a horizon high in ferric oxide from hydrolysis of jarosite. As these soils become older and better drained, the jarositic horizon and the pyritic substratum are found at progressively greater depth. The distinction between Sulfaquents and Sulfic subgroups in *Soil Taxonomy* is practical but sulfidic material and sulfuric horizon need to be redefined.

INTRODUCTION

Acid sulfate soils have, somewhere within 50 cm depth, a pH below 4 that is directly or indirectly caused by sulfuric acid formed by oxidation of pyrite (cubic FeS_2) or, rarely, of other reduced sulfur compounds. Potential acid sulfate soils are poorly drained and highly pyritic with a nearly neutral or slightly acid reaction in the field. They become acid sulfate soils if pyrite oxidizes after drainage.

Millions of hectares of acid sulfate soils and potentially acid sulfate soils occur in recent coastal plains. Probably still larger surfaces with

---

[1] Contribution from the Dep. of Soil Science and Geology, Agricultural University, Wageningen, the Netherlands.
[2] Soil scientist, Dep. of Soil Science and Geology, P.O. Box 37, Wageningen, the Netherlands.

Copyright © 1982 Soil Science Society of America, 677 S. Segoe Rd., Madison, WI 53711. *Acid Sulfate Weathering.*

older potentially acid sediments and with potentially acid sedimentary rocks occur at relatively shallow depth and are locally exposed as a result of human activities. This paper will deal mainly with acid sulfate soils developed in pyritic sediments of former Holocene tidal lands, with special reference to those of Thailand which are most familiar to the author. However, many considerations may well be applicable to other situations where acid sulfate weathering occurs. Much of the material presented in this review was based on earlier publications (van Breemen, 1973a, 1976; van Breemen and Pons, 1978).

OXIDATION OF PYRITE

The fine-grained pyrite typical for tidal sediments oxidizes readily upon exposure to the air, giving Fe(II) sulfate and sulfuric acid:

$$FeS_2 + 7/2\,O_2 + H_2O \rightarrow Fe^{2+} + 2SO_4^{2-} + 2H^+ \qquad [1]$$

Sulfite forms as an intermediate and there are indications that appreciable amounts of sulfur are sometimes removed as gaseous sulfur dioxide from young acid sulfate soils (van Breemen, 1976). Complete oxidation and hydrolysis of iron to Fe(III) oxide yields 2 moles of sulfuric acid per mole of pyrite:

$$FeS_2 + 15/4\,O_2 + 7/2\,H_2O \rightarrow Fe(OH)_3 + 2SO_4^{2-} + 4H^+ \qquad [2]$$

The term *sulfuricization* has been used to denote these acidification processes (Fanning, 1978). Oxidation of Fe(II) to Fe(III) by O_2 is slow in vitro but the reaction rate is greatly enhanced by the chemoautotrophic organism *Thiobacillus ferrooxidans*. Pyrite is oxidized more rapidly by dissolved Fe(III) than by oxygen, according to

$$FeS_2 + 14Fe^{3+} + 8H_2O \rightarrow 15Fe^{2+} + 16H^+ + 2SO_4^{2-} \qquad [3]$$

Because a low pH and the activity of *T. ferrooxidans* are required to maintain appreciable concentrations of dissolved Fe(III), these conditions also promote pyrite oxidation (Singer and Stumm, 1970). The role of other *Thiobacilli* in enhancing initial pyrite oxidation and in causing the pH drop from near neutrality to pH 4 after aeration of fresh pyritic mud (Quispel et al., 1952) has probably been overemphasized. Arkesteyn (1980) found no effect of inoculation of various *Thiobacilli*, and of extracts from acid sulfate soils, on the oxidation rate of pyrite in the range pH 7 to 4 although the *Thiobacilli* apparently benefitted and grew on sulfur released by oxidizing pyrite. Apparently, the rate of pyrite oxidation in that pH range is determined by non-biological factors. A drop in pH from 7 to 4 due to the oxidation of about 0.5% pyrite-S may take place in 1 month if pyritic soil is well exposed to the air. Pyrite oxidation in situ is usually much slower due to rate-limiting diffusion of oxygen into the wet pyritic substratum (Rasmussen, 1961; van Breemen, 1973a; de Richmond et al., 1975).

Appreciable acidification of potentially acid heavy clay soils can start only when the water table falls below the upper part of the pyritic substratum for several weeks or more. A prerequisite for such prolonged drainage is a removal or decrease of the tidal influence. This is brought about either gradually by natural processes such as a relative decrease in sea level, or more abruptly, by empoldering. In trying to curtail salt water intrusion by diking tidal land, man has inadvertently caused severe acidification of land in many areas. Locally in Southeast Asia, acidification takes place in land still under tidal influence because potentially acid material is brought to the land surface from several meters depth by the mound-building mudlobster *Thallasina anomale* (Andriesse et al., 1973).

OXIDATION PRODUCTS

Most of the iron(II), hydrogen, and sulfate ions released during pyrite oxidation undergo various further reactions in the soil. Protons are largely inactivated by ion exchange reactions and weathering reactions. Essentially all Fe(II) is further oxidized to Fe(III) in oxides, in the basic sulfate jarosite and in clay minerals. Most of the sulfate remains in solution and is removed by leaching, together with cations (mainly Mg, Ca, and Na) derived from ion exchange and mineral weathering. The remainder of the sulfate is precipitated as jarosite and as basic aluminum sulfate, and part is adsorbed, especially at low pH. Gypsum and other more readily soluble sulfates may form, often temporarily, when evaporation exceeds rainfall.

Jarosite. Pale yellow (2.5-5Y 8/3-8/6) jarosite [$KFe_3(SO_4)_2(OH)_6$] is conspicuous in most acid sulfate soils. It commonly occurs as earthy fillings of voids or as mottles in the soil matrix, and invariably gives a sharp X-ray diffraction pattern. Sodium and hydronium can substitute for potassium (K), and aluminum (Al) for iron, but almost pure (potassium) jarosite is by far most common. In synthesis experiments (Brown, 1970), and evidently also in acid sulfate soils, jarosite forms directly from solution without a precursor solid phase. Even at strong supersaturation its precipitation may be slow and seems to take place mainly at solid surfaces, generally in association with *Thiobacilli* (Nordstrom et al., 1979; Ivarson, 1973). In soils, where it is formed more readily than in pure solutions, supersaturation may be maintained for months even when the mineral is present (van Breemen, 1976).

In agreement with theoretical stability relations (Fig. 1), jarosite is formed only in acid (pH 2 to 4), oxidized (Eh > 400 mV) environments. However, in acid sulfate soils it is metastable and will eventually hydrolyse to goethite. Often, brown mottles from acid sulfate soils give a sharp jarosite X-ray diffraction pattern, so a thin coating of Fe(III) oxide may prevent recognition of jarosite in the field. Hydrolysis is enhanced by leaching and by supply of bases. Although the yellow color may turn brown within a few months by dialysis (van Breemen, 1976), yellow jarosite mottles sometimes persist for decades at pH values above 4 in limed soils (Verhoeven, 1973).

The pale yellow mottles are so characteristic that they are used, together with pH, as a diagnostic criterion for classifying acid sulfate soils (USDA, 1975). However, jarosite is lacking in some acid sulfate soils, particularly those high in organic matter (Kosaka, 1971).

Iron oxides. Eventually most of the iron from oxidized pyrite ends up in Fe(III) oxides. Fine-grained goethite may form either directly, and quickly, upon oxidation of dissolved Fe(II) sulfate released during pyrite oxidation, or, more slowly, by hydrolysis of jarosite. Both reactions are acidic and part of the sulfuricization process:

$$Fe^{2+} + SO_4^{2-} + 1/4\,O_2 + 3/2\,H_2O \rightarrow FeOOH + 2H^+ + SO_4^{2-} \quad [4]$$

$$\text{jarosite} \rightarrow 3\,FeOOH + 2\,SO_4^{2-} + K^+ + 3\,H^+$$

At pH values below 6, Fe(III) oxides are incompatible with pyrite (see Fig. 1), and indeed, most of the iron oxide is formed at an appreciable dis-

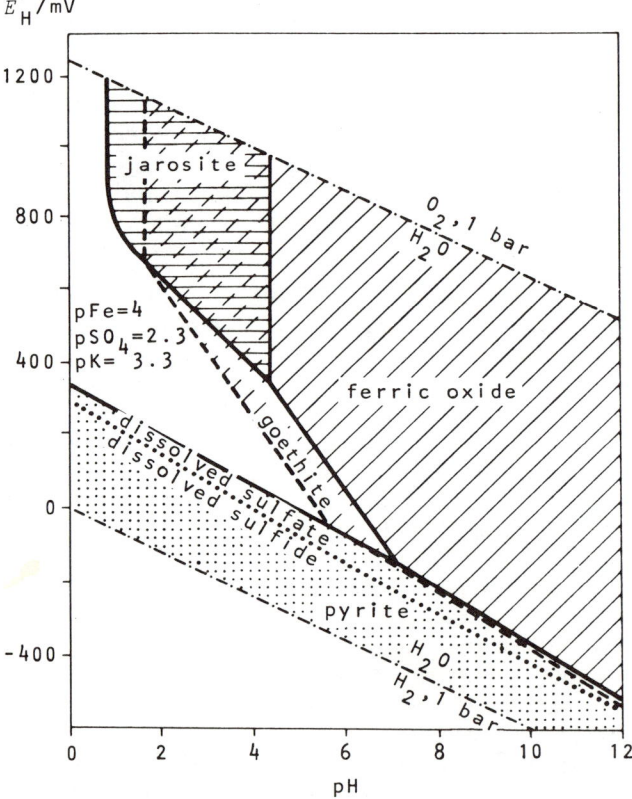

Fig. 1. E_H – pH diagram of ferric oxide (pK_{so} = 39.9), jarosite and pyrite at 25 C. Solid phases are indicated by shading. The broken hatched shading shows the extension of the Fe_2O_3 field at pK_{so} = 42.6 (goethite). After van Breemen (1976).

tance from pyrite. By contrast, at high pH, as in calcareous pyritic sediments, goethite is often pseudomorphic after pyrite (Miedema et al., 1974).

In the better drained, deeply developed acid sulfate soils, part of the Fe(III) oxides in the B horizon may occur as hematite, giving conspicuous red mottles. In Thailand, these red mottles are unknown from non-acid marine soils, and they are used as a field indicator of moderately to strongly acidic conditions (Kevie and Yenmanas, 1972). It is unlikely that the hematite is formed with ferrihydrite as a precursor (Schwertmann and Taylor, 1977), because conditions favoring ferrihydrite (rapid hydrolysis of dissolved iron at relatively high concentrations) are absent in the red mottled B horizons. Perhaps the low pH and periodically dry conditions facilitate transformation of fine-grained goethite, via a solution stage, to the thermodynamically more stable hematite (van Breemen, 1976).

Basic Aluminum Sulfate. The soil solution of acid sulfate soils is generally supersaturated with alunite, the aluminous counterpart of jarosite (van Breemen, 1973b). However, alunite has evidently not been observed in acid sulfate soils, and is more typical of rock weathering by relatively concentrated sulfuric acid in sheltered or hydrothermal environments. The activity of dissolved aluminum in acid sulfate soils in the field, as well as in more acid oxidized pyritic soil in the laboratory, behaved as if regulated by equilibrium with a basic aluminum sulfate of the stoichiometric composition $AlOHSO_4$ (van Breemen, 1973b, Vieillefon, 1974). Similar results were reported for forest soils in West Germany influenced by acid rain (Fassbender and Matzner, 1977). No phase of such a composition has been identified in any of these soils, however. The interpretation by Adams and Rawajfih (1977) that the constancy of the activity product $(Al^{3+})(OH^-)(SO_4^{2-})$ could be due to simultaneous equilibrium with gibbsite and basaluminite $[Al_4(OH)_{10}SO_4]$ is erroneous because practically all acid sulfate soil solutions are far undersaturated with those two minerals.

Watersoluble Sulfates. Gypsum has been observed in coastal marine soils over a wide pH range (3.5 to 7). The upper limit of the calcium sulfate activity product in such soils is clearly regulated by precipitation as gypsum. Due to its fairly high solubility, gypsum is confined to the dryer soils or to those with some supply of calcium carbonate. In dry periods efflorescences of still more soluble sulfate such as Na-alum, tamarugite $[NaAl(SO_4)_2(H_2O)_6]$, pickeringite $[MgAl_2(SO_4)_4(H_2O)_{22}]$ and rozenite $[FeSO_4(H_2O)_2]$, can be formed, particularly on the surface of young acid sulfate soils or of excavated pyritic soil, where pyrite oxidation is relatively rapid (van Breemen, 1976). Precipitates of melanterite $[FeSO_4(H_2O)_7]$ and efflorescences of copiapite $[Fe(II) Fe(III)_4 (SO_4)_6(OH)_2 (H_2O)_{20}]$ and coquimbite $[Fe_2(SO_4)_3(H_2O)_9]$ have been found in association with acid mine drainage (Nordstrom et al., In press). Bandy (1938) reviews the occurrence of these and other sulfates in relation to oxidized pyritic deposits from arid regions in Chile.

SILICATE WEATHERING IN ACID SULFATE SOILS

When the concentration of sulfuric acid from oxidizing pyrite or sulfur becomes very high, for instance, in sheltered but moist environments, aluminum silicates such as feldspars, mica, and smectites may rapidly weather to, e.g., alunite, jarosite, halloysite, and amorphous silica and allophane-like phases (Srebodolskiy, 1972; Keller et al., 1967). In most acid sulfate soils in coastal plains, the acidity is apparently too low to cause extensive silicate weathering. In fact, from X-ray diffraction data and elemental analyses of soil it is often difficult to find evidence for any transformation of silicate minerals. Claims that kaolinite is formed at the expense of 2:1 clay minerals (van Breemen, 1973a) could not be substantiated by later more detailed studies (van Breemen, 1976). There is strong evidence that lattice-bound Mg from saponite-like smectite components is removed as Mg sulfate and is replaced by Fe(III) from oxidizing pyrite, giving a nontronite-like smectite (van Breemen, 1976, 1980):

$$\text{saponite} + 2FeS_2 + 15/2\, O_2 + H_2O$$
$$\rightarrow \text{nontronite} + 3\,Mg^{2+} + 4SO_4^{2-} + 2H^+ \qquad [6]$$

This is the reversal of the reaction described by Drever (1971) to account for the removal of magnesium (Mg) from interstitial water in reduced, pyritic seabottom sediments. Reaction 6 does not, of course, constitute weathering in the usual sense, because the mineral lattice is preserved. Note, however, that most of the acidity normally produced during complete oxidation of pyrite (reaction 2) is inactivated in reaction 6.

In California, acid sulfate soil formation was accompanied by a distinct decrease in Mg-chlorite content and an accompanying increase in smectite, presumably due to selective removal of the Mg-hydroxide layers (Lynn and Whittig, 1966). Still other weathering processes must take place in view of the high concentrations of dissolved silica in most acid sulfate soils. Frequently, dissolved H_4SiO_4 is controlled by the solubility of amorphous SiO_2. Secondary silica has been observed indeed in acid sulfate soils in Australia (Teakle and Southern, 1937) and, as 'opal cristoballite' forming replicates of plant remains with excellent preservation of microscopic detail (Fig. 2), in Thailand, the Netherlands, and Senegal (Buurman et al., 1973).

Although jarosite constitutes a sink for dissolved K, X-ray diffraction of clay and chemical analysis of soil from jarosite and non-jarositic horizons (Fig. 3) indicates that most K in jarosite comes from elsewhere and that K bearing silicates in the immediate vicinity are unimportant as a K-source for jarosite. In fact, in the acid sulfate soils from the Bangkok Plain, weathering, including removal of potassium, is evidently more intense in the surface horizon than in the more acid B horizon (Fig. 3). Stronger weathering near the surface may be caused by ion-uptake by plants and by a seasonally fluctuating pH (4 to 6) associated with periodic flooding.

Fig. 2. Silicified plant material (probably Phragmites sp.) from an acid sulfate soil in the Netherlands. Courtesy S. Henstra, T.F.D.L., Wageningen.

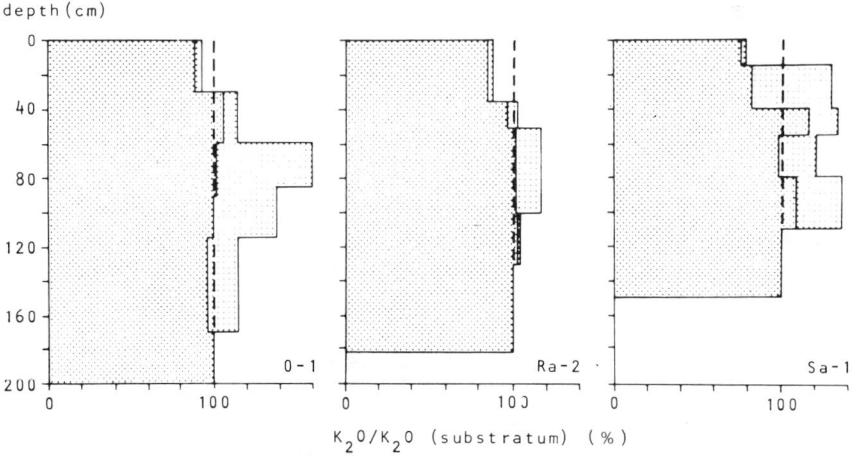

Fig. 3. Total content of K_2O relative to that in the deepest substratum in three uniformly textured acid sulfate soils from Thailand. Lighter shading indicates potassium in jarosite. After van Breemen (1976).

BUFFER REACTIONS AND pH

Calcium carbonate present in aerated pyritic sediments is rapidly dissolved by sulfuric acid. Sulfuric acid in excess of calcium carbonate reacts with clay minerals and other silicates. The reaction releases cations from exchange complex and (especially at pH < 4) magnesium, silica and Al from the mineral lattice. The various clay minerals tend to buffer the pH at 3 to 4. In B horizons of acid sulfate soils on the Bangkok Plain, the assemblage of original clay minerals and the minerals formed by acidification (ferric oxide, jarosite, basic aluminum sulfate, amorphous silica, and sometimes gypsum) make a nearly ideal pH-stat, which keeps the pH in the 3.6 to 3.8 range (van Breemen and Wielemaker, 1974). In sediments with low contents of smectite clay minerals—frequently found in equatorial regions—the equilibrium pH may be closer to 3 to 3.5. Miller (1979) calculated an equilibrium pH of 3.3 for a mine spoil assuming buffering by the assemblage kaolinite, jarosite, ferric hydroxide, and amorphous silica, and indeed found a reasonably good buffering of the pH in that range by such material in column experiments. No data are available on the buffering of pH by organic matter in highly organic acid sulfate soils, where the mineral-solution reactions are probably of lesser importance.

Of course, true equilibrium rarely if ever occurs, but the mineral assemblage of most acid sulfate soils is apparently able to pull the solution composition close to equilibrium. This is possible only at slow flow rates and slow rates of acidification (or of supply of bases) relative to rates of the buffering reactions. At very rapid acidification the buffering mechanism breaks down, which leads to lower pH values and to supersaturation with amorphous silica. For instance, in samples incubated in moist aerated conditions, pyrite oxidation is often so rapid that the pH falls below 2. The pH values between 2 and 3, or even lower, that are often reported in the literature are mostly due to such inadvertent quick oxidation and are rarely representative for soils in the field. But such extreme acidity may develop in situ as a result of sudden, deep drainage as in some reclamation schemes.

Equilibrium calculations for mineral assemblages only reflect the presence, not the amounts of the phases considered. If a phase is used up during acidification (or liming) its buffering action ceases. For instance, upon addition of lime to jarositic soils, jarosite will hydrolyse to Fe(III) oxide. Provided it hydrolyses fast enough, jarosite will keep the pH close to 4 until it is consumed in the process. This illustrates the importance of capacity factors in buffering. Theoretically, all K, Na, Ca and Mg in carbonates and silicates is available for counteracting a decrease in pH in near neutral to slightly acid conditions. Aluminum and Fe(III), whether as free oxides or bound by silica, can also provide buffering upon acidification, but these are quantitatively important only at pH values below 3. So the total amount of silicated (including exchangeable) and carbonated metals other than Al and iron (the 'bases') present in the soil is a measure for the acid-neutralizing capacity in the pH-range of most acid sulfate soils. The reduced pyritic substrate of acid sulfate soils in Thailand have an acid-neutralizing capacity that is roughly equal to, and sometimes even lower than, the potential acidity as pyrite. Yet, the overlying oxidized parts of the soil profile still contain appreciable amounts of silicated bases, reflecting the rather small degree of silicate weathering. This means either (1) that the oxidized horizons were originally lower in pyrite and/or higher in bases than the contemporary substrate; or (2) that most of the acidity formed by oxidation of pyrite formerly present in the now oxidized horizons, has not reacted with the bases present, and has left the soil in some unneutralized form; or (3) that dissolved alkalinity from the surface water or rising from deeper strata has neutralized most of the acidity. There is little chemical or sedimentological evidence for either explanation. However, the possibility that appreciable amounts of gaseous SO_2 (representing unneutralized acidity: $SO_2 + \frac{1}{2}O_2 + H_2O \rightarrow H_2SO_4$) emanate from oxidizing pyritic sediments merits further attention (van Breemen, 1976).

Most mangrove soils are typically high in organic carbon (4 to 10% or higher), which is erratically distributed with depth (Fig. 4A). In monsoon climates most of the organic matter is removed from the periodically drained part of the profile, leading to a decrease in organic

PROFILE DEVELOPMENT OF ACID SULFATE SOILS

Figure 4 shows a hypothetical chronosequence of seasonally flooded acid sulfate soils (Harmsen and van Breemen, 1975). Stage A represents an undrained mangrove soil, and B and C illustrate increasingly older and deeper developed acid sulfate soils. When pyritic soil is drained periodically, e.g., during a dry season, pyrite within 50 cm of the surface can be removed completely in a few decades. Jarosite is formed by oxidation of ferrous sulfate that diffuses upward from the zone where pyrite oxidizes, and accumulates at shallower depth as yellow mottles along pores and cracks (stabe B). Jarosite is slowly hydrolyzed to fine-grained goethite in the upper part of the yellow mottled horizon, causing a residue of brown mottles (stage C). In the surface soil, reduction during flooding mobilizes iron, part of which migrates downward. This eventually leads to a lowering of ferric oxide near the soil surface (stage C). Thus as the soils become older (in terms of pedological maturity, not absolute age) and better drained, the different horizons are found at progressively greater depth. For example, while the acid sulfate soils from inland parts of the Bangkok Plain and the Mekong Delta, especially those in the Plain of Reeds (Moormann, 1959), developed in sediments of probably the same age, most of the soils from Vietnam are more poorly drained and hence less developed and younger than those from Thailand.

The pyritic substratum is generally dark gray or, if highly organic, dark brown. The yellow and brown mottled B horizon may have a grayish brown (Bangkok Plain), light gray (Kalimantan, Malaysia), or dark brown (Kalimantan, Senegal, Philippines) matrix color. Light gray colors may reflect a high kaolinite content, and dark brown colors may be due to high organic matter. Fingers of the relatively more oxidized B horizon often penetrate the pyritic substratum along cracks and old root channels.

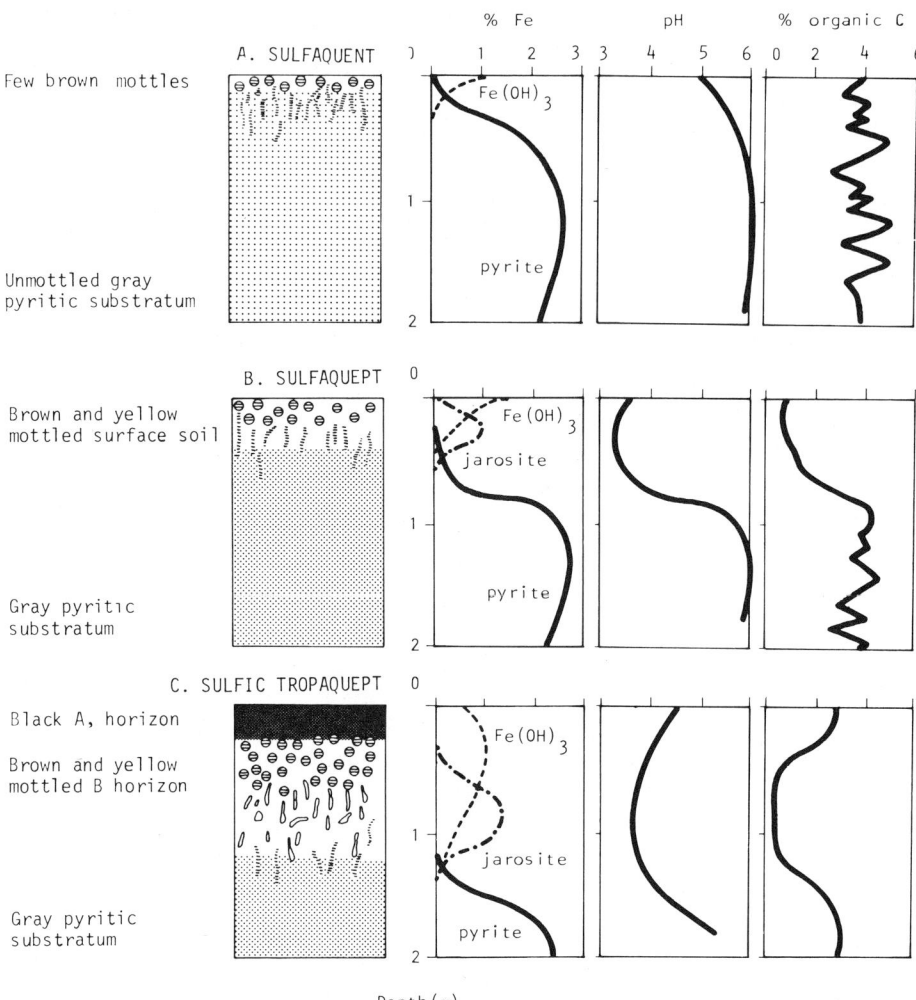

Fig. 4. Profile development (A → C) of acid sulfate soils in the Bangkok Plain, Thailand. A. Sulfaquent; B. Sulfaquept; C. Sulfic Tropaquept.

matter except near the surface where the postmangrove vegetation supplies plant residues (Fig. 4C). Even so, due to low decomposition rates under acidic conditions, acid sulfate soils often have a distinctly higher organic carbon content in the surface horizon than comparable non-acid marine soils (Kawaguchi and Kyuma, 1969). Still lower rates of organic matter decomposition prevail in acid sulfate soils in continuously wet equatorial climates. Such soils often have a peaty surface layer as in Western Malaysia (Chow and Mg, 1969) and southern Kalimantan (Driessen and Soepraptohardjo, 1974).

Peaty surface horizons are also found in poorly drained acid sulfate soils in a monsoon climate as in Kerala state, India (Rao et al., 1975), and the Plain of Reeds, Vietnam (Moormann, 1959). Acid sulfate soils developed in highly organic pyritic sediments sometimes lack the conspicuous yellow mottles, probably because they are generally too reduced to permit the persistence of jarosite (Kosaka, 1971). Often these soils have a typical dark brown matrix color and a buttery consistence (Marius, 1973).

Young acid sulfate soils, especially if they are high in organic matter, undergo rapid reduction after submergence and many again revert to near-neutral potential acid sulfate soils. This reversion may take 1 to 2 years for very young soils under tidal influence, as was found in Sierra Leone (Bloomfield and Coulter, 1973), but takes longer for older soils.

As drainage improves, the soil changes physically—water is lost more or less irreversibly—resulting in a higher bulk density and a lower n-value (Pons and Zonneveld, 1965). In the absence of drainage by gravity as in most coastal plain soils, evapotranspiration is mainly responsible for physical ripening. Because root development and plant growth are hampered in acid sulfate soils, they generally show retarded physical ripening, and a slow decrease of the water table compared with non-acid marine soils. This in turn limits the oxidation of pyrite in the substratum and hence slows their profile development. However, given enough time, complete sulfuricization will eventually take place under favorable hydrologic conditions. Ultimately, this leads to removal of sulfuric acid (of reactions 4 and 5) and the resulting soils cannot be considered as acid sulfate soils any more (post-sulfuricization stage).

CLASSIFICATION OF ACID SULFATE SOILS

Central to the classification of actual and potential acid sulfate soils in *Soil Taxonomy* (USDA, 1975) are the sulfuric horizon and sulfidic material. A sulfuric horizon is composed of mineral or organic soil material with a pH < 3.5 and yellow jarosite mottles. Sulfidic materials are waterlogged mineral or organic soil materials with ≥ 0.75% sulfur and less than three times as much carbonate ($CaCO_3$ equivalent) as sulfide sulfur. In the definition of sulfidic material, the acid neutralizing capacity of clay minerals and the exchange complex is implicitly taken as equivalent to at most 0.75% S. Although this is probably a reasonable limit for many heavy textured coastal plain soils, the actual buffering capacity may vary considerably depending on texture and clay mineralogy. Because some factors that play a role in acid buffering are difficult to quantify (e.g., the effect of easily weatherable silicates) it is probably better to let the soil "speak for itself" by allowing it to oxidize by aeration and by measuring the pH drop. Because the acidification proceeds rather quickly such a method is practically feasible.

The following definition of sulfidic material is based on such a method:
"sulfidic material is waterlogged mineral, organic, or mixed soil material with a pH of 3.5 or higher, containing oxidizable sulfur compounds, which, if incubated as a 1 cm thick layer under moist, aerobic conditions (field capacity) at room temperature, shows a drop in pH of at least 0.5 unit to a pH below 3.5 within 4 weeks".

In view of the highly transient character of sulfidic material, it is essential that the initial pH is determined immediately after sampling.

A drawback of the definition of the sulfuric horizon is that yellow jarosite mottles are absent in some acid sulfate soils. Therefore, another characteristic pointing to sulfuric acid as the acidifying agent should be used in combination with pH to define the sulfuric horizon. Data available so far indicate that the presence of at least 0.05% of watersoluble SO_4 is a better criterion than the presence of jarosite. Therefore, the following definition is proposed:

"a sulfuric horizon" is composed of mineral, organic, or mixed soil material, generally containing yellow jarosite mottles with hue 2.5Y or yellower and chroma 6 or more, that has a pH ≤ 3.5 (1:1 in water) and contains at least 0.05% water soluble sulfate.

In practice, the pH criterion alone will be sufficient because very few soils with a pH (in water) below 3.5 were not influenced by sulfuric acid.

Soils influenced by sulfuricization but not sufficiently acid to be classified as sulfic subgroups have been loosely termed "para" or "pseudo acid sulfate soils" (Pons, 1973). They may either develop due to relatively small amounts of pyrite in the parent material or represent the post-sulfuricization stage of soils that were once acid sulfate soils. They can normally be classified in other taxons, often as Tropaquepts or Haplaquepts.

Potential acid sulfate soils are either Sulfaquents (Aquents with sulfidic material within 50 cm of the mineral soil surface), Sulfic Fluvaquents (Fluvaquents with sulfidic material between 50 and 100-cm depth), or Sulfihemists (Histosols with sulfidic material within the 100-cm depth).

Acid sulfate soils can be classified as Sulfaquepts (Aquepts with a sulfuric horizon that has its upper boundary within 50 cm of the soil surface), Sulfic Tropaquepts (Tropaquepts with jarosite mottles and a pH 3.5 to 4 somewhere within the 50-cm depth, or with jarosite mottles and a pH < 4 in some part between 50 to 150 cm depth), or Sulfic Haplaquepts (comparable to Sulfic Tropaquept but under a more temperate climate). The distinction between Sulfaquepts and Sulfic Subgroups is very useful agronomically in that the former are generally unsuitable for agriculture without costly amendment measures, whereas the latter can often be made productive easily.

Acid sulfate soils that are dominantly organic may be Sulfohemists (Histosols with a sulfuric horizon that has its upper boundary within 50 cm of the surface). Unfortunately, acid sulfate soils with a peaty top soil, which may well be different agronomically from those lower in organic matter, are not separated in histic subgroups.

LITERATURE CITED

1. Adams, F., and Z. Rawajfih. 1977. Basaluminite and alunite: a possible cause of sulfate retention by acid soils. Soil Sci. Soc. Am. J. 41:686–692.
2. Andriesse, J. P., N. van Breemen, and W. A. Blokhuis. 1973. The influence of mudlobsters (*Thallasina anomale*) on the development of acid sulfate soils in mangrove swamps in Sarawak (East Malaysia). p. 11–39. *In* H. Dost (ed.) Acid sulphate soils. ILRI Publ. 18, Vol. II, Wageningen.

3. Arkesteyn, G. J. M. W. 1980. Pyritic oxidation in acid sulphate soils: The role of microorganisms. Plant and Soil 54:119–134.
4. Bandy, M. C. 1938. Mineralogy of three sulfate deposits of Northern Chile. Am. Mineral. 23:669–760.
5. Bloomfield, C., and J. K. Coulter. 1973. Genesis and management of acid sulfate soils. Adv. Agron. 25:265–326.
6. Brown, J. B. 1970. A chemical study of some synthetic potassium-hydromium jarosites. Can. Mineral. 10:696–703.
7. Buurman, P., N. van Breemen, and S. Henstra. 1973. Recent silification of plant remains in acid sulphate soils. Neues Jahrb. Mineral. Monatsh. 1973:117–124.
8. Chow, W. T., and S. K. Ng. 1969. Preliminary study of acid sulfate soils in W. Malaysia. Malay. Agric. J. 47:253–267.
9. De Richmond, T., J. Williams, and Uma Datt. 1975. Influence of drainage on pH, sulphate content and mechanical strength of a potential acid sulphate soil in Fiji. Trop. Agric. (Trinidad) 52:325–334.
10. Drever, J. I. 1971. Magnesium-iron replacement in clay minerals in anoxic marine sediments. Science 172:1334–1336.
11. Driessen, P. M., and M. Soepraptohardjo. 1974. Soils for agricultural expansion in Indonesia. Bull. 1. Soil Res. Inst., Bogor. 63 p.
12. Fanning, D. S. 1978. Soil morphology, genesis, classification and geography. Publ. by the author, Dep. of Agronomy, Univ. of Maryland, College Park, MD 20742.
13. Fassbender, H. W., and E. Matzner. 1977. Zur Bildung von basischen Aluminiumsulfaten im Boden. Mitt. Dtsch. Bodenkundl. Ges. 25:175–182.
14. Harmsen, K., and N. van Breemen. 1975. Translocation of iron in acid sulfate soils. II. Production and diffusion of dissolved ferrous iron. Proc. Soil Sci. Soc. Am. 39:1148–1153.
15. Ivarson, K. C. 1973. Microbiological formation of basic ferric sulfates. Can. J. Soil Sci. 53:315–323.
16. Kawaguchi, K., and K. Kyuma. 1969. Lowland rice soils in Thailand. Natural Science Series N-4. Center for Southeast Asian Studies, Kyoto University. 270 p.
17. Keller, W. D., R. J. Gentille, and A. L. Reesman. 1967. Allophane and Na-rich alunite from kaolinite nodules in shale. J. Sediment Petrol. 37:215–220.
18. Kevie, W. van der, and B. Yenmanas. 1972. Detailed reconnaissance soil survey of Southern Central Plain area. Report SSR-89. Soil Survey Div., Land Dev. Dep., Bangkok. 187 p.
19. Kosaka, J. 1971. Report on acid sulfate soils in the Muda irrigation area. (Japanese, English summary) Trop. Agric. Res. Center (Japan) Bull. 17:29 p.
20. Lynn, W. C., and L. D. Whittig. 1966. Alteration and transformation of clay minerals during catclay development. p. 241–248. In Clays and Clay Miner. Proc. 14th Natl. Conf. Pergamon.
21. Marius, C. 1973. Les sols de l'estuaire du Saloum. ORSTOM, Centre de Dakar. 25 p.
22. Miedema, R., A. G. Jongmans, and S. Slager. 1974. Micromorphological observations on pyrite and its oxidation products in four Holocene alluvial soils in the Netherlands. p. 772–794. In G. K. Rutherford (ed.) Soil Microscopy. Proc. 4th Int. Working meeting soil micromorphology. Limestone Press, Kingston Press, Kingston, Ontario.
23. Miller, S. D. 1979. Chemistry of a pyritic strip-mine spoil. Ph.D. Thesis. Yale University. 201 p. Univ. Microfilms Int., JPK 79-26660, London.
24. Moormann, F. R. 1959. Notes sur les conditions pedologiques et la genese de la plaine des joncs. Secretariat d'etat a l'agriculture, Saigon, Vietnam. 34 p.
25. Nordstrom, D. K., E. A. Jenne, and J. W. Ball. 1981. Redox equilibria of iron in acid mine waters. p. 51–79. In E. A. Jenne (ed.) Chemical modeling in aqueous systems. Speciation, sorption, solubility, and kinetics. Symp. Ser. Am. Chem. Soc. 93.
26. Pons, L. J. 1973. Outline of the genesis, characteristics and classification of acid sulphate soils. p. 3–27. In H. Dost (ed.) Acid sulphate soils. ILRI Publ. 18, Vol. I, Wageningen, Netherlands.

27. Pons, L. J., and I. S. Zonneveld. 1965. Soil ripening and soil classification. Initial soil formation in alluvial deposits and a classification of the resulting soils. ILRI Publ. 13, Wageningen, Netherlands.
28. Quispel, A., G. W. Harmsen, and D. Otzen. 1952. Contribution to the chemical and bacteriological oxidation of pyrite in soil. Plant and Soil 4:43–55.
29. Rao, V., K. P. Rajaram, R. Siddaramappa, and N. Sethunathan. 1975. Phenols in rice soils. Soil Biol. Biochem. 7:227–229.
30. Rasmussen, K. 1961. Uorganske svovlforbindelsers omsaetninger i jordbunden. Thesis. Veterinaer og Landbohøskole, Copenhagen. 176 p.
31. Schwertmann, U., and R. M. Taylor. 1977. Iron oxides. p. 145–180. In Minerals in soil environments. Soil Sci. Soc. Am., Madison, Wis.
32. Singer, P. C., and W. Stumm. 1970. Acidic mine drainage: the rate determining step. Science 167:1121–1123.
33. Srebodolskiy, B. I. 1972. Sulfuric acid weathering of hydromica and montmorillonite. Dokl. Akad. Sci. USSR. Earth Sci. Sect. 211:201–202.
34. Teakle, L. J. H., and B. L. Southern. 1937. The peat soils and related soils of Western Australia. II. Soil survey of Herdsman Lake. J. Agric. W. Austr. 14:404–424.
35. USDA, Soil Conservation Service, Soil Survey Staff. 1975. Soil taxonomy: a basic system of soil classification for making and interpreting soil surveys. USDA Agric. Handb. 436. U.S. Government Printing Office, Washington, DC. 754 p.
36. Van Breemen, N. 1973a. Soil forming processes in acid sulfate soils. p. 66–130. In H. Dost (ed.) Acid sulphate soils. ILRI Publ. 18, Vol. I. Wageningen.
37. ————. 1973b. Dissolved aluminum in acid sulfate soils and in acid mine waters. Proc. Soil Sci. Soc. Am. 37:694–697.
38. ————. 1976. Genesis and solution chemistry of acid sulfate soils in Thailand. Agric. Res. Rep. 848. PUDOC, Wageningen. 263 p.
39. Van Breemen, N. 1980. Magnesium-ferric iron replacement in smectite during aeration of pyritic sediments. Clay Mineral. 15:101–110.
40. ————, and L. J. Pons. 1978. Acid sulfate soils and rice. p. 739–761. In Soils and Rice. International Rice Research Institute, Manila, Philippines.
41. ————, and W. G. Wielemaker. 1974. Buffer intensities and equilibrium pH of minerals and soils. II. Theoretical and actual pH of minerals and soils. Proc. Soil Sci. Soc. Am. 38:61–66.
42. Verhoeven, B. 1973. Acid sulphate soils in the Wieringermeerpolder, a good 40 years after reclamation. p. 114–128. In H. Dost (ed.) Acid sulphate soils. ILRI Publ. 18, Vol. II, Wageningen.
43. Vieillefon, J. 1974. Contribution a l'etude de la pedogenese dans le domaine fluvio-marin en climat tropical d'Afrique de l'Ouest. These. ORSTOM et CRG. Paris et Thonon. 361 p.

Chapter 7

Morphological and Mineralogical Features Related to Sulfide Oxidation under Natural and Disturbed Land Surfaces in Maryland[1]

D. P. WAGNER, D. S. FANNING, J. E. FOSS, M. S. PATTERSON, AND P. A. SNOW[2]

ABSTRACT

Soils and geologic materials exhibiting properties related to sulfide oxidation in upland positions in the Coastal Plain and Appalachian surface mine areas of Maryland were investigated. Features attributable to acid sulfate processes were observed both in recently excavated sulfide-bearing materials as well as in soils on undisturbed land surfaces. Relative to underlying unoxidized strata, near-surface zones of active or recently active sulfide oxidation are characterized by higher contents of free iron oxides, lower contents of total and sulfide-S, and higher contents of sulfate-S. In a profile developed in recently exposed sulfidic sediments, ratios of sulfate-S to total S range from almost 1 near the surface to less than 0.2 below 80 cm. Sulfate minerals associated with recent sulfide oxidation include jarosite, gypsum, copiapite, rozenite, and szomolnokite. These minerals were observed as mottles or efflorescences along ped faces, in soil pores, and on rock surfaces. Jarosite and gypsum were the only sulfate minerals identified under natural land surfaces. The subsoil and substrata of an undisturbed Ultisol developed from glauconitic sediments were found to contain up to 0.3% S as jarosite.

INTRODUCTION

The excessive release of sulfuric acid is a well known problem in areas where sulfide-rich materials are exposed to an oxidizing environment at or near the earth's surface. Even though sulfide occurrence is not

---

[1] Contribution No. 5679 and Scientific Article No. A2638 of the Maryland Agric. Exp. Stn., Dep. of Agronomy, College Park, MD 20742.
[2] Graduate student, professor of soil science, professor of soil science, former research asst., and research asst., respectively, Dep. of Agronomy, Univ. of Maryland, College Park.

Copyright © 1982 Soil Science Society of America, 677 S. Segoe Rd., Madison, WI 53711. *Acid Sulfate Weathering.*

restricted to recent tidal environments, most pedological studies of acid sulfate phenomena have concentrated on the effects of sulfide oxidation in tidal landforms. Having received little scrutiny, the pedogenic importance of acid sulfate processes in upland positions has not been fully appreciated.

There are numerous examples of sulfide oxidation in upland areas. Among the most widely known are those of acid generation in excavated overburdens associated with Paleozoic coal deposits. Less known, but of recently growing interest, have been studies of upland Coastal Plain sediments that exhibit acid sulfate features (6, 13). Additional reports (2, 3, 4, 5, 9, 12, 14, 15, 17, 19, 25) have described acid sulfate soils or sulfide-rich materials not associated with modern tidal environments.

Most investigations of acid sulfate processes in upland positions have dealt with materials artificially exposed at the land surface by man's activities. However, some observations (5, 6, 16) have provided evidence of sulfide weathering under natural landscapes relatively undisturbed by man. Such observations suggest that the processes associated with sulfide weathering may have an important role in the genesis of many soils on natural land surfaces.

Before the role of acid sulfate processes in the genesis of upland soils can be fully defined, two questions must be answered. These are 1) What are the long lasting effects of acid sulfate processes on soil properties? and 2) How may soils that have undergone acid sulfate processes be recognized after these processes are no longer active?

The purpose of this paper is to characterize some of the morphological, mineralogical, and chemical properties of acid sulfate affected soils on upland land surfaces. It is hoped that, as a report of some recent observations on acid sulfate weathering under both disturbed and undisturbed surfaces in Maryland, the findings presented may be applied toward answering the two questions posed above.

MATERIALS AND METHODS

Field

Sites selected for study were chosen on the basis of observed characteristics resulting from sulfide oxidation, or because geologic strata known to contain sulfides were present. Characteristics useful in identifying sites of active sulfide oxidation were lack of vegetation, gray to black soil or sediment colors, jarosite and iron oxide mottling, efflorescences of sulfate minerals, or occasionally a sulfur (probably SO_2) odor. Soil and mineral samples were obtained from roadcuts, surface mine excavations, hand borings, or hand-dug pits.

Two Coastal Plain soils were selected for detailed profile sampling. Both soils were developed from glauconitic sediments. One soil, henceforth designated Profile 1, was developed in recently exposed sulfide-rich and calcareous sediments. The other soil (Profile 2) was a relatively undisturbed soil identified as Collington fine sandy loam (a fine loamy, mixed, mesic Typic Hapludult).

Laboratory

All soil samples were air-dried. Particle size was determined by the pipette method. Soil pH was determined using a 1:1 mixture of air-dried soil and water. The pH of samples from one profile was determined prior to air-drying. Mineral identification was determined by X-ray diffraction of ground powder specimens using a Philips diffractometer equipped with a 2-theta compensating silt and graphite crystal monochrometer. Scanning electron microscopy was also employed in mineral identification. Analyses of dithionite-extractable iron were made using a Philips X-ray spectrograph following the X-ray spectroscopic procedure of Fanning et al. (8).

Total iron was determined by X-ray spectroscopic analysis of digested soil residue. Soil decomposition was accomplished using the HF–HClO$_4$–HNO$_3$ digestion procedure outlined by Pratt (18). After digestion, the residue was dissolved with water, mixed with cellulose, freeze-dried, and pelletized.

Sulfur analyses were made by X-ray spectroscopic analysis of extractant-impregnated cellulose, also freeze-dried and pelletized. Sulfur forms were extracted using a modification of the American Standards for Testing Materials (ASTM) procedure for sulfur determination (1). Sulfate-S was extracted by heating 1.0 g of soil in 25 ml of 3 N HCl at 60 C for a minimum of 3 hours. Extraction of total S was accomplished by heating a separate 1.0 g soil sample in 25 ml of 4 N HNO$_3$ at 60 C for a minimum of 3 hours.

RESULTS AND DISCUSSION

Background Soil-Geology Relationships

Most of the soils and geologic materials studied were in positions where sulfide-rich strata had been exposed by various construction or mining excavations. The man-influenced exposures of the sulfide-rich materials have led to accelerated oxidation of sulfides that were previously preserved under anaerobic conditions. Sites studied were distributed throughout the western portion of the Coastal Plain and in coal surface-mining areas of the Appalachian Province. Figure 1 illustrates the general distribution of the sulfide-bearing formations studied.

The coal mining areas of Maryland are located in the extreme western portion of the state. Geologically, the coal and associated sulfide-rich strata occur within sedimentary formations of Pennsylvanian or Permian Age (24). The geological formations are identified as the Washington, Monongahela, Conemaugh, Allegheny, and Pottsville, and these include the Waynesburg, Pittsburgh, Barton, Upper Freeport, and Cookville coal seams. Coal-mining areas were not examined as extensively as Coastal Plain sites in this study and were investigated principally for the purpose of obtaining sulfate mineral samples.

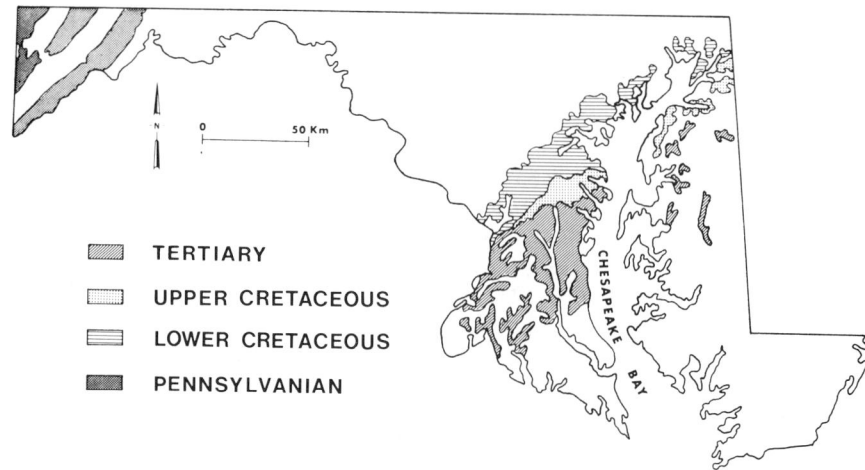

Fig. 1. Map of Maryland showing the distribution of geologic outcrops studied.

Upland acid sulfate soils of the Coastal Plain may be classified into three groups on the basis of geology, morphology, and mineralogy. These groups include soils developed from sulfide-rich sediments of lower Cretaceous, upper Cretaceous, and Tertiary age. Although sulfidic sediments have also been identified in Maryland's tidal marshes of recent age (7), only sediments occurring in upland positions are discussed in this report.

Lower Cretaceous formations are the basal sediments of the Maryland Coastal Plain and form the westernmost outcrops of the Coastal Plain sediments. Lower Cretaceous formations are members of the Potomac Group of sediments. These sediments were deposited by predominantly fluvial processes including in-channel deposition by braided and meandering rivers as well as backswamp deposition on a low, deltaic plain (10). Acid sulfate soils developed from Lower Cretaceous sediments are mostly limited to fine-textured strata containing pyritic lignite. Textures for these strata center in the clay, silty clay, and silty clay loam textural classes (Fig. 2). Lower Cretaceous outcrops range in elevation from nearly 100 m to sea level, and colors range from black (10YR 2.5/1) to gray (10YR 5/1). No jarosite mottles have been observed in what would otherwise be sulfuric horizons formed in Lower Cretaceous sediments.

Upper Cretaceous outcrops occur in positions east of Lower Cretaceous outcrops. A comparison with the Geologic Map of Maryland (26) indicated that the Upper Cretaceous sulfide-rich deposits studied were mapped as the Monmouth, Matawan, and Magothy formations. Upper Cretaceous sediments are of marine origin and are more coarse-textured than sulfide-rich Lower Cretaceous strata. Acid sulfate soils formed from Upper Cretaceous sediments center in the fine sandy loam and sandy clay loam textural classes (Fig. 2). These materials are glauconitic, and substrata below oxidized horizons commonly contain calcareous fossil shell fragments. Outcrops range in elevation from 65 m to sea level. These

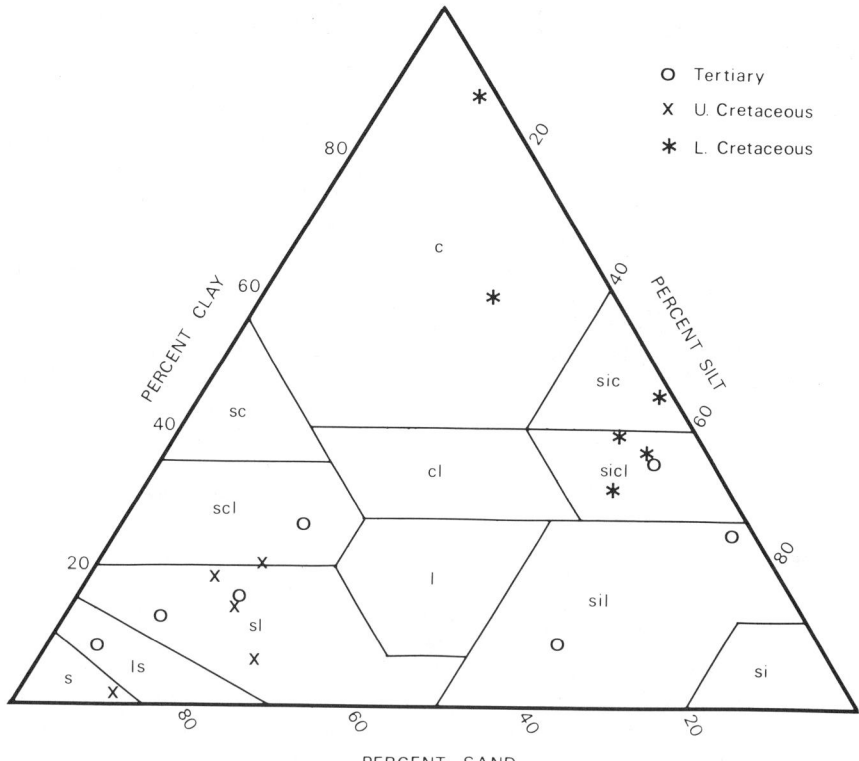

Fig. 2. Textural distribution of sulfide-rich Coastal Plain sediments.

sediments have black (N 2.5 or 5Y 2.5/1) to dark olive gray (5Y 3/2) colors, and pale yellow (5Y 7/4) to olive 5Y 5/6) jarosite mottles are usually present in upper horizons.

Tertiary formations from which acid sulfate soils have developed are also of marine origin, but may or may not contain glauconite or shell fossils. Sulfide-rich outcrops are identified as the Nanjemoy Formation of Eocene Age, the Aquia Formation of Paleocene Age, and the Calvert Formation of Miocene Age. Outcrops of Tertiary sediments lie to the south and east of Cretaceous outcrops. The sediments have variable textures, ranging from silty clay loam to loamy sand (Fig. 2). Outcrop elevations are between 50 m and sea level. Sediments are generally dark grayish brown (2.5Y 4/2) to dark olive gray (5Y 3/2), and commonly have yellowish jarosite mottles near the surface and along cracks.

Sulfate Mineralogy

Several sulfate minerals were identified in the soils and geologic materials. Rozenite ($FeSO_4 \cdot 4H_2O$) was the most common sulfate mineral found on the surfaces of recently exposed coal seams, lignite, and associated strata. This mineral appears as a white powdery precipitate. Individual crystals were usually not discernible without the aid of magnification, but crusts approached one cm in thickness. Rozenite apparently forms by migration of solutions to points of dessication where aqueous sulfate and ferrous ions contained in solution precipitate as rozenite. During the air-drying of samples in the lab, rozenite formed on the surfaces of samples which previously contained no visible sulfate minerals. The mineral is readily soluble in water and was observed in the field only on recently exposed sulfide-rich materials.

Szomolnokite ($FeSO_4 \cdot H_2O$) is megascopically similar in appearance to rozenite and apparently forms as a product of the dehydration of rozenite. Mineral specimens identified as rozenite by X-ray diffraction and then subsequently re-analyzed following several months of storage were found to have transformed to szomolnokite. Szomolnokite samples were never obtained directly from the field, but presumably under sufficiently low humidity this mineral may exist in the field.

Copiapite, giving an X-ray diffraction pattern similar to the ASTM pattern (Card No. 20-659) for alumino-copiapite ($Mg_{0.59}$ $Al_{0.30}$) ($Fe_{3.56}$ $Al_{0.44}$) $(SO_4)_6$ $(OH)_2 \cdot 19.7H_2O$, was occasionally identified in association with rozenite on the surfaces of recently exposed coal and lignite. The mineral has an intense lemon-yellow color (e.g., Plate 4C) and, like rozenite, is readily soluble in water. The association of copiapite with rozenite suggests the two minerals may form by similar mechanisms. Unlike rozenite, however, the iron contained in copiapite is oxidized (Fe^{3+}).

Gypsum ($CaSO_4 \cdot 2 H_2O$) was identified in the sulfuric horizons of several Coastal Plain soils in which the unoxidized parent materials contain both sulfides and calcium carbonate. Where gypsum was present in sulfuric horizons formed from recently excavated sediments, the mineral generally occurred as small (< 1 mm) clusters of needle-like crystals. Figure 3 shows a typical cluster of gypsum crystals. Gypsum typically was concentrated along ped faces or sediment fracture planes in near-surface horizons. Large (up to 5 cm) gypsum crystals have also been observed in the lower substratum of an undisturbed Coastal Plain soil exposed in a bluff along the Potomac River. These crystals were observed at a depth of 3 m and were situated within a transitional zone between oxidized sulfate-bearing horizons and unoxidized sulfide-bearing, calcareous sediment. The gypsiferous horizon was sufficiently acidic to be classified as a sulfuric horizon, and appears to have developed as a result of natural weathering in the geologic column. Gypsum was not observed in any of the coal surface-mining sites investigated; however, the occurrence of this mineral is apparently contingent only on a suitable combination of sulfides and calcium in parent materials being exposed to conditions under which the sulfides oxidize.

Jarosite [$KFe_3(SO_4)_2(OH)_6$] was observed in most of the soil and geologic materials studied. including disturbed materials as well as natural weathering columns. However, acid sulfate soils developed from freshly excavated Lower Cretaceous sediments were distinctly lacking in jarosite. Jarosite was present in nearly all other upland acid sulfate soils of the Coastal Plain and, although not commonly observed in surface mine spoils, jarosite was identified in one "mine-soil".

Fig. 3. Scanning electron micrograph of gypsum crystals on the surface of a plate face from a sulfuric horizon.

Typically, jarosite occurred as yellowish (2.5Y 6/6, 5Y 6/6, 5Y 7/4) mottles along ped faces or in soil pores. Colors are often lower in chroma than the chroma of 6 specified for jarosite by Soil Taxonomy (20) in the definition of the sulfuric horizon. Where present in the substrata of undisturbed Coastal Plain soils, jarosite occurred almost exclusively along the plate faces of the sediment structure. In excavated exposures of sulfide-rich sediments, such as road cuts, jarosite mottles also appeared to have formed along vertical faces and fracture planes. Jarosite was commonly associated with strong brown (7.5YR 5/6) and dark reddish brown (5YR 3/4) iron "oxide" mottles, and was generally present in horizons where gypsum had formed. Figure 4 shows the scanning electron microscopy (SEM) appearance of jarosite crystals along a ped face.

Jarosite was observed both in sulfuric and nonsulfuric (pH > 3.5) horizons. In freshly-exposed materials, or in natural weathering columns where zones of active sulfide oxidation occur, jarosite was present in sulfuric horizons. The mineral also occurred in nonsulfuric subsoil and substrata horizons of undisturbed Ultisols developed from glauconitic sediments. In these soils, jarosite was first encountered within 70 cm of the soil surface, and below this was present to depths of up to 3 m. The pH levels of the nonsulfuric jarosite-rich horizons ranged from 3.6 to 4.6.

The identification of jarosite in the subsoil and deeply weathered substrata of Ultisols suggests that jarosite may have long persistence in soils. This is consistent with the reported low solubility of jarosite (23),

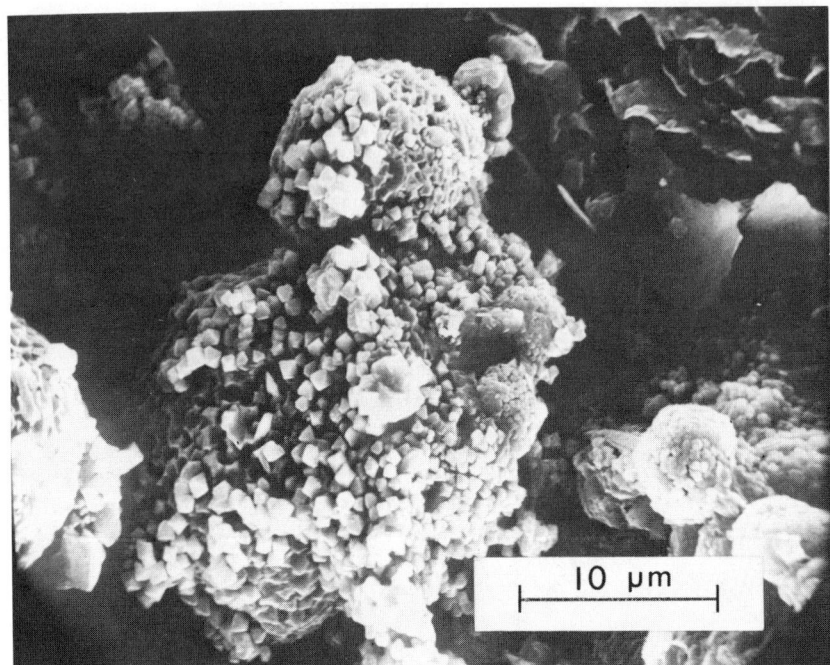

Fig. 4A. Scanning electron micrographs of jarosite crystals in a pale yellow (5Y 7/4) mottle from a sulfuric horizon.

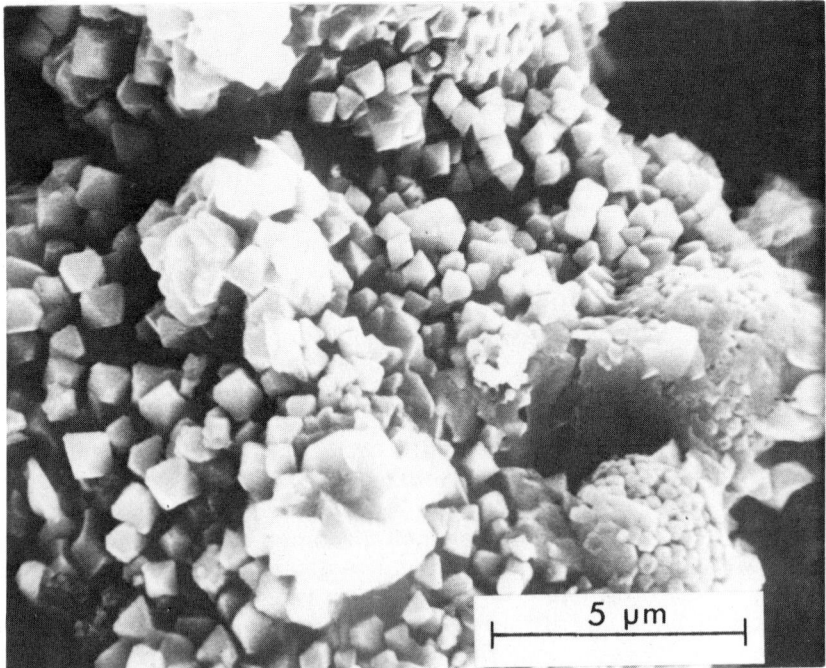

Fig. 4B. Enlargement of the central portion of Fig. 4A.

and is supported by other findings of probable old jarosite (3, 5). Brinkman and Pons (3) have suggested the term "pseudo-acid sulfate soil" be applied to soils having jarosite mottles but pH values above 4. At high pH (>4.4) and in an oxidizing environment, jarosite stability apparently is regulated by the slow transformation of jarosite to ferric oxide and/or goethite (22). The relative stability of jarosite together with its unique field appearance and presumed conditions of formation, make jarosite an important indicator mineral for identifying soils or soil materials that have gone through acid sulfate weathering but no longer have excessively low pH. A soil containing jarosite but in which no horizon is sufficiently acidic to qualify as sulfuric is shown in color Plates 3A and B. Although not studied in detail it appears that this soil would be classified as a fine loamy, mixed, mesic Hapludult.

In addition to the sulfate minerals previously described, other sulfate minerals were also observed but have not yet been identified. These sulfate minerals occurred as crusts on soil or rock surfaces and were present only in freshly exposed materials undergoing active sulfide oxidation. They were mostly olive yellow (5Y 6/6) in color, apparently hydrated as evidenced by a glistening luster, and readily soluble in water. Because of high solubility, such minerals are very short-lived, and are present only during initial stages of sulfide oxidation.

Sulfur and Iron Forms in Soil Profiles

The two soil profiles sampled for the purpose of characterizing iron and sulfur forms represent two time-contrasting stages in soil development. Profile 1 is in recently exposed sulfide-rich sediments, whereas Profile 2 is that of a relatively undisturbed Ultisol developed in deeply weathered sediments. Both soils have formed from glauconitic Coastal Plain sediments.

Active Acid Sulfate Soil. Profile 1 was situated on gently sloping terrain modified by excavation activities in the construction of a nearby highway. A sulfuric horizon was present in the upper 30 cm of the soil, and the acid sulfate conditions have deveoped within approximately one decade, subsequent to construction of the highway. Excavation and removal of an estimated 2 to 3 m of overburden appear to have caused the abrupt exposure of sulfide-rich sediments. Because of the suddenness of sulfide exposure at this site, accelerated processes active in forming profile features probably cannot be construed as completely paralleling early stages of sulfide oxidation in natural weathering columns. Observations of deep sulfide weathering in natural columns indicate that substrata normally undergo oxidation well before incorporation into the sola of soil profiles. This would suggest that oxidation of sulfides deep within undisturbed columns probably proceeds at considerably slower rates than the oxidation of sulfides in freshly exposed materials. Nevertheless, characterization of this profile is important in demonstrating both the pedological implications of man-induced sulfide oxidation as well as the possible direction of parent material transformation in natural weathering columns.

Profile 1 consisted of unstratified material having a very fine sandy loam to sandy clay loam texture. Profile colors were dark olive gray (5Y 3/2) in the upper 30 cm and black (5Y 2.5/1) below this depth. Many pale olive (5Y 7/4) jarosite and strong brown (7/5YR 5/6) or dark reddish brown (5YR 3/4) iron oxide mottles were present on plate faces in the sulfuric horizon. Gypsum clusters were abundant in the upper 30 cm and below 35 cm calcareous fossil shell fragments were present in variable amounts. Pyrite was the only sulfide mineral identified by X-ray diffraction in a sample from the reduced substratum. The water table was not encountered in a boring to 2.4 m. Time of sampling was early autumn. Profile 1 is shown in Plate 3C.

The distribution of pH and total sulfur content with depth for Profile 1 is given in Fig. 5. The pH of samples was determined prior to air-drying and represents field conditions. Lowest pH values are in the sulfuric horizon occupying the upper 30 cm of the profile. Near the surface, sulfuric acid has been released in quantities sufficient to neutralize all carbonates and bring about the acidic conditions indicated. At greater depths pH values are indicative of substantially unoxidized sediments in which $CaCO_3$ from shell fragments appears to control pH.

Concentrations and forms of sulfur in Profile 1 are related to near-surface weathering. The sulfuric horizon has the lowest sulfur content of the profile, whereas below the sulfuric horizon, sulfur content remains

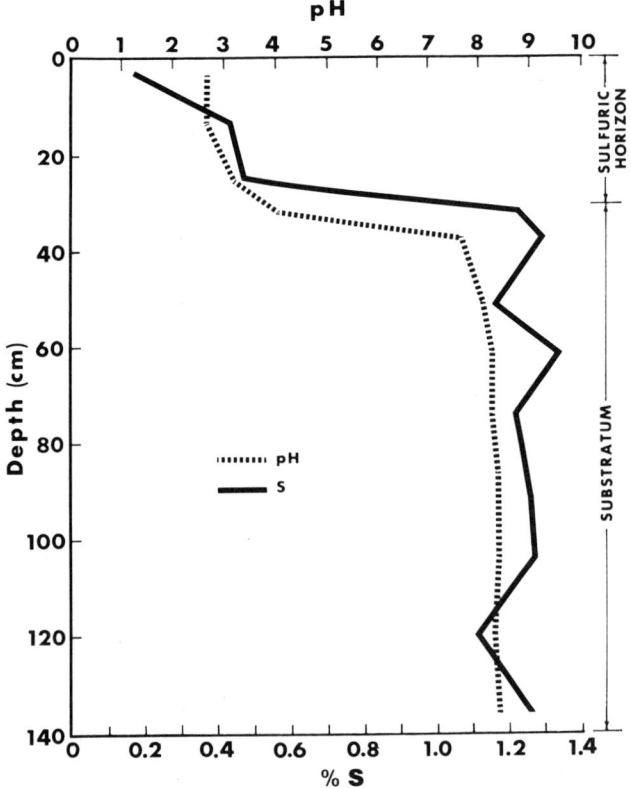

Fig. 5. Sulfur distribution and pH with depth in Profile 1.

relatively constant with depth. Lower sulfur contents near the surface likely represent a loss of sulfur by lateral or vertical leaching of soluble sulfates formed in the oxidation of pyrite. Calculations of sulfate-S to total S ratios from data presented in Fig. 6 give values of nearly 1 within the sulfuric horizon, indicating the complete absence of sulfides. Below the sulfuric horizon, sulfate content gradually decreases at the expense of sulfide content, and sulfate-S to total S ratios become less than 0.2 below 80 cm. The increase in sulfate-S in the 30 to 80 cm depth increment may indicate an illuvial accumulation of sulfate-S, however, the nearly constant distribution of total S strongly suggests in situ conversion of sulfides to sulfates. Relatively high levels of sulfate-S in the 30 to 80 cm increment probably indicate progressive deepening of the oxidized zone in this profile, and as indicated by sulfate-S concentrations of about 0.2%, conditions suitable for oxidation of pyrite may be present to 140 cm. Subsequent studies have shown, however, that most of the sulfates extracted from samples below 80 cm were formed as a result of partial pyrite oxidation during air-drying and storage of samples.

Distribution of dithionite-extractable iron is also shown in Fig. 6. At depths well below the sulfuric horizon concentrations of free iron closely parallel those of sulfate-S. This probably indicates that both the sulfates

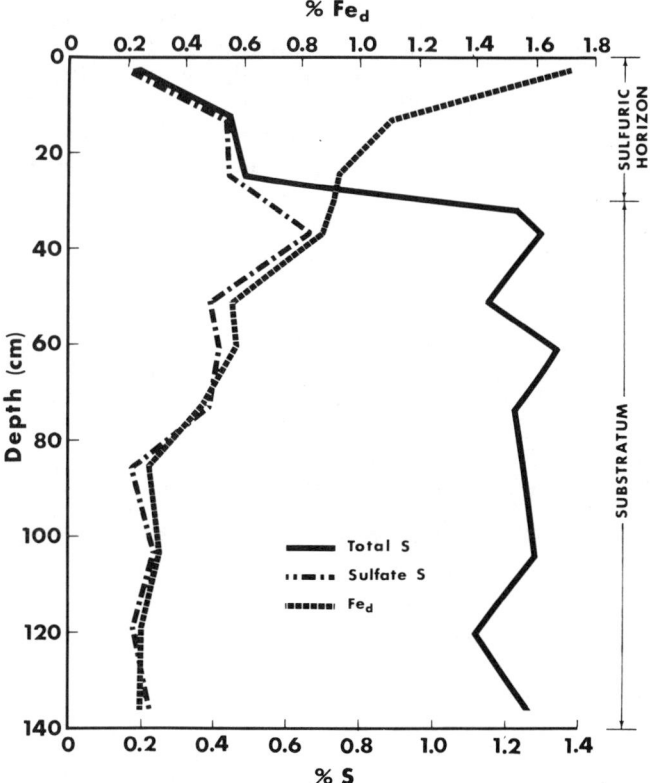

Fig. 6. Distribution of sulfate-S (HCl-extractable), total S (HNO_4-extractable), and Fe_d (free iron), dithionite-extractable) with depth in Profile 1.

and free iron present at lower depths have formed in situ from the oxidation of pyrite. As the surface is approached, the distributions of sulfate sulfate-S and free iron become strongly divergent, and within the sulfuric horizon free iron content increases as sulfate-S decreases. The upward decrease in sulfate-S within the sulfuric horizon can probably be attributed to greater leaching losses near the surface. Possible mechanisms responsible for the upward increase in free iron within the sulfuric horizon are not as readily apparent. It should be noted here that dithionite-extractable iron includes iron from jarosite and ferrous sulfate as well as from iron oxides. As evidenced by high chroma mottling patterns, the bulk of the iron extracted from the sulfuric horizon is probably derived from oxidized forms (free iron oxides and jarosite).

In discussing the distribution of free iron in the sulfuric horizon considerations must be given to both the sources and means of retention of free iron in this soil. Iron released from pyrite oxidation, together with iron released from acid attack on other iron-bearing minerals, likely accounts for most of the iron presently recoverable in oxide form. Because of this, substantial importance must be attached to the fact that no pyrite remains in the sulfuric horizon. The absence of pyrite throughout the sulfuric horizon suggests all portions of the sulfuric horizon have undergong comparable degrees of weathering. This further suggests that equal supplies of iron for conversion to free iron forms would have been available throughout the sulfuric horizon. If original in situ supply is assumed to have been equal, then the uneven distribution of free iron must be credited to relative variations in iron retention in different portions of the sulfuric horizon, or to mechanisms of iron mobility. This presents two apparent possibilities that allow for increasing free iron concentrations toward the surface. Either there has been greater loss of iron from the lower part of the sulfuric horizon relative to the surface, thus implying more favorable conditions for precipitation and retention of iron oxides at the surface; or upward translocation of iron, perhaps initially in ferrous sulfate form, has occurred.

Data shown in Table 1 offer little conclusive evidence to preferentially support either of the above possibilities. The upward increase in iron content in the sulfuric horizon is reflected in values both for dithionite-extractable as well as total iron. Since total iron increases toward the surface, it seems unlikely that higher free iron contents at the surface represent more intense weathering and resultant increased iron transformation at this level. Rather, it appears iron distribution must be accounted for in terms of additions or losses within the sulfuric horizon. Existing laboratory data do little to direct the discussion beyond this point.

Additional evidence giving some insight into the understanding of iron distribution in Profile 1 has been provided by field observations. Efflorescences of rozenite in mixture with other sulfate minerals have been observed and identified (X-ray diffraction) on the surface of this soil. These efflorescences appeared in the form of powdery crusts along the edges of coarse fragments or on ped faces. The question arises as to where the ferrous sulfate comprising the readily soluble rozenite has originated. Since there is no remaining pyrite in the sulfuric horizon, it is unlikely that ferrous sulfate is actively being generated in this horizon. Nor does it seem likely that the surface rozenite has formed from residual ferrous sulfate surviving since the period when pyrite was actively weathering in the surface zone. The high solubility of rozenite and general tendency to lose sulfates from the surface argue against this. Most probably, ferrous sulfate found in the sulfuric horizon, and perhaps much of the acid which sustains the low pH of the sulfuric horizon, have originated from below the sulfuric horizon where active oxidation of pyrite is still progressing. This would require an upward migration of at least 30 cm.

Low sulfate concentrations in the sulfuric horizon indicate that at any given time the amount of free iron extracted directly from ferrous sulfate is not appreciable. Because of this, any model for free iron accumulation which initially involves transport of iron as ferrous sulfate must also include as a subsequent stage some mechanism for the conversion of ferrous sulfate iron to a more insoluble form. The exact mechanism by which such a conversion may occur is not known. Upward translocation of iron has been reported in acid sulfate soils of Thailand (11, 23), where the principal mechanism believed responsible for upward migration and accumulation of iron was the seasonal migration of Fe^{2+} in solution in response to evapotranspiration of the water followed by dessication and Fe^{3+} oxide precipitation. This model does not account for an intermediate Fe^{2+} holding stage, which to some extent appears to occur in this soil; namely Fe^{2+} held in rozenite in the presence of atmospheric O_2. The low pH of the sulfuric horizon apparently allows for the stability of Fe^{2+} under such high E_H conditions. It is suggested that dissolution of rozenite by rainwater and the attendant increase in soil pH by dilution, may provide a means for conversion of rozenite-iron to iron oxide-iron. While much of the iron would precipitate from solution, soluble sulfates would be subject to removal from the soil by leaching, thus accounting for the observed sulfur losses and iron increase as the surface is approached (Fig. 6 and Table 1).

Table 1. Distribution of iron forms with depth in Profile 1.

Depth (cm)	Fe (Total)	Fe (Dithionite)	Fe† (Pyrite)	Fe† (Silicate)
		%		
0–6	5.09	1.72	0.01	3.36
6–20	4.03	1.11	0.00	2.92
20–30	3.36	0.92	0.03	2.41
30–35	4.25	0.91	0.59	2.75
35–46	4.04	0.87	0.55	2.62
46–56	3.64	0.55	0.66	2.43
56–66	4.10	0.58	0.80	2.72
66–82	4.37	0.45	0.73	3.19
82–97	4.32	0.27	0.93	3.12
97–112	4.22	0.30	0.90	3.02
112–128	3.96	0.24	0.81	2.91
128–143	4.21	0.25	0.90	3.06

† Calculated values. Fe (Pyrite) was calculated from the Formula FeS_2, using S (pyrite) as the difference between S (Total) and S (SO_4). Fe (Silicate) was calculated as the difference between Fe (Total) and the combined sum of Fe (Dithionite) and Fe (Pyrite).

Inactive (Pseudo or Fossil) Acid Sulfate Soil. Profile 2 was situated on a nearly level, undisturbed land surface. This soil was developed from coarse-textured glauconitic sediments and was identified as Collington fine sandy loam, a Typic Hapludult. The profile did not contain a sulfuric horizon, but pale yellow (5Y 7/4) jarosite mottles were common in the lower solum and upper C horizons. Horizons with jarosite ranged in pH from 4.6 to 4.3. An argillic horizon occurred between the depths of 50 and 70 cm, and jarosite mottles were present in the lower part of the argillic horizon. Neither the solum nor substrata were calcareous; however, a 10 cm thick silica-cemented layer of fossil shell casts marked the base of the argillic horizon.

The geologic sediments below Profile 2 were deeply weathered. High concentrations of sulfur were limited to the upper 150 cm of the profile, and analyses of samples from a deep boring indicated little or no sulfur in deeper layers to a depth of over 8 m. Investigation of materials at greater depth was prevented by caving of the auger hole. Deep substrata had loamy sand textures and were predominantly very dark grayish brown (2.5Y 3/2) in color. Common-to-many reddish brown (5YR 4/4) iron oxide mottles were encountered throughout the column, and several 1 to 2 cm thick iron oxide cemented layers were penetrated. The geologic column is evidently oxidized sufficiently that any sulfides that may originally have been present have been oxidized.

Maximum sulfur concentration in Profile 2 was between the depths of 60 and 100 cm (Fig. 7). This depth increment corresponds with the B2t and B3 horizons, and also comprises the zone of maximum iron oxide accumulation. Since the dithionite treatment extracts jarosite iron as well as iron oxide iron, a portion of the iron concentrations shown in Fig. 7 must be attributed to jarosite. Using the jarosite formula of $KFe_3(SO_4)_2(OH)_6$ and assuming that all sulfur is in jarosite form, a value of 0.73% jarosite iron is obtained for the argillic horizon. This indicates that as much as 33% of the dithionite-extractable iron in the argillic horizon is from jarosite.

The retention of sulfur as jarosite in the solum of Profile 2 allows for some speculation concerning the weathering history of this soil. It is not certain whether sulfides were ever present throughout the entire geologic column; however, a supposition of originally ubiquitous sulfide distribution is not without evidence. The presumed acid sulfate origin of the jarosite indicates that at least part of the column originally contained sulfides, and except for the present pattern of sulfur distribution, the deposit was otherwise quite uniform. Also, the association of pyrite with unweathered glauconitic sediments seems common in Maryland and has been reported in other parts of the world as well (2, 5, 14, 21, 25). Assuming an initial distribution of sulfides throughout the geologic column, conditions for jarosite formation and/or stability must have existed only in relatively close proximity to the land surface. Elsewhere in the column, oxidation rates or the necessary activity of constituent ions in the soil solution may have been too low for jarosite to have formed or persisted. The apparent great depth of sulfide weathering and the presence of jarosite in the argillic horizon suggest that the jarosite in the profile is very old. Thus

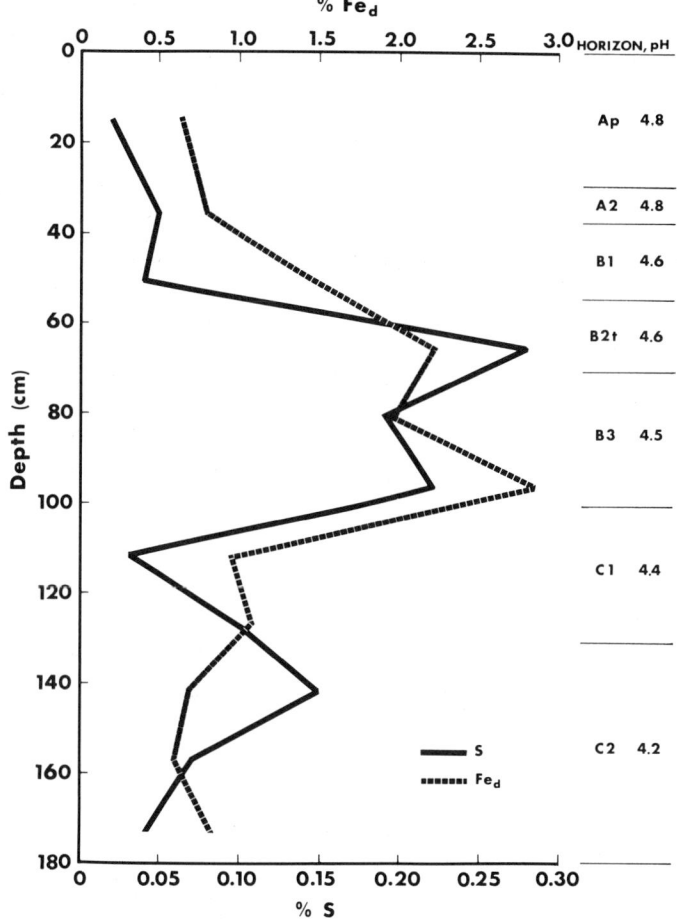

Fig. 7. Distribution of sulfur and Fe_d (free iron) in Profile 2 with depth. Positions of horizons and soil pH are shown on the right.

this soil should represent the post-sulfuricization stage of soil development described in the paper by Carson et al. (6) in this publication, and this soil may be considered to be a pseudo-acid sulfate soil by the terminology of Brinkman and Pons (3).

Subsequent studies are showing that many of the soils developed in glauconitic sediments in Maryland, most of which are Hapludults, contain jarosite in their lower B and in C horizons.

CONCLUSIONS

Upland soils and geologic materials having acid sulfate properties occur in the Appalachian coal mining and Coastal Plain regions of Maryland. These sulfide-bearing formations range in age from Pennsylvanian or Permian in the Appalachian coal basins to Lower Cretaceous through Tertiary in the Coastal Plain. Many soils having acid sulfate properties occur where sulfide-rich strata have been recently exposed by excavation. Other soils with acid sulfate weathering features are situated on relatively undisturbed landscapes where processes of sulfide oxidation have not been artificially accelerated by man.

Sulfate minerals often are indicators of active or previously active sulfide oxidation. Water soluble, hydrated sulfates such as rozenite form rather quickly in newly exposed sulfide-bearing materials. Less soluble minerals such as gypsum and particularly jarosite are found both in recently disturbed materials as well as in undisturbed weathering profiles.

Distributions of iron and sulfur forms in soil profiles are influenced by duration of weathering and transport of ions by evaporation or leaching. Soils developed in recently weathered sediments have the highest sulfate and iron oxide contents in near-surface, sulfuric horizons. The sulfuric horizon of one profile was found to be lower in sulfur than the underlying substrata, indicating the apparent leaching of soluble sulfates from the surface. In the same soil, a progressive increase in dithionite extractable free iron toward the soil surface was identified. This suggests that upward translocation of iron as ferrous sulfate may be an important process in acid sulfate soils.

In undisturbed Coastal Plain soils developed from glauconitic sediments, jarosite has been observed in subsoil and substrata horizons. The presence of jarosite in the argillic horizon of Ultisols indicates that this mineral may persist long after processes of sulfide oxidation have ceased. Jarosite appears to be a useful indicator mineral for identifying soils that have undergone acid sulfate weathering but which no longer exhibit excessively low pH.

LITERATURE CITED

1. American Society for Testing and Materials. 1974. Forms of sulfur in coal. D3177 p. 507–511. *In* 1974 Book of ASTM Standards: Part 26, Gaseous fuels; coal and coke. Philadelphis, Pa. 1974.
2. Briggs, L. I. 1951. Jarosite from the California tertiary. Am. Mineral. 36:902–906.
3. Brinkman, R., and L. J. Pons. 1973. Recognition and prediction of acid sulphate soil conditions. p. 169–203. *In* H. Dost (ed.) Acid sulphate soils. ILRI Pub. 18. Vol. 1. Wageningen, Netherlands.
4. Brophy, G. P., and M. F. Sheridan. 1965. Sulfate studies IV: The jarosite-natrojarosite-hydronium jarosite solid solution series. Am. Mineral. 50:1595–1607.
5. Buurman, P., N. van Breemen, and A. G. Jongmans. 1973. A fossil acid sulphate soil in ice-pushed Tertiary deposits near Uelson (Krels Nordhorn), Germany. p. 52–75. *In* H. Dost (ed.) Acid sulphate soils. ILRI Pub. 18. Vol. 11. Wageningen, Netherlands.
6. Carson, C. D., D. S. Fanning, and J. B. Dixon. 1982. Alfisols and Ultisols with acid sulfate weathering features in Texas. p. 127–146. *In* J. A. Kittrick, D. S. Fanning, and L. R. Hossner (ed.) Acid sulfate weathering. SSSA Spec. Pub. no. 10. Madison, Wis.
7. Darmody, R. G., and J. E. Foss. 1979. Soil-landscape relationships of the tidal marshes of Maryland. Soil Sci. Soc. Am. J. 43:534–541.
8. Fanning, D. S., R. F. Korcak, and C. B. Coffman. 1970. Free iron oxides: rapid determination utilizing X-ray spectroscopy to determine iron in solution. Soil Sci. Soc. Am. Proc. 34:941–946.
9. Furbish, W. J. 1963. Geologic implications of jarosite, pseudomorphic after pyrite. Am. Mineral. 48:703–706.

10. Glaser, J. D. 1969. Petrology and origin of Potomac and Magothy (Cretaceous) sediments, Middle Atlantic Coastal Plain. Maryland Geol. Survey Rep. Inv. 11. 101 p.
11. Harmsen, K., and N. van Breemen. 1975. Translocation of iron in acid sulfate soils: II. Production and diffusion of dissolved ferrous iron. Soil Sci. Soc. Am. Proc. 39:1148–1153.
12. Ivarson, K. C. 1973. Microbiological formation of basic ferric sulfates. Can. J. Soil Sci. 53:315–323.
13. Miller, W. L., C. L. Godfrey, W. G. McCully, and G. W. Thomas. 1976. Formation of soil acidity in carbonaceous soil materials exposed by highway excavations in East Texas. Soil Sci. 121:162–169.
14. Mitchell, R. S. 1962. New occurrences of jarosite in Virginia. Am. Mineral. 47:788–789.
15. Moorman, F. R. 1963. Acid sulphate soils (cat-clays) of the tropics. Soil Sci. 95:271–275.
16. Pawluk, S., and M. Dudas. 1978. Reorganization of soil materials in the genesis of an acid Luvisolic soil of the Peace River Region, Alberta. Can. J. Soil Sci. 58:209–220.
17. Poelman, J. N. B. 1973. Soil material rich in pyrite in noncoastal areas. p. 197–204. In H. Dost (ed.) Acid sulphate soils. ILRI Pub. 18. Vol. II. Wageningen, Netherlands.
18. Pratt, P. F. 1965. Digestion with hydrofluoric and perchloric acids for total potassium and sodium. In C. A. Black (ed.) Methods of soil analysis, Part 2. Agronomy 9:1019–1021. Am. Soc. of Agronomy, Madison, Wis.
19. Rimsait, J. 1975. Natural alteration of mica and reactions between released ions in mineral deposits. Clays and Clay Miner. 23:247–255.
20. Soil Survey Staff. 1975. Soil taxonomy. USDA Handb. 436, U.S. Government Printing Office, Washington, DC.
21. Tyler, S. A. 1936. Heavy minerals of the St. Peter sandstone in Wisconsin. J. Sediment. Petrol. 6:55–84.
22. van Breemen, N., and K. Harmsen. 1975. Translocation of iron in acid sulfate soils: I. Soil morphology, and the chemistry and mineralogy of iron in a chronosequence of acid sulfate soils. Soil Sci. Soc. Am. Proc. 39:1140–1148.
23. Vlek, P. L. G., Th. J. M. Blom, J. Beek, and W. L. Lindsay. 1974. Determination of solubility product of various iron hydroxides and jarosite by the chelation method. Soil Sci. Am. Proc. 38:429–432.
24. Waage, K. M. 1959. Refractory clays of the Maryland coal measures. Maryland Dep. Geol. Mines and Water Resources. Bull. 9. 182 p.
25. Warshaw, C. M. 1956. The occurrence of jarosite in underclays Am. Mineral. 41:288–296.
26. Weaver, K. N., E. T. Cleaves, J. Edwards, and J. D. Glaser. 1968. Geologic Map of Maryland. Maryland Geol. Survey. Baltimore, Maryland.

Chapter 8

Alfisols and Ultisols with Acid Sulfate Weathering Features in Texas[1]

C. D. CARSON, D. S. FANNING, AND J. B. DIXON[2]

ABSTRACT

Jarosite has been identified in soils classified as Ultisols (e.g., Aubrey soil series) and/or Alfisols (e.g., Lufkin soil series) developed on the Woodbine, Catahoula, and Yegua geologic formations in Texas. The Woodbine is Upper Cretaceous, the Catahoula is Oligocene, and the Yegua is Eocene. The extent of these geologic formations suggests that there may be many other upland soils that contain jarosite. Jarosite has been identified in very acid soil horizons and in other soil horizons with near neutral pH. Some horizons containing jarosite have as much as 8.1 meq/100 g exchangeable aluminum (Al) and 23.8 meq/100 g nonexchangeable titratable acidity. Appreciable amounts of acidity determined in soils with a near neutral pH may reflect an earlier acid sulfate weathering regime. None of these upland soils have pH values low enough to have any horizons that qualify as sulfuric horizons today, but the presence of jarosite is interpreted as indicating that these soils underwent active sulfuricization at some time in the past.

Jarosite has been identified in these soils by noting mottles with yellow (5Y) hues and with verification in the laboratory by X-ray diffraction or differential thermal analysis on relatively pure hand picked specimens. Barite, gypsum, and calcite have been identified in some horizons that contain jarosite by employing a similar combination of field and laboratory techniques. The mineralogy of the matrix soil materials associated with the presence of jarosite is usually mixed in the Woodbine derived soils and smectitic in the Yegua and Catahoula derived soils. Pyrite weathering has apparently been the source of sulfate ions and acidity necessary for jarosite formation. Jarosite seems to be preserved in some Claypan soils due in part to their dense clayey subsoil.

---

[1] Contribution from Texas Agric. Exp. Stn., College Station; Southwest Texas State Univ., San Marcos; and Univ. of Maryland, College Park.

[2] Associate professor of agriculture at Southwest Texas State Univ., San Marcos, TX 78666; professor of soil science, Univ. of Maryland, College Park, MD 20742; and professor of soil mineralogy, Texas A&M Univ., College Station, TX 77843.

Copyright © 1982 Soil Science Society of America, 677 S. Segoe Rd., Madison, WI 53711. *Acid Sulfate Weathering.*

INTRODUCTION

This paper deals mainly with soils that show relict features of acid sulfate weathering. The bulk of this weathering, referred to below as sulfuricization, probably occurred at an early stage in the genesis of these soils or as a kind of parent material pre-weathering. The soils today have been classified as Alfisols and Ultisols (Table 1). These soils are envisioned as fitting into the post-sulfuricization stage of the idealized scheme of acid sulfate weathering presented below. Sulfuricization has been defined (Fanning, D. S., J. B. Dixon, D. P. Wagner, and L. R. Hossner. 1979. Sulfuricization: acid sulfate weathering of soils, sediments and sedimentary rocks. Abstracts for Clay Minerals Society meetings, in Macon, Ga., p. 29; also Fanning, 1978) as the overall process by which sulfide bearing materials are oxidized, minerals are weathered by the sulfuric acid produced and new mineral phases are formed from the dissolution products. The dissolved products of the sulfuric acid weathering are conceived, in some instances, as moving by diffusion or mass flow (e.g., by leaching, or upwards or laterally by capillary flow that may be enhanced by evapotranspiration) before they are precipitated in response to Eh or concentration factors. Among the minerals that appear to weather are sulfides, carbonates, and silicates. Among new minerals formed are sulfates (e.g., jarosite and gypsum), iron oxides (e.g., goethite), and probably also silicates (e.g., smectite).

The idealized stages in the sulfuricization of a material are:

Stage 1. Pre-Sulfuricization. Sulfide bearing material is present with essentially no sulfates because oxidation has been prevented (e.g., by continuous saturation with water). The pH may be neutral or above. Material representing this stage may be a sulfidic material as defined in *Soil Taxonomy* (Soil Survey Staff, 1975) if 0.75% S or more is present mainly in the form of sulfides and there is less than three times as much $CaCO_3$ equivalent as S and the material is waterlogged. The intent here, however, is to encompass any sulfide bearing material that is not undergoing a net oxidation. Sulfide minerals, trioctahedral chlorites, siderite, and other minerals with iron (Fe), manganese (Mn), and sulfur (S) in reduced states are considered to be quite stable under the conditions represented by this stage and may actually be forming.[3]

Stage 2. Actively Sulfuricizing. The pH may be very low (<3.5) if carbonates are not present and sulfuric acid is being produced more rapidly than it reacts with silicate minerals. If pH is <3.5 (1:1 in H_2O), and jarosite is present, material representing this stage may qualify as a sulfuric horizon as defined in *Soil Taxonomy* (Soil Survey Staff, 1975). Rate of O_2 supply is probably the main control on the rate of H_2SO_4 production. Many silicates and other minerals are decomposed. New minerals form from reaction products, but this may take place in another location (e.g., jarosite precipitation in structural partings where Eh is high enough for Fe^{+3} to be stable).

---

[3] The process or processes by which sulfide minerals form (especially in tidal marshes) have been described by Rickard (1973) and Goldhaber and Kaplan (1982). This process has been called sulfidation or pyritization (Rickard, 1973) or sulfidization (Fanning, 1978).

Table 1. Classification and location (county) of pedons of jarosite containing soils from Texas with name and age of geologic formations upon which the pedons occurred.

Soil series†	Taxonomic name	County	Geologic formation	Geologic age
Aubrey	clayey, mixed thermic Typic Haplustult	Denton	Woodbine	Upper Cretaceous
Birome	fine, mixed, thermic Ultic Paleustalf	Hill	Woodbine	Upper Cretaceous
Shalba	fine, montmorillonitic, thermic Typic Albaqualf	Washington	Catahoula	Oligocene
Lufkin	fine, montmorillonitic, thermic Vertic Albaqualf	Brazos	Yegua	Eocene

† Soil series into which the pedon would be classified. The Aubrey and Birome soils are found predominantly in the East Cross Timbers land resource area and the Shalba and Lufkin are predominantly in the Claypan land resource area.

Stage 3. Post-Sulfuricization. Sulfides are completely oxidized. The pH is normally 4 or above. Evidence that a material has been sulfuricized may include the following: a) the presence of jarosite or other sulfate minerals (e.g., gypsum and barite) with underlying material representing Stages 1 and/or 2, b) underlying materials representing Stages 1 and/or 2 and evidence that the column was lithologically continuous. The material may be acid (pH 4 to 5) and have high color chromas from iron oxides and lack sulfates under certain oxidizing and leaching conditions.

Our experience, reported here, indicates that materials representing Stage 3 with jarosite mottles may have a low pH or a high pH. High pH (e.g., secondary carbonates present) in the presence of jarosite may indicate that the material has had a low pH (probably 4 or below when the jarosite formed) and has subsequently developed a high pH. The mechanism for the apparent pH rise beyond what an aluminum (Al) saturated material would have (4 to 5) is not fully understood. It may involve a) re-saturation of the material with bases from gypsum and those released during silicate weathering (acidity may slowly release bases from feldspars and other minerals). This may raise the pH if H_2SO_4 production stops or is slow, b) bases and alkalinity brought in from surrounding materials by lateral or vertical leaching or capillary rise (perhaps in response to evapotranspiration), c) CO_2 from the biosphere converting gypsum to calcite with some leaching to get rid of H_2SO_4 generated in this process, or d) $CaCO_3$ or other bases added to the soil as dust (natural or artificial liming).

Not considered above in these idealized sulfuricization stages is the possibility of reversals because of saturation or alternating oxidizing and reducing conditions. At least two different reduction situations would seem to exist. Under weak reducing conditions Fe^{+3} may be reduced (to Fe^{+2}), but not sulfate; since Fe^{+3} is reduced at a higher Eh than $SO_4^=$ (e.g., see Patrick and Mikkelsen, 1971). This should cause dissolution of jarosite and its constituent ions would be mobilized for possible leaching. Under stronger reducing conditions, sulfate S of jarosite and other sulfate minerals could be reduced to sulfide as well as having Fe^{+3} (of jarosite and iron "oxides") reduced to Fe^{+2}. Iron sulfides should then form. This probably would require the presence of sulfur reducing bacteria. This possibility has been demonstrated in laboratory experiments, in which jarosite was converted to iron sulfides in the presence of Desulfovibrio desulfuricans, by Ivarson et al. (1976) and Ivarson and Hallberg (1976).

Another aspect of reactions in which Fe^{+3} and sulfate S are reduced, is that some alkalinity may be produced; e.g., from the formation of HCO_3^- when organic matter is oxidized during some of these reactions (van Breemen and Brinkman, 1976, p. 162). This may also contribute to the rise in pH noted above in some materials representing the post-sulfuricization stage.

Materials representing the various stages are often present in the same geologic column and/or landscape. Weathering products, especially from Stage 2, may sometimes move into materials representing the other stages and affect them. Thus the various stages are mainly conceptual.

Some problems arise concerning nomenclature for "acid sulfate soils" that represent the various stages of sulfuricization noted above. This problem has been dealt with by Brinkman and Pons (1973), although the term sulfuricization was not used by them. Corresponding to Stage 1 of the process, Brinkman and Pons defined *potential acid sulfate soil*:

"A *potential acid sulfate soil or material* is a soil or reduced parent material which is expected by the person identifying it to become an acid sulfate soil or material upon drainage and oxidation under certain defined future field conditions."

It apparently was this concept that the authors of *Soil Taxonomy* had in mind when they defined *sulfidic materials* (Soil Survey Staff, 1975).

Corresponding to Stage 2 of the process, Brinkman and Pons defined *actual acid sulfate soils*:

"An *(actual) acid sulfate soil* is a soil with one or more horizons consisting of (actual) acid sulfate materials, i.e., materials containing soluble acid aluminum and ferric sulphates in concentration toxic to most common dry-land crops. Such materials have high proportions of exchangeable aluminum, pH (water) below 4, and may have characteristically pale yellow mottles by acid sulphates of iron, potassium iron or sodium iron. Some have white mottles of aluminum sulphates."

Soil Taxonomy authors must have had the actual acid sulfate soil concept in mind when they defined the *sulfuric horizon*.

Corresponding to Stage 3 of the sulfuricization process, Brinkman and Pons (1973) defined *pseudo-acid sulfate soil*:

"A *pseudo-acid sulphate soil* contains one or more horizons with characteristic yellow mottling (basic iron sulfates) commonly associated with acid sulfate conditions, but does not have a pH below 4 and does not contain free acids or more than about 60% exchangeable aluminum."

Elsewhere a soil that apparently represented Stage 3 of the process has been referred to as a "fossil acid sulfate soil" (Buurman et al., 1973).

For purposes of the symposium on acid sulfate soils held in the Netherlands in 1972 (Dost, 1973), Pons apparently intended that all three kinds of acid sulfate soils recognized by Brinkman and Pons (1973) be included under the blanket term "acid sulfate soils" (quotation marks added by us) as indicated in the following quotation from Pons (1973):

"The term acid sulfate soils is employed to pertain to all materials and soils in which as a result of processes of soil formation, sulfuric acids either will be produced, are being produced, or have been produced in amounts that have a lasting effect on main soil characteristics."

It is this broad use of the term acid sulfate soils that is employed in this paper. However, as indicated earlier, the soils from Texas that we shall describe in some detail, appear to represent Stage 3 of the sulfuricization process and would technically be considered pseudo-acid sulfate soils by the definitions of Brinkman and Pons (1973).

Although a number of others have apparently reported upland soils that might also be considered to be pseudo-acid sulfate soils (see later section of this paper on upland acid sulfate soil distribution), this may be the first paper to report soils that apparently have become acid enough for jarosite to form and to have subsequently developed a relatively high pH (near or above neutral) to the point that secondary calcite has, at least locally, formed in the presence of jarosite.

We have observed and described several soil pedons with the relict features of acid sulfate weathering. However, only two have been selected for somewhat detailed description in this paper. These pedons will be referred to by the soil series names, Aubrey and Lufkin, that would probably be given to these soils if they were encountered in soil surveys. Pedons of two other soil series (Birome and Shalba, Table 1) with acid sulfate weathering features will also be mentioned. We do not mean to imply that any of these soil series typically have relic features of acid sulfate weathering, such as jarosite, exchangeable Al, etc., however, many other pedons of these and other soil series probably do. The main sulfuricization feature that we confidently feel has been imprinted on the pedons presented here is the presence of jarosite in lower B and C horizons. However, we are inclined to think that many of the other physical, chemical, and mineralogical properties of these soils have also been affected by sulfuricization.

UPLAND ACID SULFATE SOIL DISTRIBUTION

Acid sulfate soils have been reported worldwide by several reviewers such as Kawalec (1973), Bloomfield and Coulter (1973), and van Breemen (1973). Most of the locations cited by these authors involve soils in deltas or coastal areas that are currently, or recently have been, under the influence of seawater.

Upland acid sulfate soils have been found in the Netherlands (Poelman, 1973; van Wallenburg, 1973), Java (in a review by Chenery, 1954), Russia (Chenery, 1954), Uganda (Chenery, 1954), Germany (Buurman et al., 1973), British Columbia, Canada (Clark et al., 1961), Alberta, Canada (Pawluk and Dudas, 1978), North Carolina, USA

(Furbish, 1963), Virginia, USA (Mitchell, 1962), Maryland, USA (Wagner et al., 1982), California, USA (Briggs, 1951) and Texas, USA (Carson et al., 1982—this paper). Many of these appear to be examples of soils representing the post-sulfuricization stage (see Introduction).

Other examples of upland acid sulfate soils or soil-like materials include: 1) exposed road cuts in Texas, USA (Miller et al., 1976) and Ontario, Canada (Ivarson, 1973) and 2) acid mine spoil materials in South Africa, Germany, England, and the Netherlands (Poelman, 1973), Venezuela (Hedberg et al., 1947), Pennsylvania, USA, (Warshaw, 1956) and West Virginia USA (Bloomfield and Coulter, 1973). Most of these apparently are examples of materials undergoing active sulfuricization.

Soil pedons in Texas that have relict features of acid sulfate weathering, or at least with jarosite present under pH conditions too high for a sulfuric horizon to be recognized, have been classified in the Aubrey and Lufkin series as well as in the Birome and Shalba soil series (Table 1). The Aubrey and Birome soils are formed on outcroppings of the Woodbine formation (Fig. 1 and Table 1), which is of Cretaceous age. The Woodbine is composed of geologic deposits that are considered to have been deposited in shallow-water, brackish or marine environments relatively near shore. The Aubrey pedon reported here is from Denton County,

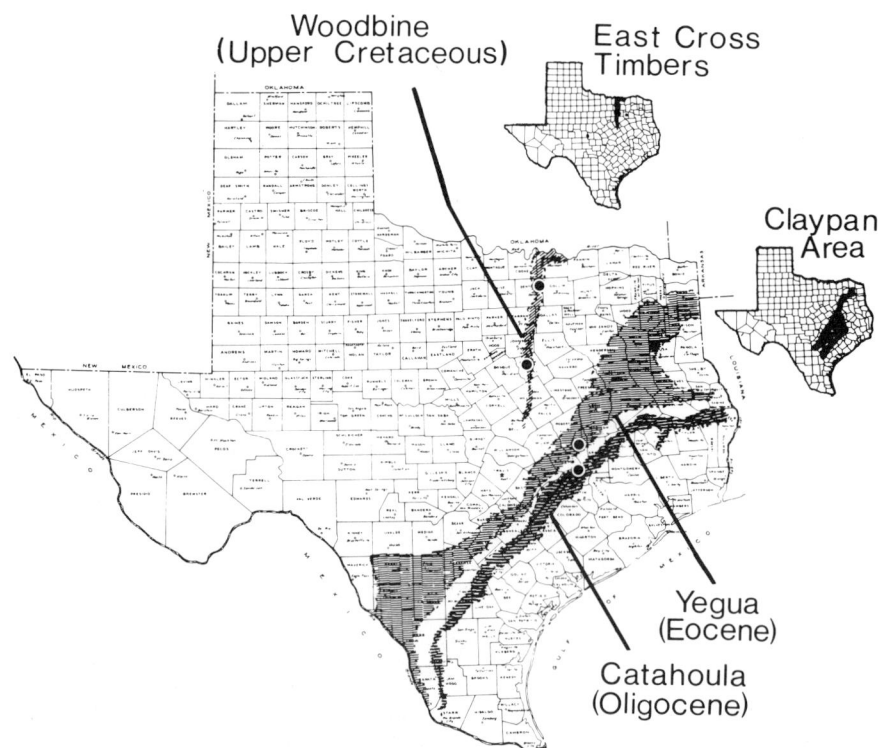

Fig. 1. Location and distribution of jarosite containing soils.

noted by the most northerly dot in Fig. 1, while the Birome pedon is from almost due south of the Aubrey location in Hill County. Both soils occur in the East Cross Timbers land resource area (Godfrey et al., 1973) (Fig. 1). The Aubrey and Birome soils are characterized by having jarosite present in the lower horizons with acid pH values throughout the profiles.

The Lufkin pedon reported here is developed on the Eocene age Yegua Formation in the Claypan land resource area outlined in Fig. 1. This soil has jarosite and small amounts of barite, gypsum, and calcite in the lower B and C horizons (Table 2). This soil is slightly above to near neutral pH in the surface and also in the lower horizons (Table 4). Other pedons of the same soil located within a few meters of this one were acid (pH about 5.7 in water) in the surface horizons, but were otherwise similar with depth. Jarosite was identified in the lower 3 horizons (Table 2). The Shalba soil is formed on the Catahoula Formation (Fig. 1, Table 1) which is of Oligocene age. This soil, like the Lufkin, is also in the Claypan land resource area (Fig. 1). Jarosite was identified in this soil but no other data are available.

PROFILE CHARACTERISTICS

Soils (Aubrey and Birome) from the East Cross Timbers area (Fig. 1) generally have a sandy textured A horizon over a clay textured B horizon and the C horizon is sandy clay, shaly clay, or shaly clay intermixed with weakly cemented sandstone. The Claypan Area (Fig. 1) soils (Lufkin and Shalba) have B horizons that are very compact with little structural development. The abundance of clay in the B horizon of these soils with jarosite present agrees with the observation of van Wallenburg (1973) that the subsoil of some cat-clay soils of inland polders contain more than 53% clay. Field descriptions of texture as well as several other properties are given in Tables 2 and 3 for the Aubrey and Lufkin soils. The Aubrey soil is characterized by having visible jarosite and being acid throughout (Tables 2 and 4). The Lufkin soil has jarosite at near neutral pH as well as acid pH levels (Tables 2 and 4).

Pale yellow jarosite coatings on shale fragments and bedding planes were easily visible in the C1 and C2 horizons of the Aubrey soil (Table 2). Jarosite was not readily visible in the B3 horizon but was detected when this material was analyzed by X-ray diffraction (XRD) and differential thermal analysis. Dark red colors in the B2 horizons of the Aubrey suggest ferric oxide and/or hydroxide compounds are accumulating in the upper profile. The Lufkin pedon has jarosite visible in the B23t, B3, and C1 horizons (Table 2). The jarosite is present as mottles and coatings of peds. The jarosite mottles usually have a 5Y7/6 color. A unique feature of the Lufkin pedon (Table 2) is the presence of white barite mottles in the B3 and C1 and also the presence of small amounts of calcite. Calcite in the presence of jarosite suggests the pH has been lower in the past and has now risen to the near neutral levels observed in the B3 and C1. Fine texture and poor structure have likely retarded leaching and weathering and have probably helped to preserve jarosite in this environment. Van Bree-

Table 2. Field descriptions of Aubrey and Lufkin pedons.

Horizon	Depth	Color	Texture†	Structure†	Consistence†	Boundary†	Special features†
				Aubrey Pedon			
A1	0-8	7.5YR4/4	fsl	2fsbk	mfr, dvh	cs	
A2			Barely distinguishable and not thick enough to sample.				
B21t	8-18	7.5YR4/4 to 5YR5/1	c	2mbk	mvfi, dvh	s	
B22t	18-58	2.5YR3/6	c	2mbk	mvfi, dvh	gs	
B3	58-74	2.5YR4/6 and 5YR5/1	c	2mbk	mvfi, dvh	cw	gray shale fragments
C1	74-86	5YR6/1	c	0m	mvfi, deh	cw	5Y 7/8 jarosite coatings on bedding planes
C2	86-152	5YR6/1	c	0m	mvfi, deh		common distinct 5Y 7/8 jarosite mottles as coatings on bedding planes
				Lufkin Pedon			
A11	0-4	10YR3/1	vfsl	1mpl	mfr	w	
A12	4-8	10YR5/3	lfs	0msg	mvfr, dsh	aw	
A13	8-23	7.5YR3/2	vfsl	0msg	mvfr, dh	aw	
B21tg	23-36	7.5YR3/2, some 5YR3/4 ped faces	sic	2fpr/2msbk	mfi	gs	apparent thin organo-clay skins
B22tg	36-56	10YR5/2, some 10YR3/2 and 4/4 ped faces	sic	1mpr/2msbk	mfi	cw	
B23tg	56-86	10YR5/3, some 10YR 3/2 ped faces	sicl	1f-msbk	mfi	gs	few 5Y7/6 jarosite mottles; cld 7.5YR 5/8 mottles
B3	86-114	2.5Y5/2	sicl	1cpr/1mbk	mfr	s	few 5Y7/6 jarosite and few, fine, distinct white barite mottles, also a few fine calcite nodules. Also m 1 & 2 d 10YR 5/4 and few 7.5YR 5/8 mottles.
C1	114-152	2.5Y5/2	sicl	1vcpl	mfr		common, distinct jarosite and barite mottles; other mottles similar to B3

† Abbreviations follow the Soil Survey Manual (Soil Survey Staff, 1951, p. 139.)

Table 3. Additional description information for Aubrey and Lufkin soil pedons.

Physiography vegetation and slope	Root activity	Climate	
		Aubrey Pedon	
Upland, 4–5% SEE	Many fine and medium roots in A; few fine and medium roots in B2 horizons; few fine roots in B3 and no roots in C1, C2.	Moist subhumid 33 to 38 inches (842 to 970 mm) mean annual precipitation; mean annual temperature is 66 to 68 F (17 to 20 C)	Scrub post oak with Florida pospalum, eastern gramma, panicums, three awn, and common bermudagrass
		Lufkin Pedon	
Upland, 4% W	Common fine roots in A horizons; a few coarse tree roots in B2 horizons; common fine roots in B21t and a few fine and medium roots in B and in C horizons—less with depth.	Moist subhumid 39 inches (990 mm) mean annual precipitation; mean annual temperature is 69 F (20 C)	Scrub post oak and live oak

men (1973) noted that jarosite can persist at pH values in excess of 4 in cases of minimal leaching or slow hydrolysis to ferric oxides. These soils with jarosite present occur in a moist, subhumid climate (Table 3).

ACIDITY

The acidity of acid sulfate soils is generally attributed to pyrite oxidation at some stage of weathering. As much as 4 moles of H^+ can be produced if 1 mole of pyrite is completely oxidized and the oxidized Fe is hydrolyzed to ferric oxides or hydroxides (van Breemen, 1973) as shown by Eq. [1]:

$$FeS_2 + 15/4\ O_2 + 7/2\ H_2O \rightarrow Fe(OH)_3 + 2\ H_2SO_4. \qquad [1]$$

Most of the geologic sediments mentioned in this report that contain pyrite also are rich in buffering materials such as clay minerals like montmorillonite and may contain varying amounts of carbonates. The acidity from pyrite oxidation whether complete (as in Eq. [1]) or partial can furnish H^+ to attack silicate clay minerals to liberate Al^{3+} and Fe^{3+}, which in turn may produce more acidity by hydrolysis as seen in Eq. [2] and [3]:

$$Al^{3+} + 3H_2O \rightarrow Al(OH)_3 + 3H^+ \qquad [2]$$

$$Fe^{3+} + 3H_2O \rightarrow Fe(OH)_3 + 3H^+ \qquad [3]$$

When incomplete hydrolysis of the Fe^{+3} from pyrite or other weathering occurs, and if the soil solutions contain sufficient K and sulfate, then jarosite can form, which is or has been the case in many acid sulfate soils.

Table 4. Some properties of Aubrey and Lufkin soil pedons.

Soil horizon	Depth	pH (H₂O) 1:2		pH (KCl) 1:2		Exchangeable acidity‡ (Al)	Nonexchangeable titratable acidity‡	Clay‡	Content of Organic carbon‡	Sulfur‡	Jarosite† Max.	Jarosite† Cor.
	cm	1 hour	24 hour	1 hour	24 hour	meq/100 g	meq/100 g		%	%		
							Aubrey pedon					
A1	0–8	5.3	5.5	4.3	4.3	0.1	7.2	21	1.2	0.07	0.6	ND
B21†	8–18	5.0	5.0	3.7	3.7	2.9	17.3	68	1.0	0.06	0.5	ND
B22†	18–58	4.7	4.7	3.5	3.5	5.6	21.1	69	0.8	0.04	0.4	ND
B3	58–74	4.7	4.5	3.4	3.3	7.4	23.5	75	0.6	0.23	1.8	1.6
C1	74–86	4.6	4.3	3.3	3.2	8.1	23.8	73	0.4	0.72	5.6	5.5
C2	86–152	4.4	4.0	3.1	3.1	7.6	22.1	73	0.3	0.89	6.9	6.7
							Lufkin pedon					
A11	0–4	7.5	6.9	6.9	7.2	0.0	2.0	16	1.4	0.13	1.0	ND
A12	4–8	7.6	7.1	7.0	7.3	0.0	1.5	12	0.9	0.14	1.1	ND
A13	8–23	7.0	6.7	6.8	7.0	0.0	1.6	14	0.7	0.13	1.0	ND
B21tg	23–36	5.3	4.6	3.9	3.8	2.2	12.4	52	0.5	0.13	1.0	ND
B22tg	36–56	4.8	4.4	3.8	3.7	1.9	8.1	46	0.5	0.15	1.2	ND
B23tg	56–86	5.2	4.9	4.2	4.1	0.6	7.7	46	0.4	0.18	1.4	0.9
B3	86–114	6.7	6.2	5.7	5.4	0.9	5.1	44	0.1	0.16	1.3	1.2
C1	114–152	6.9	6.4	5.9	5.6	0.0	5.9	44	0.1	0.19	1.5	1.4

† Jarosite maximum (Max.) calculated from total sulfur content assuming 12.8% S = 100% jarosite and corrected (Cor.) by subtracting sulfur assumed to be organic matter S. ND = not detected by visual observation or XRD.
‡ Exchangeable acidity (Al) was extracted with 1 N KCl and measured colorimetrically; nonexchangeable titratable acidity was taken as the difference between acidity measured at pH 8.0, BaCl₂-triethanol amine (total acidity)., and exchangeable acidity; clay was measured by fractionation and weighing; organic carbon by wet digestion—carbon train; and total sulfur by Leco apparatus.

As jarosite weathers it can undergo hydrolysis to ferric oxides, oxyhydroxides, or hydroxides producing 3 moles H⁺ for every mole of jarosite as shown in Eq. [4]:

$$2KFe_3(SO_4)_2(OH)_6 + 6H_2O \rightarrow 6Fe(OH)_3 + K_2SO_4 + 3H_2SO_4 \quad [4]$$

Since jarosite is fairly insoluble it probably would not be a large contributor to the present total acidity. The presence of low pH and large amounts of total acidity are probably largely the result of Al^{3+} and Fe^{3+} liberated from clay minerals during periods of pyrite oxidation thus giving rise to adsorbed Al^{3+} and Fe^{3+} and hydroxy forms of these elements. However, during active sulfuricization (see introduction) some amount of free sulfuric acid presumably is also present.

pH

Many acid sulfate soils may be buffered between pH 3 and 4 when pyrite is actively being oxidized, but how low the pH drops is also influenced by the rate at which H_2SO_4 is produced compared to the rate at which buffering reactions occur. Kaolinite and montmorillonite are buffered in this range (van Breemen and Wielemaker, 1974). Many of the acid sulfate soils related to modern tidal marshes have had fairly high levels of pyrite (2 to 5% pyrite-S, van Breemen, 1973), whereas the soils mentioned in this paper are likely derived from sediments that contained less pyrite. Sediments more diluted with respect to pyrite along with the presence of varying amounts of carbonates may partially explain why the jarosite containing soil horizons mentioned here are in a pH (water) range from 4 to 5 (Aubrey) or 5 to 7 (Lufkin) (Table 4). On the other hand, since it is normally considered that a pH of about 4 or below is required for jarosite formation (e.g., van Breemen and Harmsen, 1975), it seems more likely that the pH of the jarosite bearing horizons of the Aubrey and Lufkin soil was this low when the jarosite formed during an active sulfuricization stage and that the pH has subsequently risen. Some possible mechanisms for this pH rise are mentioned in the introduction to this paper.

Aubrey soil pH measured in 1 N KCl ranged downward to 3.1 and changed little after 24 hours, whereas pH measured in water showed a marked decline in 24 hours, perhaps indicating that hydrolysis was occurring. In those horizons where jarosite was identified it may have been partially responsible for a pH drop by hydrolysis.

Total Acidity

The summation of exchangeable Al and nonexchangeable, titratable acidity will be considered here as total acidity. Exchangeable hydrogen (H) has been neglected because it would likely account for a small fraction of the total in the pH ranges of the soils studied. In the Aubrey profile 0.1 to 0.7 meq H per 100 g was determined (Carson, 1977). A positive relationship between clay content and total acidity seems evident in all soils studied where each has a coarser textured surface above more clayey B and C horizons. Acidity from organic matter probably is negligible below the upper horizons (Table 4). This acidity vs. clay content relationship may relate to the fact that matrix clay minerals such as montmorillonite were H sinks during pyrite oxidation resulting in mineral deterioration and Al release. The Al freed would normally exist as exchangeable if the pH was low and as nonexchangeable if the pH was somewhat higher. This relationship is borne out as shown in Table 4 where exchangeable Al is not a significant part of the total acidity until the soil pH (water) drops below about 5.

Exchangeable Al accounts for no more than about 25% of the total acidity with the most occurring in strongly acid B and C horizons. The presence of as much as 8.1 meq/100 g (Aubrey, Table 4) exchangeable Al relfects the current pH and the influence of acid weathering on aluminosilicate clay minerals. Van Breemen (1973) noted when a non-calcareous pyritic soil was oxidized for 47 days, 7.2 meq/100 g exchangeable Al was produced.

Most of the acidity in upland acid sulfate soils is nonexchangeable, titratable acidity. This acidity includes organic matter complexed Al, hydroxy Al, or hydroxy Fe associated with clay minerals and dissociatable H from either organic matter or clay minerals. About one-third of the organic matter associated acidity in the Aubrey surface soil was organic matter complexed Al (Carson, 1977). The upper horizons have relatively more organic matter and less clay, therefore organic matter contributes significantly to nonexchangeable acidity in those layers. The B and C horizons are most likely more influenced by clay mineral associated acidity than the loamy upper horizons since they contain more clay and less organic matter (Table 4).

The contribution of jarosite to nonexchangeable acidity was considered. As shown in Eq. [4], complete hydrolysis of jarosite can produce 3 moles of H for every mole of jarosite. Complete hydrolysis of jarosite in a soil containing 5% jarosite could yield approximately 30 meq/100 g of H. The Aubrey soil materials contain between 1.6 and 6.7% jarosite, while the Lufkin soil has at most 0.6 to 1.4% jarosite (Table 4). However, jarosite is quite insoluble (Ksp = $10^{-98.6}$; Vlek et al., 1974) and may not undergo much hydrolysis. Jarosite picked from the Aubrey soil (87% pure) was analyzed and found to contain 18.7 meq/100 g total acidity by the Ba-triethanolamine method. This low figure compared to a theoretical value of 522 meq/100 g, if the jarosite completely hydrolyzed, probably reflects the low solubility of jarosite.

MINERALOGY

Overall Mineralogy

The mineralogy of the upland acid sulfate soils studied reflects the parent material from geologic deposits and effects of weathering. The soils located in the East Cross Timbers area of Texas are associated mainly with the outcroppings of the Woodbine formation (Oliver, 1971) as shown by Fig. 1. This formation has sand interbedded with shale, with the shale being the most prominent material in which jarosite has been identified. Subsurface geologic material associated with the Woodbine in Hill and Johnson Counties in Texas (most southerly outcropping area of the Woodbine where a jarosite-containing soil has been identified, see Fig. 1) has been described as having clay ironstone concretions, jarosite crusts, and selenite crystals in cracks and along bedding planes, and minor amounts of lignitic material and pyrite crystals. The shale portion of the southern outcrop is known as the Pepper shale (Sellards et al., 1931). The more northern outcrop of the Woodbine that gave rise to the Aubrey soil described in this paper is the Lewisville Member. This shale material is in a subsurface position in Dallas County (the next county south of Denton County) and has been described as having pyrite, lignites, and glauconite or micaceous materials present (Foster, 1965).

Although pyrite has been found in subsurface positions of the Woodbine geologic material, none has been identified in soil profiles. The East Cross Timber acid sulfate soils have mixed mineralogies in the soil pedons (Table 1). The Aubrey soil was found to have almost equal amounts of smectite and kaolinite with slightly less mica and small amounts of vermiculite, quartz, and iron oxides present in clay fractions (Carson, 1977). The three lowest horizons had detectable jarosite. Matrix soil clay minerals such as smectite have shown some signs of acid weathering in the upper horizons of the Aubrey soil. Broad, weak 18 Å X-ray diffraction peaks in the upper three horizons indicate the smectite is more weathered than in the lower horizons (Carson, 1977).

Upland acid sulfate soil pedons of the Claypan Area (Table 1) occur in two types of formations: one, the Catahoula formation (Oligocene) which follows a contour roughly paralleling the Gulf of Mexico, about 75 to 125 miles inland, from east-northeast to south-southwest, and; two, the Yegua (Eocene) and associated sediments in the Claiborn group that occur inland from the Gulf of Mexico about 100 to 150 miles (Sellards et al., 1931) (Fig. 1). The Catahoula beds in east, east-central Texas are light colored, tuffaceous, gritty, noncalcareous volcanic-type sediments sometimes interbedded with clays and ash beds. Glass shards, glass coated quartz, and other volcanic ejecta as well as barite have been identified in soils on the Catahoula (Lynn et al., 1971). Laboratory data of a soil developed on the Catahoula (Falba soil from Walker Co.), which is similar to the Shalba that has jarosite, indicated the clay contained 41 to 58% smectite; 27 to 31% kaolinite; and small amounts of vermiculite, mica, quartz, and iron oxides (Carson, 1977). Barite has been identified in the B horizon of the Falba soil and jarosite has been found in the upper C horizon of the Shalba soil in Washington County, both developed on the Catahoula formation. These data plus other information (McKee and Brown, 1977) indicated the acid soil pedons on the Catahoula often contain barite as well as jarosite and the matrix mineralogy is mostly smectite, especially in the argillic B horizons.

The marine clays and shales of the Yegua (Eocene) formation are sometimes interspersed with some lenses of volcanic-type materials similar to the Catahoula formation. The clays in the parent materials developed on the Yegua formation are high in smectite. X-ray diffraction and differential thermal analysis of another acid claypan soil pedon (Lufkin no. 2 of Kunze and Oakes, 1957) indicated the clay fraction was mostly smectite (montmorillonite) with lesser amounts of kaolinite and quartz. Gypsum was present in the B and C horizons of the Lufkin no. 2 of Kunze and Oakes with pH values of 4.6 and 4.7.

The Lufkin pedon of the present study contained smectite as the predominant clay mineral (unpublished XRD patterns). This smectite appears to have high quantities of hydroxy Al interlayers based on poor collapse with potassium (K) saturation and heat. The soil also contained quartz and some feldspars. Jarosite and barite, as well as small quantities of calcite were present in the B3 and C1 horizons and the soil solution of these horizons is apparently occasionally saturated with respect to gypsum as evidenced by gypsum crystals that formed on an exposed profile face during a dry period. In a nearby gully and in a core taken with a power coring apparatus, jarosite was found to also occur in a similar manner to that in the C1 horizon to depths of over 5 m. Other cores in this area showed that sulfides were not present until a depth of about 10 m.

Weathering in the Lufkin soil has apparently progressed from pyrite in localized areas of stratified marine sediments to jarosite in isolated spots of the lower B and C horizons. Barite and gypsum have also been formed during the weathering process. The fact that jarosite, barite, and gypsum are sometimes found in close proximity to each other suggests the leaching and weathering occurring in these soils may be modified to a great extent by the extremely high amount of smectite present that causes the permeability to be very slow.

Sulfate Minerals

The presence of jarosite, barite, or gypsum appears to depend upon several interrelated soil properties and the nature of the original sulfur containing parent sediment. The weathering sequence that includes oxidizing conditions can cause pyrite and/or other sulfide minerals to yield large amounts of acids and certain sulfate minerals as described in the various sulfuricization stages discussed in the introduction.

The formation of jarosite, $KFe_3(SO_4)_2(OH)_6$, appears to take place when oxidizing conditions exist and sufficient Fe, K, and sulfates are present as well as acid pH values. Sulfate and Fe may be present mainly as weathering products of pyrite. The basic cations needed for jarosite formation are apparently available as acid weathering products of alumino-silicates or other base-containing minerals or rocks. The Fe and K can also be available as exchangeable cations in equilibrium with the soil solution.

Barite ($BaSO_4$) is conceived to form in acid sulfate soils during the sulfuricization process when minimal Al is present and both barium and sulfates are available. Barite seems to be found in soils with relatively high base saturations even though pH values may be extremely acid (Lynn et al., 1971). Barium is probably present in the original parent sediment and may not have moved very far during the weathering-dissolution processes due to its immobility, however, its exact mineralogical source (primary mineral) in these Texas sediments and soils is not known. Sulfate on the other hand could move as part of the soil solution from weathering pyrite or other sulfur sources to precipitate with the barium. Van Breemen (1973) reports that the pH at equilibrium for amorphous silica, kaolinite, and montmorillonite in typical acid sulfate soils is similar to published values (Kittrick, 1971a and 1971b). Earlier work on a Texas smectite (Carson et al., 1976) showed the equilibrium pH to be near 3 for similar conditions. Barite formation seems to have occurred in a matrix that is highly smectitic, high base, high silica, and low pH.

Gypsum ($CaSO_4 \cdot 2H_2O$) has been shown to occur over a wide range of pH values (van Breemen, 1973). At pH values near 4 gypsum may occur in association with jarosite. The calcium (Ca) may come from Ca bearing minerals or the exchange complex, whereas sulfur comes from the weathering of pyrite or other sulfur bearing minerals. Gypsum commonly occurs where $CaCO_3$ is present when pyrite oxidation is proceeding. Calcite neutralizes the acidity produced and liberates calcium to react with sulfate. Even though gypsum is fairly soluble it can persist in an environment of minimum leaching. The dense lower horizons of Claypan soils may have gypsum present due to a minimal amount of leaching. Jarosite can replace gypsum under conditions of low pH, low solution Ca, or high solution K (van Breemen, 1973).

Field Identification of Sulfate Minerals. Recognition of sulfate minerals in the field can be accomplished once the observer has had some experience with observation and verification. Distinct characteristics are usually associated with jarosite, barite, and gypsum but care has to be exercised not to confuse them with minerals of similar color, morphology, or hardness.

Jarosite is usually characterized by yellow colors such as 5Y hue with a value of 6 to 8 and chroma of 4 to 6 or more. A 2.5Y 8/4 color has been reported for jarosite in a Thailand soil (van Breemen and Harmsen, 1975). The yellow color may occur as flecks of relatively pure material (87.6% jarosite in particles picked from Aubrey soil, Carson, 1977) or as mottles occurring as coating on ped faces (e.g., see color Plate 4A) or in old root channels. The yellow mottles apparently may also occur as small masses inbedded in the matrix of soil (Clark et al., 1961). It is not uncommon for jarosite to be in association with iron oxides (see color Plate 4A), which may cause more browns and reds to be admixed with the yellow.

Barite that occurs in geologic bentonitic clays or in soils is usually a fine, white powder on structural surfaces or soft, white nodular masses (see color Plate 3D). Barite mottles occurring as "nodular" masses found in the Lufkin and in other soil materials in the vicinity of College Station, Tex., have colors whiter than N8 both moist and dry. One problem with barite identification is possible confusion with calcite. Small white calcite crystals can be distinguished from barite because the calcite will effervesce when treated with HCl.

Gypsum in acid sulfate soils of Texas is nearly white to colorless and is sometimes vitreous. If the crystals and cleavable masses are large enough to test (Fig. 2), the gypsum will be much softer than either quartz

Fig. 2. Gypsum as it occurs in a large crystalline mass.

or calcite (gypsum may be scratched with a fingernail). Also, gypsum often forms visible lathes or needles in contrast to the more equidimensional shape of quartz and calcite.

Jarosite occurs as very fine crystals (Fig. 5A of Dixon et al., 1982). The microcrystalline particles are weakly compacted into small flecks (few mm diam) or as thin coatings on peds, fracture planes, and cracks or in voids such as in old root channels. These coatings when scrapped off have a consistence of silt. Barite occurs as small white crystals (0.025 to 0.080 mm, diam) that may sparkle in light, especially when viewed in sunlight with a hand lens. The crystals of barite may vary from one location to another. A barite sample from the Lufkin has flat hexagonal shaped particles while samples found in Louisiana are tabular and clear and grains from Oklahoma are euhedral with hexagonal, octagonal, or rhombic outlines (Lynn et al., 1971). Gypsum found in acid sulfate soils may be clusters of crystals, or as loose, weakly attached masses of crystals (Fig. 2) or as flat sheets that are almost clear. Gypsum particles vary from silt size to crystals (often sheets) that are several centimeters across (see color Plate 4B).

Laboratory Identification and Verification of Sulfate Minerals. Field recognition of sulfate minerals in acid sulfate soils is sometimes difficult due to the small amounts present and masking by the matrix soil material. Also, many soil profiles are only sampled to a depth of 100 to 150 cm which might exclude some of the sulfate minerals. It is important to know the overall soil mineralogy as well as recognizing the presence of

Fig. 3. Scanning electron micrograph of jarosite in the Aubrey soil.

individual sulfate minerals. Extensive prefractionation chemical treatments hinder the identification of sulfate minerals in the whole soil clay. Therefore, identification must be made on untreated or mildly treated whole samples or handpicked specimens. X-ray diffraction was done on finely ground whole samples of the Aubrey soil to establish the presence of jarosite. Diagnostic peaks used for identification of jarosite were the 3.11 and 3.08 Å peaks. Hand picked specimens from the other jarosite containing soils were also vertified by XRD. Barite presence may be confirmed by XRD of powder from suspected material that has been manually separated. Diffraction peaks at 3.57 and 3.31 Å indicate barite. Gypsum crystals can be verified by the XRD peak at 7.56 Å. Verification can be made by comparisons with patterns of these minerals given in the American Standards for Testing Materials (ASTM) or other XRD pattern catalogs.

Thermal analysis corrobarates XRD results for the normal suite of soil minerals such as smectite, kaolinite, mica, and quartz. This tool is especially useful in verifying jarosite. An endotherm at 416 C is a diagnostic feature that makes jarosite detectable down to approximately 2% of the whole soil (Carson, 1977). Jarosite was detectable by differential thermal analyses (DTA) in the untreated, Aubrey soil in the lowest three horizons. These results show that DTA is a sensitive tool for jarosite identification because the B3 horizon has only 1.6% jarosite present.

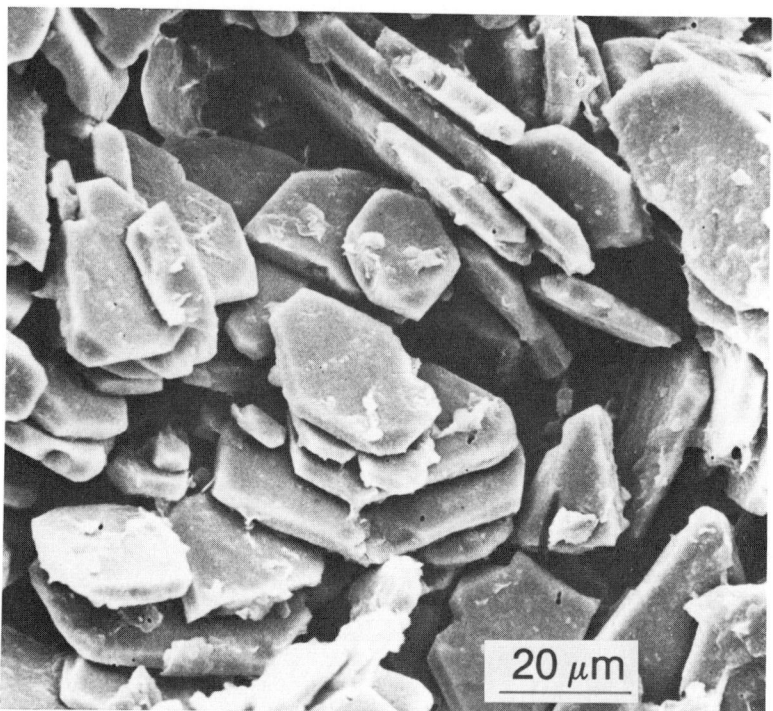

Fig. 4. Scanning electron micrograph of barite in the Lufkin soil.

The electron microscopic investigation of crystal morphology is useful in identification, genesis, and weathering studies of sulfide and sulfate minerals. Scanning electron microscopy (SEM) of jarosite (Fig. 3) picked from an acid sulfate soil (Aubrey) shows cubic particles with octahedral faces. These morphological forms are like those expected for pyrite suggesting that jarosite may have crystallized as a pseudomorph of pyrite. A scanning electron micrograph of barite from the Lufkin soil is shown in Fig. 4.

Quantitative Jarosite Analyses Based on Sulfur Content. The quantity of total sulfur present in a soil can be used as a measure of jarosite if no other inorganic sulfur sources are present. Sometimes organic matter may be present in enough quantity to need a correction for sulfur in it. This may be difficult unless some horizons contain organic matter but no jarosite (as determined by other tests) so that an organic matter sulfur content can be estimated. Ideally, jarosite should have 12.8% sulfur. Using an organic matter sulfur correction, the Aubrey soil was found to have 1.6% jarosite in the B3 horizon, 5.5% in the C1, and 6.7% in the C2 as measured by total sulfur analysis, whereas Lufkin was found to have 0.6, 0.9, 1.2, and 1.4% jarosite in the B22, B23, B3, and C1 horizons respectively (Table 4). However, since some other sulfate minerals, notably barite, also occur in the Lufkin soil materials these jarosite estimates must still be too high for the Lufkin. Also, as is noted in the pedon descriptions, the jarosite present was highly localized into mottles associated with structural partings like that shown in color Plate 4A.

LITERATURE CITED

1. Bloomfield, C., and J. K. Coulter. 1973. Genesis and management of acid sulfate soils. Adv. Agron. 25:265–326.
2. Briggs, L. I., Jr. 1951. Jarosite from the California Tertiary. Am. Mineral. 36:902–906.
3. Brinkman, R., and L. J. Pons. 1973. Recognition and prediction of acid sulfate soil conditions. p. 169–203. In H. Dost (ed.) Acid sulfate soils. Vol. I. ILRI Publ. 18, Wageningen, The Netherlands.
4. Buurman, P., N. van Breemen, and A. G. Jongmaus. 1973. A fossil acid sulfate soil in ice-pushed Tertiary deposits near Uelsen (Kreis Nordhorn), Germany. p. 52–55. In H. Dost (ed.) Acid sulfate soils. Vol. II. ILRI Publ. 18. Wageningen, The Netherlands.
5. Carson, C. D. 1977. The nature of acidity in selected acid soils. Ph. D. Thesis. Texas A&M Univ. Univ. Microfilms. Ann Arbor, Mich. (Diss. Abstr. 38:3498–3499).
6. Carson, C. D., D. S. Fanning, J. B. Dixon. 1982. Alfisols and Ultisols with acid sulfate weathering features in Texas. p. 127–146. In J. A. Kittrick, D. S. Fanning, and L. R. Hossner (ed.) Acid sulfate weathering. SSSA Spec. Pub. no. 10. Madison, Wis.
7. ————, J. A. Kittrick, J. B. Dixon, and T. R. Mckee. 1976. Stability of soil smectite from a Houston Black Clay. Clays Clay Miner. 24:151–155.
8. Chenery, E. M. 1954. Acid sulfate soils in Central Africa. Int. Congr. Soil Sci. Trans. 5th (Leopoldville) IV:195–198.
9. Clark, J. S., C. A. Gobin, and P. N. Sprout. 1961. Yellow mottles in some poorly drained soils of the lower Fraser Valley, British Columbia. Can. J. Soil Sci. 41:218–227.
10. Dixon, J. B., L. R. Hossner, A. L. Senkayi, and K. Egashira. 1982. Mineralogical properties of lignite overburden as they relate to mine spoil reclamation. p. 169–191. In J. A. Kittrick, D. S. Fanning, and L. R. Hossner (ed.) Acid sulfate weathering. SSSA Spec. Pub. no. 10. Madison, Wis.
11. Dost, H. (ed.). 1973. Acid sulfate soils. ILRI Publ. 18, Vols. I and II. Int. Institute for Land Reclamation and Improvement. P. O. Box 45, Wageningen, The Netherlands.
12. Fanning, D. S. 1978. Soil morphology, genesis, classification and geography. Published by the author, Dep. of Agronomy, Univ. of Maryland, College Park, Maryland. 133 p.
13. Foster, P. W. 1965. Subsurface geology of Dallas County. p. 126–189. In H. L. Schaeffer (ed.) The geology of Dallas County. The Dallas Geol. Soc., Dallas, Tex.
14. Furbish, W. J. 1963. Geologic implications of jarosite, pseudomorphic after pyrite. Am. Mineral. 48:703–706.
15. Godfrey, C. L., G. S. McKee, and H. Oakes. 1973. General soil map of Texas. Texas Agric. Exp. Stn. Texas A&M Univ., College Station, Tex.
16. Goldhaber, M. B., and I. R. Kaplan. 1982. Controls and consequences of sulfate reduction rates in recent marine sediments. p. 19–36. In J. A. Kittrick, D. S. Fanning, L. R. Hossner (ed.) Acid sulfate weathering. SSSA Spec. Pub. no. 10. Madison, Wis.
17. Hedberg, H. D., L. C. Sass, and H. F. Funkhouser. 1947. Oil fields of Greater Oficina Area, Central Anzoategui, Venezuela. Am. Assoc. Pet. Geol. Bull. 31:2108–2110.
18. Ivarson, K. C. 1973. Microbiological formation of basic ferric sulfates. Can. J. Soil Sci. 53:315–323.
19. ————, and R. O. Hallberg. 1976. Formation of mackinawite by the microbial reduction of jarosite and its application to tidal sediments. Geoderma. 16:1–7.
20. ————, R. O. Hallberg, and T. Wadsten. 1976. The pyritization of basic ferric sulfates in acid sulfate soils: a microbiological interpretation. Can. J. Soil Sci. 56:393–406.
21. Kawalec, A. 1973. World distribution of acid sulfate soils. References and Map. p. 292–295. In H. Dost (ed.) Acid sulfate soils. ILRI Publ. 18, Vol. 1. Wageningen, The Netherlands.
22. Kittrick, J. A. 1971a. Stability of montmorillonites: I. Belle Fourche and Clay Spur montmorillonites. Soil Sci. Soc. Am. Proc. 35:140–145.
23. ————. 1971b. Stability of montmorillonites: II. Aberdeen montmorillonite. Soil Sci. Soc. Am. Proc. 35:820–823.
24. Kunze, G. W., and H. Oakes. 1957. Field and laboratory studies of the Lufkin soil, a Planasol. Soil Sci. Soc. Am. Proc. 21:330–335.
25. Lynn, W. C., H. Y. Tu, and D. P. Franzmeier. 1971. Authigenic barite in soils. Soil Sci. Soc. Am. Proc. 35:160–161.

26. McKee, T. R., and J. L. Brown. 1977. Preparation of specimens for electron microscopic examination. p. 809–846. *In* J. B. Dixon and S. B. Weed (ed.) Minerals in soil environments. Soil Sci. Soc. of Am., Madison, Wis.
27. Miller, W. L., C. L. Godfrey, W. G. McCully, and G. W. Thomas. 1976. Formation of soil acidity in carbonaceous soil materials exposed by highway excavations in East Texas. Soil Sci. 121:162–169.
28. Mitchell, R. S. 1962. New occurrence of jarosite in Virginia. Am. Mineral. 47:788–789.
29. Oliver, W. B. 1971. Depositional systems in the Woodbine Formation (Upper Cretaceous), Northeast Texas. Univ. Texas Bur. Econ. Geol. Rep. No. 73.
30. Patrick, W. H., Jr., and D. S. Mikkelsen. 1971. Plant nutrient behavior in flooded soil. p. 187–216. *In* R. A. Olson (ed.) Fertilizer technology and use, 2nd ed. Soil Sci. Soc. Am., Madison, Wis.
31. Pawluk, S., and M. Dudas. 1978. Reorganization of soil materials in the genesis of an acid sulfate Luvisolic soil of the Peach River region, Alberta. Can. J. Soil Sci. 58:209–220.
32. Poelman, J. N. B. 1973. Soil material rich in pyrite in non-coastal areas. p. 197–207. *In* H. Dost (ed.) Acid sulfate soils, Vol. I. ILRI Publ. 18. Wageningen, The Netherlands.
33. Pons, L. J. 1973. Outline of the genesis, characteristics, classification and improvement of acid sulfate soils. p. 3–27. *In* H. Dost (ed.) Acid sulfate soils. Vol. I. ILRI Publ. 18, Wageningen, The Netherlands.
34. Rickard, D. T. 1973. Sedimentary iron sulphide formation. p. 28–65. *In* H. Dost (ed.) Acid sulfate soils. Vol. I. ILRI Publ. 18, Wageningen, The Netherlands.
35. Sellards, E. H., W. S. Adkins, and R. B. Plummer. 1931. The geology of Texas. Univ. Texas, Bur. Econ. Geol. Bull. No. 3232.
36. Soil Survey Staff. 1951. Soil survey manual. U.S. Government Printing Office, Washington, DC.
37. ————. 1975. Soil taxonomy: a basic system of soil classification for making and interpreting soil surveys. USDA Handb. 436. U.S. Government Printing Office, Washington, DC.
38. ————. 1976. Erosion and sediment control guidelines for developing areas in Texas. USDA. U.S. Government Printing Office, Washington, DC.
39. van Breemen, N. 1973. Soil forming processes in acid sulfate soils. p. 66–131. *In* H. Dost (ed.) Acid sulfate soils, Vol. I. ILRI Publ. 18, Wageningen, The Netherlands.
40. van Breemen, N., and R. Brinkman. 1976. Chemical equilibria and soil formation. p. 141–170. *In* G. H. Bolt, and M. G. M. Bruggenwert (ed.) Soil chemistry. Part A. Basic elements. Elsevier Scientific Publishing Co., Amsterdam.
41. ————, and K. Harmsen. 1975. Translocation of iron in acid sulfate soils: I. Soil morphology, and the chemistry and mineralogy of iron in a chronosequence of acid sulfate soils. Soil Sci. Soc. Am. J. 39:1140–1148.
42. ————, and W. G. Wielemaker. 1974. Buffer intensities and equilibrium pH of minerals and soils: II. Theoretical and actual pH of minerals and soils. Soil Sci. Soc. Am. Proc. 38:61–66.
43. van Wallenburg, C. 1973. Cat-clay soils and potential cat-clays in inland polders in the western part of The Netherlands. p. 264–283. *In* H. Dost (ed.) Acid sulfate soils, Vol. I. ILRI Publ. 18. Wageningen, The Netherlands.
44. Vlek, P. L. G., Th. J. M. Blom, J. Beek, and W. L. Lindsay. 1974. Determination of solubility product of various iron hydroxides and jarosite by the chelation method. Soil Sci. Soc. Am. Proc. 38:429–432.
45. Wagner, D. P., D. S. Fanning, J. E. Foss, M. S. Patterson, and P. A. Snow. 1982. Morphological and mineralogical features related to sulfide oxidation under natural and disturbed upland surfaces in Maryland. p. 109–125. *In* J. A. Kittrick, D. S. Fanning, and L. R. Hossner (ed.) Acid sulfate weathering. SSSA Spec. Pub. no. 10. Madison, Wis.
46. Warshaw, H. D. 1956. The occurrence of jarosite in underclays. Am. Mineral. 41:288–296.

Chapter 9

Gypsiferous Soils in the Western United States

W. D. NETTLETON, R. E. NELSON, B. R. BRASHER, AND P. S. DERR[1]

ABSTRACT

Gypsum ($CaSO_4 \cdot 2H_2O$) is the only pedogenic calcium sulfate mineral that has been found in soils with ustic, xeric, and aridic moisture regimes. It has been found in soils in 14 of the 17 conterminous western states by the National Soil Survey Laboratory and likely will be found in the other three. In these arid to subhumid soils, parent material differences in large part control the occurrence of gypsum. But gypsum in soils can be from other sources too. For example, drainage of coastal wetlands oxidizes sulfides to acid sulfates, and in their reclamation, if not before, the acid sulfates are neutralized by carbonate to form gypsum. Gypsum is also formed in minesoils by neutralization of the acid sulfates released by oxidation of sulfides. Pedogenic gypsum, in contrast to allogenic gypsum, accumulates in subsurface horizons relative to surface and underlying horizons mostly as euhedral to subhedral spindle-shaped crystals in pores and veins. As the s-matrix in a developing gypsic horizon becomes plugged, the pore volume decreases and the restricted hydraulic conductivity keeps the soil moist longer allowing the growing gypsum crystals to interlock and indurate the horizon. Subsidence of soils through solution and removal of gypsum can crack building foundations, break irrigation canals, and make roads uneven. Concrete in slab structures, irrigation canals, and building foundations deteriorates and cracks as extremely high pressures develop during formation of highly hydrated ettringite [$Ca_6 \cdot Al_2(SO_4)_3(OH)_{12} \cdot 26H_2O$] or during conversion of thenardite (Na_2SO_4) to mirabilite ($Na_2SO_4 \cdot 10H_2O$) if the temperature in the concrete and the soil drops low enough.

INTRODUCTION

Calcium sulfate is a common constituent of many sedimentary rocks and dryland soils. Anhydrite ($CaSO_4$) is the form found in many marine evaporites, gypsum ($CaSO_4 \cdot 2H_2O$) is the form found in many Quaternary

---

[1] Research soil scientists, National Soil Survey Lab., Soil Conservation Service, Lincoln, NE 68501; Soil scientist, Soil Conservation Service, Casper, Wyoming.

Copyright © 1982 Soil Science Society of America, 677 S. Segoe Rd., Madison, WI 53711. *Acid Sulfate Weathering.*

deposits. Hemihydrate ($CaSO_4 \cdot \frac{1}{2}H_2O$) is a transition form. Surface or near surface gypsum is replaced by anhydrite in geologic formations as a result of burial and it, in turn, is replaced by gypsum if the anhydrite is thereafter brought close to the surface (Murray, 1964).

In this paper, we will show that gypsum is a major, if not the only, calcium sulfate mineral in soils; that it is most common in arid lands but also occurs in minesoils and in acid sulfate soils of coastal marshes; that its recognition is possible by a study of soil morphology; and that although it improves soil permeability, it also causes concrete to corrode, and its removal can cause soils to subside and undergo failure.

METHODS AND MATERIALS

We analyzed a number of soils (Table 1) for calcium sulfate by the water extract procedure (6F1A and 6F1B, Soil Conservation Service, 1972). The soils were sampled and analyzed in support of the National Cooperative Soil Survey. Sample collection and preparation are described by codes 1A1, 1A2, and 1B1 (Soil Conservation Service, 1972). The collection, preparation, and storage of samples would have maintained them within the stability range for gypsum and hemihydrate (Palache et al., 1951, p. 484 and 485). Anhydrite converts to gypsum at room temperature, but the process apparently is slow under the conditions of our laboratory because $CaSO_4$ has remained in the anhydrite crystal form there for several years.

Table 1. Listing of the location, classification, and parent materials of some of the soils studied.

Series	Location	Classification	Parent material
Calcross (S72UT-21-9)	Iron Co, Utah	Xerric Torrifluvent, fine-silty, mixed (calcareous), mesic	Mixed alluvium from sedimentary rocks
Holloman (S74NV-3-5)	Clark Co, Nev.	Typic Torriorthent, loamy, gypsic, thermic, shallow	Alluvium from Muddy Creek frm.
Rampart Tax. (S74AK-999-4)	Interior, Alaska	Typic Cryoboralf, coarse-silty, mixed	Loess
Unnamed (S74NM-35-1)	Tularosa Basin, N.M.	Ustollic Paleorthid, coarse-loamy, gypsic, thermic	Wind and water laid sediments from the Yeso frm.
Unnamed (S74NM-35-2)	Tularosa Basin, N.M.	Typic Gypsiorthid, coarse-loamy, gypsic, thermic	Same
Russler (S74NM-35-7)	Tularosa Basin, N.M.	Typic Camborthid, fine-silty, mixed, thermic	Same
Unnamed (S74NM-35-9)	Tularosa Basin, N.M.	Petrogypsic Gypsiorthid, coarse-loamy, gypsic, thermic	Same
Unnamed (S74NV-3-1)	Clark Co, Nev.	Petrogypsic Gypsiorthid, loamy, gypsic, thermic	Alluvium from the Muddy Creek frm.

Thin sections were prepared by the method of Innes and Pluth (1970) except that the water used during fine grinding was saturated with gypsum to prevent pitting of the sections. Descriptive terms are mostly those of Brewer (1964).

Identification of the Calcium Sulfate Mineral

A simple experiment was conducted to gain an understanding of the stability of the forms of calcium sulfate. Gypsum and commercial plaster of Paris were heated to 105 C, converting both to anhydrite. Then with as few as three wetting and air-drying cycles, the anhydrite was converted back to gypsum. Because of this, we expected gypsum to be the dominant, if not the sole form of calcium sulfate in soils. At least calcium sulfate in the surface horizons of soils in an aridic moisture regime would be expected to be in the form of gypsum, since these horizons become moist occasionally. Calcium sulfate below the leaching zone in these soils could be either hemihydrate or anhydrite since the soils are almost always dry.

Next we studied 23 soil samples containing measurable amounts of calcium sulfate using X-ray diffraction (XRD) and chemical methods to identify the major calcium sulfate mineral. The samples were from the soils from Alaska, Nevada, and Utah that are listed in Table 1. The samples were crushed to less than 60 mesh for the analyses. Gypsum was the only calcium sulfate mineral identified. Three typical examples are listed in Table 2. The XRD peaks from these examples of soil gypsum are near those reported for gypsum in the literature (Brown, 1961).

In the chemistry experiment, 1 to 10-g samples were weighed in a tared filter crucible and washed with increments of ethanol to remove the soluble salts. Washings were continued until the electrical conductivity of the final ethanol wash was less than 0.04 mmhos per cm. The ethanol washed samples were then ovendried, weighed, and extracted with water to remove the gypsum. Aliquots of the extracts were analyzed for sulfate (Soil Conservation Service, 1972; method 6L1b). We quantitatively transferred the soil samples back to the tared filter crucibles and then ovendried and weighed them. The difference in ovendried sample weights after an ethanol wash to remove the salts and after water extraction to remove the gypsum was considered to be the weight of calcium sulfate.

All of the 23 samples from the three pedons were calcareous with calcium carbonate contents ranging from 8 to 40%. Presumably, some calcium carbonate was dissolved by the water and contributed along with calcium sulfate to the weight loss. Small amounts of organic matter, sodium sulfate, and magnesium sulfate could also have been dissolved by the water. Apparently, these other losses are small in comparison to gypsum since gypsum contents calculated from water dissolution of the samples are highly correlated ($r = 0.999$) with the gypsum measured from the total sulfate (Nelson et al., 1978).

Table 2. X-ray diffraction data for examples of the CaSO₄ mineral in three of the soils and for gypsum, hemihydrate, and anhydrite.

S72UT-21-9 Torrifluvent C3, 91–110 cm		S74NM-35-9 Gypsiorthid C2csm, 48–66 cm		S74NV-3-5 Torriorthent C3, 46–88 cm		Gypsum†			Hemihydrate†		Anhydrite†		
d	I	d	I	d	I	d	I	hkl	d	I	d	I	hkl
7.63	56	7.66	93	7.62	100	7.56	100	020	6.07	10	3.498	100	020;002
4.27	100	4.30	100	4.29	15	4.27	51		3.50	9	2.849	33	210
3.80	31	3.82	22	3.80	19	3.79	21	031;040	3.02	10	2.328	22	202;220
3.07	44	3.08	83	3.07	16	3.059	57		2.818	10	2.208	20	212

† Brown, G., 1961, p. 477.

RESULTS

Occurrence

Gypsum, then, is the pedogenic calcium sulfate mineral in soils with ustic, xeric, and aridic moisture regimes. Many of the Aridisols have gypsum, a few have gypsic horizons[2], and a very few are in gypsic families[3]. Of these, there are more series recognized in the Cambic Gypsiorthids[4] than in the Typic Gypsiorthids[4], Calcic Gypsiorthids[4], or Petrogypsic Gypsiorthids[4] (USDA Soil Survey Staff, 1979).

The National Soil Survey Laboratory has had several hundred soil characterization projects in the 17 conterminous western states in the last 25 years. We have found gypsum in all of the states (Fig. 1) except Washington, Kansas, and Oklahoma, and since in the drier regions soil gypsum is mostly inherited from the soil parent material (see the section on the effect of parent material), in future studies we expect to find gypsum in soils in these states also because they do have some gypsum mines (Fig. 2). Kansas and Oklahoma, in particular, have extensive gypsum deposits. Even river water in these states has the calcium magnesium sulfate-chloride chemical types of water (Fig. 3) which one would expect from gypsiferous soils.

Sulfate is more concentrated than chloride in the rain water in the more arid parts of the western states. This is largely a result of the higher concentration of sulfate relative to chloride at ground level (Junge, 1954) and at altitudes of 1,270 m (5,000 ft) (Byers et al., 1955). The high con-

---

[2] The *gypsic horizon* is a noncemented or weakly cemented horizon of enrichment with secondary sulfates that is 15 cm or more thick, has at least 5% more gypsum than the C horizon, or the underlying stratum, and in which the product of the thickness in cm and the percent gypsum is ≥ 150. A *petrogypsic horizon* is a gypsic horizon that is strongly enough cemented with gypsum that dry fragments do not slake in water and that roots cannot enter.

[3] The gypsic family includes soils having more than 40% by weight of carbonates plus gypsum and the gypsum is > 35% of the sum of carbonates and gypsum.

[4] Gypsiorthids are the soils that lack an argillic or natric horizon unless it is a buried horizon and that have a gypsic or petrogypsic horizon whose upper boundary is within 1 m of the soil surface. Further differentiation is made into subgroups depending on whether the soil has a calcic or petrogypsic horizon, or a large amount of gypsum.

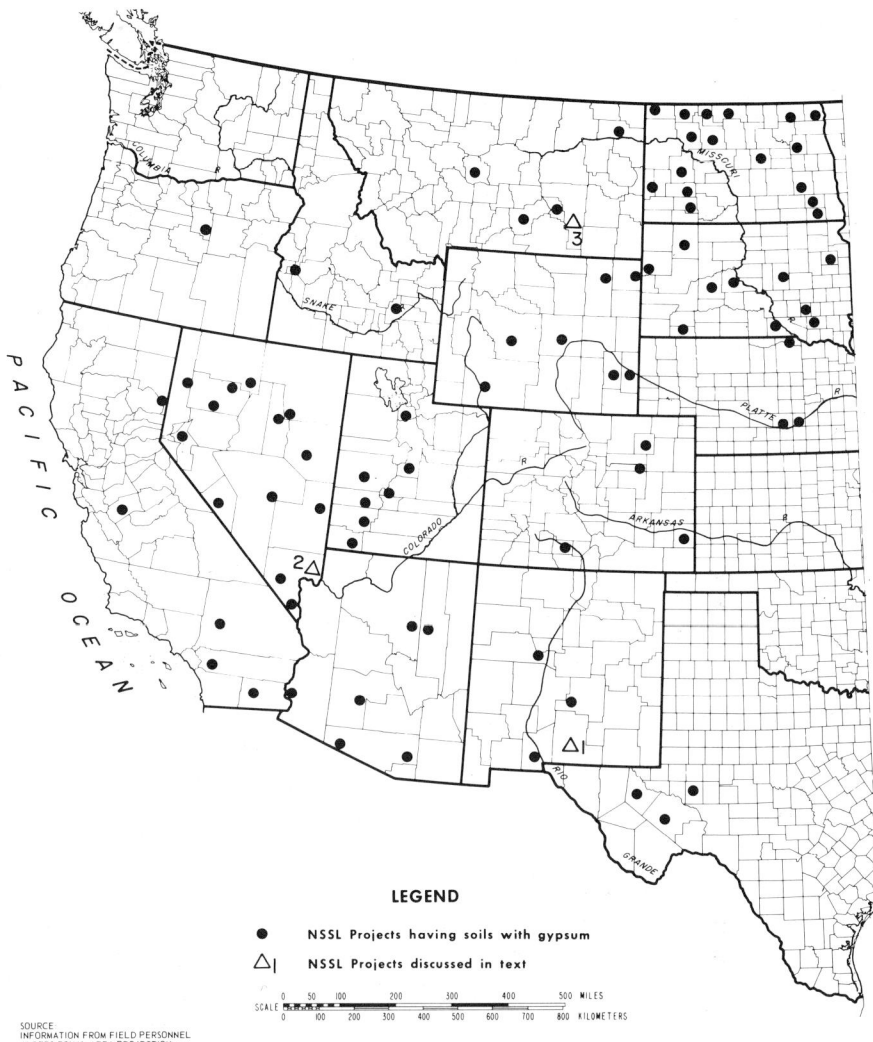

Fig. 1. Counties in the western United States in which the NSSL has found soils which contain gypsum.

centrations of sulfate in the Midwest is consistent with the high consumption of sulfur bearing fuels while nearer the ocean, chloride is more important because of the salt spray effect (Lodge et al., 1968).

Effect of Climate

The distribution of gypsum is limited to soils in which modern precipitation and leaching have not been sufficient to remove it. A Petrogypsic Gypsiorthid from Clark County, Nevada, (study area 2 of Fig. 1) is an example. This pedon received an estimated 14 cm of precipitation in

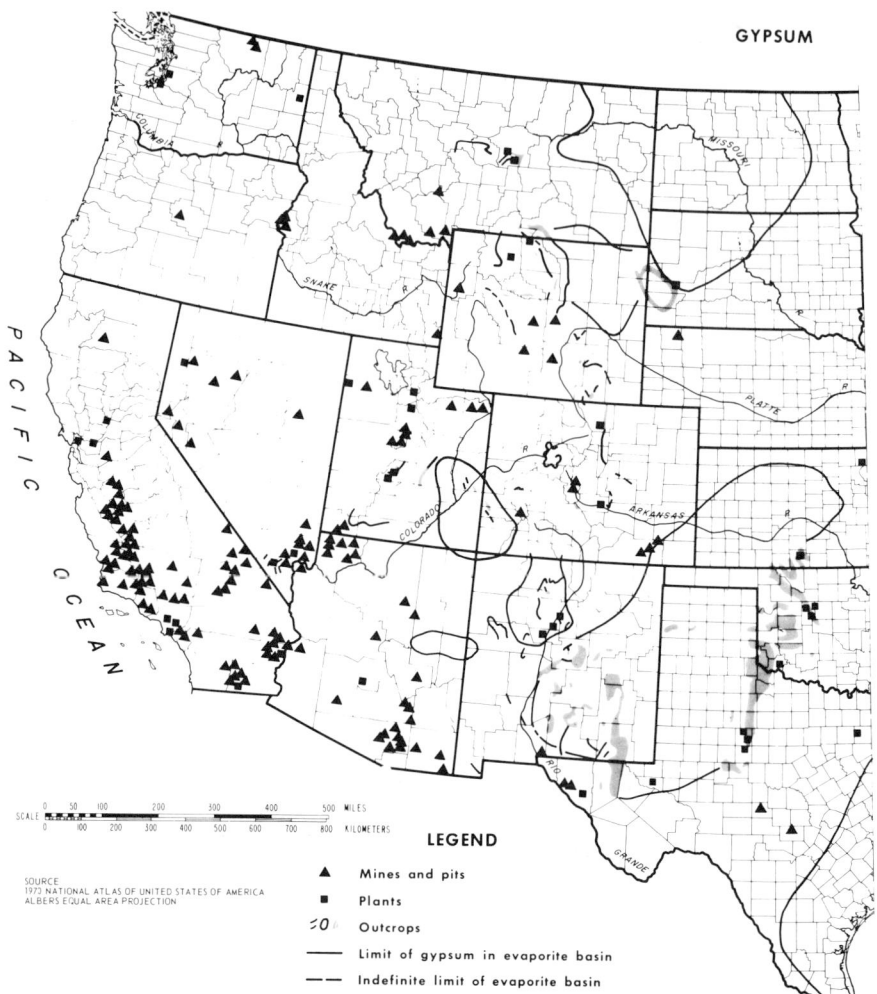

Fig. 2. Location of gypsum and anhydrite mines in the western United States.

an average year. After evapotranspiration losses, enough water is left to wet the soil to a depth of 32 cm[5]. Depth to the petrogypsic horizon is 12 cm. Gypsum content reaches its maximum, however, between 28 and 58 cm. One year in five, precipitation exceeds 2½ cm in 2 weeks (Gifford et al., 1967) and would be expected to wet the soil to a depth of 52 cm. Unpublished studies by us also show that gypsum is precipitated near the wetting front in many other dryland soils.

---

[5] United States Weather Bureau, 30-year normal monthly precipitation temperature data 1931–1960, Card Deck 490, San Diego, California. Data for Boulder City, Nev. Mean annual precipitation, 14 cm. Mean annual temperature, 19.4 C. Mean annual potential evaporation, 115.2 cm. Precipitation exceeds potential evapotranspiration in December, January, and February reaching an average of 1.7 cm in February. The water holding capacity of the soil, 1/3 bar less 15 bar water, is 0.04 cm/cm for the A1, 0.08 cm/cm for the B2, and 0.04 cm/cm for the underlying horizons. The soil in February in an average year then wets to a depth of about 32 cm.

GYPSIFEROUS SOILS

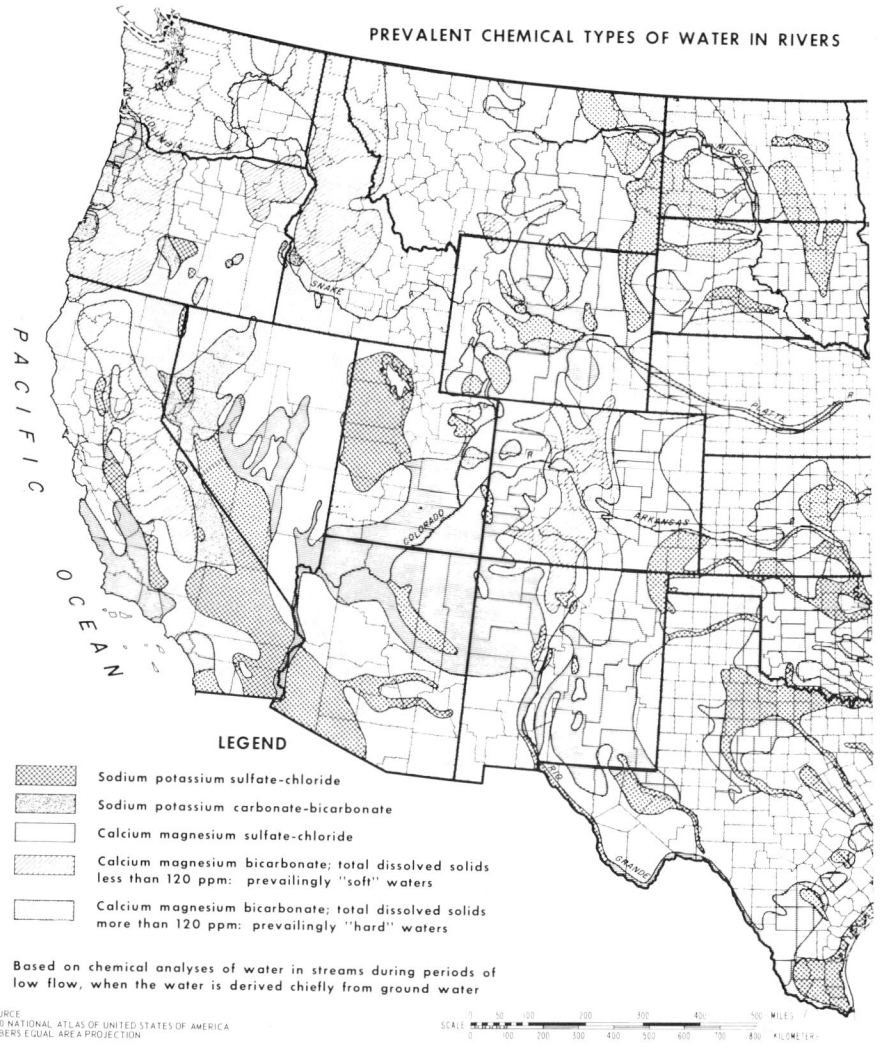

Fig. 3. Chemical types of waters in the major rivers of the western United States.

Effect of Parent Material

In soils with shallow depths of wetting, parent material differences in large part control the occurrence of gypsum. The soils in which we have found gypsum closely correspond to the distribution of anhydrite and gypsum in underlying geologic formations (Fig. 1 and 2). Anhydrite is most common in the Permian-age formations but does occur in some younger deposits derived from Permian-age formations (U.S. Geological Survey, 1932).

The Quaternary-age sand dunes in the Tularosa Basin (study area 1 on Fig. 1) were mostly derived from the Yeso formation (Spanish for gypsum). The Yeso is one of the more extensive gypsiferous formations in New Mexico (Weber and Kottlowski, 1959). It is exposed along the western frontal escarpment of the Sacramento Mountains (Fig. 4 and 5) where there are bedded gypsum deposits ranging in thickness from 1.2 to 16 m (4 to 54 ft). These exposures occur at elevations of 1,585 to 2,591 m (5,200 to 8,500 ft). Outcrops of similar Permian calcium sulfate beds in southeastern New Mexico are largely gypsum within 152 m (500 ft) of the surface, and anhydrite below 152 m (500 ft) (Adams, 1944).

The White Sands in the central part of the Tularosa Basin near Alamogordo (Fig. 4) are gypsum sand dunes derived from dry lake bed efflorescences. The present day ephemeral lakes at the southwestern point of the dunes are relics of the large perennial Pleistocene lake, Lake Otero, formed during the glacial pluvial intervals. The largest contemporary

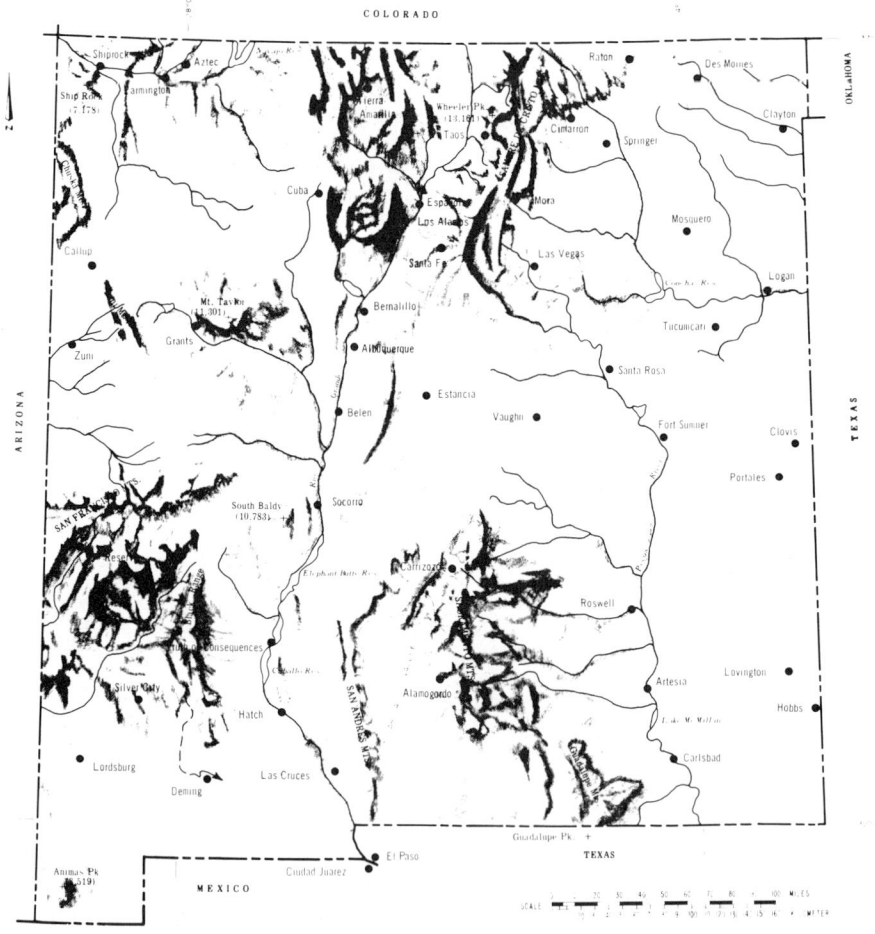

Fig. 4. Map of New Mexico showing some major mountains and valleys.

lake, Lake Lucero, may occupy an area of up to 25 km² (9.7 mi²), however, relic shorelines and evaporite deposits at about 1,219 m (4,000 ft, the dashed line in Fig. 5) indicate that Lake Lucero's Wisconsin predecessor's area reached approximately 1,800 km² (700 mi²) during at least one glacial pluvial substage. Lake Lucero is the present source of gypsum crystals which now form over 70,000 ha (730 km²) of active gypsum dune lands. The bedded gypsum surrounding the dune lands, however, cover over 200,000 ha (2,000 km²). These beds are comprised of two parts, those that are lacustrine salt deposits and those that are stabilized eolian uplands. During high winds, gypsum in the form of salts and very fine sands is being deposited over much of the area.

In a northeasterly direction from Lake Lucero, most soils within 32 to 40 km (20 to 25 mi) are in gypsic families. Those in dune areas near the lake are Typic Torriorthents. The soils in interdunal and surrounding areas are Typic Gypsiorthids. Typic Gypsiorthids predominate in the next belt of soils, though there are some Cambic Gypsiorthids and Petrogypsic Gypsiorthids on the outer part of this belt. The town of Alamogordo is in the next area of soils. These have formed in gypsiferous materials but have received less eolian gypsum. Most are Ustollic Camborthids, Torriorthents, and Fluvents and have mixed mineralogy.

Some gypsic families of Typic Gypsiorthids have formed on steep ridges and hills at the foot of Sierra Blanca, an igneous dike on the northeast rim of the basin. Since the igneous dike and the alluvium from it did not originally contain gypsum, we take this as further evidence of the influence of eolian gypsum in the basin.

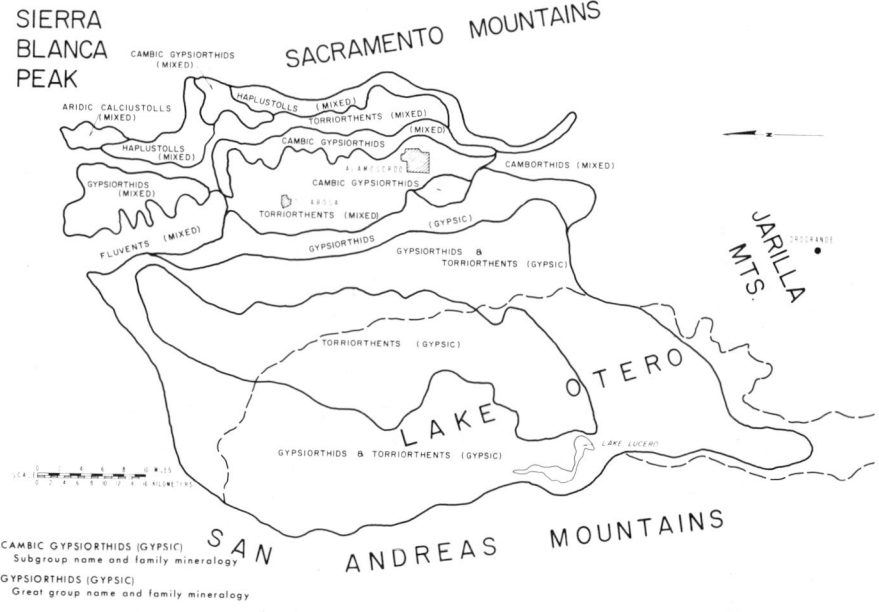

Fig. 5. Soils map of the Tularosa Basin in southern New Mexico. The family mineralogy of each map unit is in parentheses. Either the subgroup or great group name of each map unit is given.

Beyond Alamogordo to the east, alluvium from the Yeso fm has had more influence than eolian gypsum has had on the parent materials of the soils. First are the Cambic Gypsiorthids and then some Typic Haplustolls and Aridic Calciustolls on higher alluvial fans adjacent to the escarpment of the Sacramento Mountains. All these soils have formed in gypsiferous parent materials, but have less than 40% gypsum or gypsum and carbonate (Soil Survey Staff, 1979), hence, none are in gypsic families. The soils closer to the mountains have undergone leaching due to the higher precipitation there and hence do not have any gypsic horizons within 1 m of the surface. Gypsic horizons do occur on surfaces in the basin that have artifacts known to be older than 5,000 years. Some of the surfaces date from the last pluvial of the Pleistocene. Gypsic horizons in the soils on these surfaces, however, may be much younger because the area now receives annual deposits of gypsum dust. We conclude that none of the gypsic horizons are older than the last pluvial of the Pleistocene (about 12,000 years ago, Martin and Mehringer, 1965).

Further evidence of the influence of parent material on the occurrence of gypsum in soils follows. Allogenic gypsum is the source of sulfate in the arid region of the United States where "nonacid sulfate" soils occur and where there is no evidence that there has been any sulfate formed there via an "acid sulfate" stage.

One of the major Permian gypsiferous formations, in Texas and Oklahoma, is the Cloud Chief formation. It is associated with outcropping redbeds in north-central Texas, central Oklahoma, and northeastern Kansas (Keroher et al., 1966).

In southeastern Nebraska, gypsum has been found in Johnson shale (Burchett, 1970). Johnson shale is also of Permian age. Anhydrite beds of Mississippian to Jurassic age are found in the subsurface throughout most of western South Dakota. The Minnelusa and Spearfish are two formations which contain anhydrite beds (Cox, 1964).

In eastern South Dakota, allogenic gypsum is common in soils formed in glacial till. In the subsurface, it occurs in the upper Cretaceous Carlile shale. This formation underlies glacial till in some areas and may account for the physical incorporation of gypsum in the till. Rocks in the Graneros shale, Greenhorn limestone, and the Dakota group, immediately below the Carlile shale, contain either pyrite or limestone or both (Hedges, 1968). The pyrite and limestone can combine to form gypsum under near surface conditions.

Mancos shale is an important source of allogenic gypsum in soils of the states of Wyoming, Colorado, Utah, New Mexico, and Arizona (Keroher et al., 1966). The formation is of Lower and Upper Cretaceous age. The Colorado River drains much of the area. Consequently, a number of soils formed in Colorado River sediment contain allogenic gypsum.

In California, gypsiferous deposits are mostly in Lower Triassic series rocks and in alluvial deposits of younger age; however, there are some deposits in Tertiary formations in Imperial and Ventura Counties (VerPlanck, 1954) and in Quaternary deposits of impure gypsum that occur along the southwestern margin of the San Joaquin Valley (Heart, 1966).

Morphology

Gypsum crystallizes in the monoclinic system as tabular, prismatic, or acicular crystals including twinned varieties. A massive, opaque, granular variety, rock gypsum, is the most abundant form present in geologic formations. The fine-grained form is alabaster. Crystalline varieties of gypsum include selenite, the rhombic, or tabular platy cleavable form; and satinspar an aggregate of fine, parallel, acicular crystals with a silky luster and pearly opalescence. We have seen all of these forms except satinspar in soils. Hence, in a single sample it is difficult to identify pedogenic gypsum if allogenic gypsum is also present.

Pedogenic gypsum can be identified, however, by study of its distribution to depths to which the soil becomes wet. The upper part of the gypsic horizon, or the horizon of accumulation (Ccs), for the most part contains euhedral to subhedral spindle-shaped grains of gypsum (Fig. 6). These crystals are mostly randomly oriented throughout the s-matrix.

In the middle parts of some gypsic horizons, euhedral gypsum crystals also line channel voids (Fig. 7). In some fabrics, granular gypsum grains completely fill some channels. The s-matrix in these parts also contain some spindle-shaped crystals of gypsum similar to those found in the upper part.

In petrogypsic horizons, the crystals of gypsum grow together into an

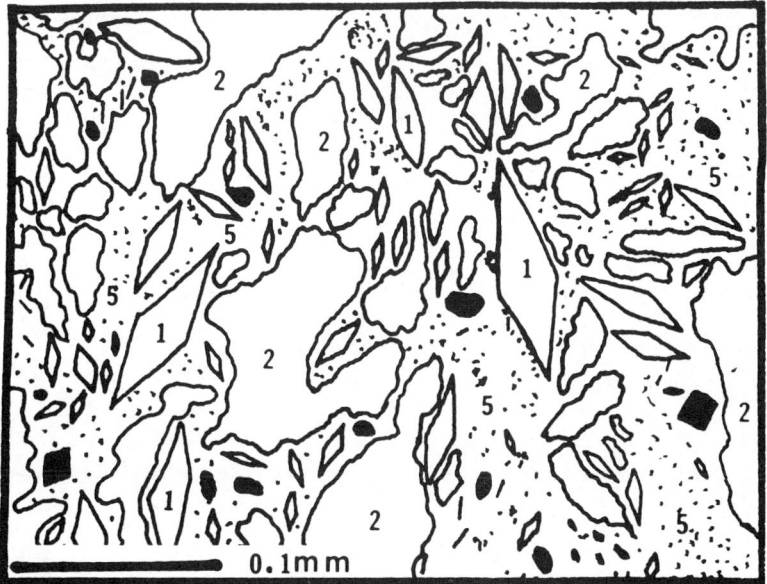

Fig. 6. Drawing of a thin section of the upper part of the s-matrix of a gypsic horizon. Most of the gypsum occurs as euhedral to subhedral grains of silt and sand size. The grains are spindle shaped (1) and randomly oriented. There are also many sand size aggregates having irregular shape and diffuse boundaries (2). These are composed of silt size carbonate, silicate clay and fine grained gypsum. The s-matrix also includes nongypsum skeleton grains (5) all shown in black.

interlocking network of subhedral to granular grains (Fig. 8). Below these petrogypsic horizons, the fabrics will be similar to the lower parts of the gypsic horizons. Some examples of gypsic horizons follow.

Rhombohedral, or spindle-shaped, euhedral crystals of coarse-silt size or coarser occur in the gypsic horizon of a Typic Gypsiorthid (S74NM-35-7) from the Tularosa Basin (study area 1 of Fig. 1). This pedon has a maximum of 24% gypsum (Table 3), an intertextic basic fabric, and a silasepic plasmic fabric. This assumes that the uniformally distributed silt-size crystals of gypsum within the s-matrix are plasma and that the gypsum crystals in voids and cracks are pedogenic features (Fig. 9). Most of the gypsum is pedogenic and occurs as gypsans in voids and cracks.

A second gypsiferous soil (S74NM-35-2) from the Tularosa Basin has a gypsic horizon with 75% gypsum. This Typic Gypsiorthid also has a gypsic horizon with an intertextic basic structure and a silasepic plasmic structure (Table 4, Fig. 10). Again, part of the gypsum is silt size and uniformly distributed throughout the s-matrix. A far greater part of the gypsum, however, occurs as euhedral rhombs and spindles of fine sand, very fine sand, and coarse silt size in voids and cracks.

An Ustollic Paleorthid from the Tularosa Basin contains as much as 92% gypsum in its gypsic horizon, but like the gypsic horizons in pedons 35-2 and 7, it remains soft (Table 5). It too has an intertextic basic and a

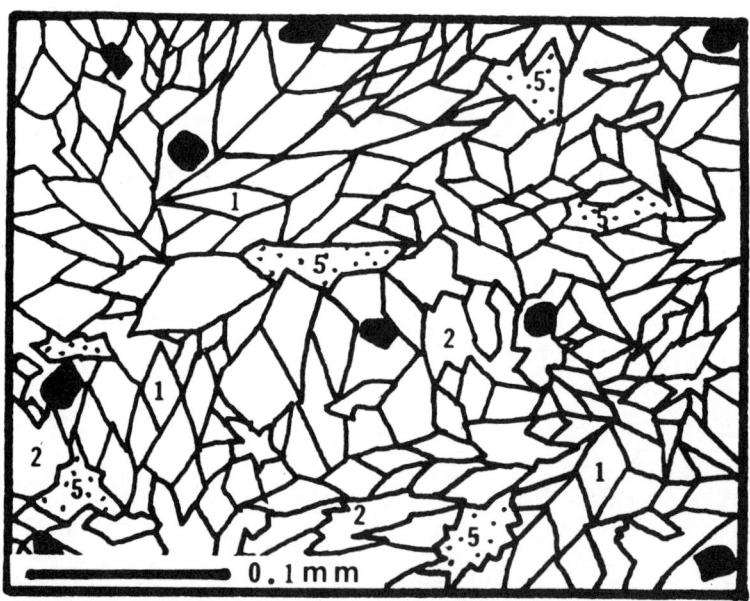

Fig. 7. Drawing of a thin section of the middle part of a gypsic horizon. The gypsum occurs as euhedral, subhedral, and granular grains of sand and silt size within the s-matrix and lining channel voids. Some of the gypsum grains are spindle shaped (1). Some of these line channel voids (4). Others are randomly oriented within the s-matrix. Some are granular (3) interlocking grains partly filling voids. The s-matrix also includes nongypsum skeleton grains (5) all shown in black.

GYPSIFEROUS SOILS

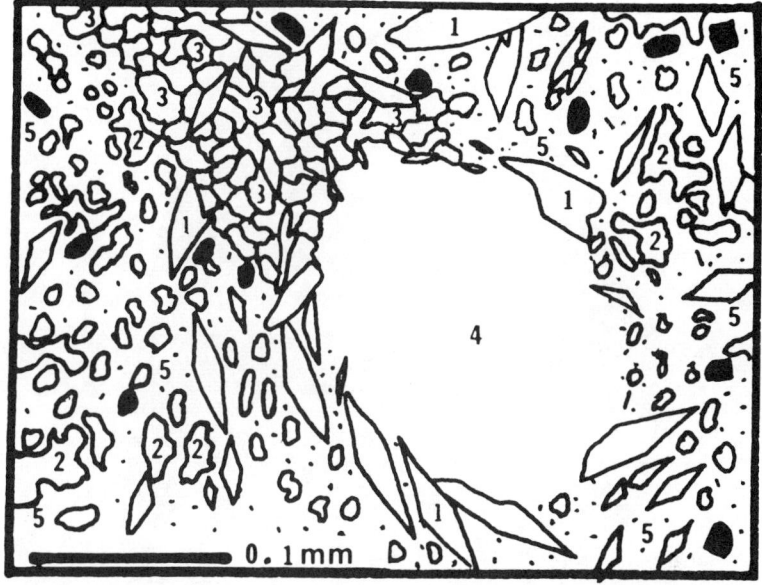

Fig. 8. Drawing of a thin section of a petrogypsic horizon. Interlocking subhedral, euhedral, and granular gypsum grains (1) provide the horizon's strength. There are some remnants of the irregular shaped sand size aggregates of carbonate, silicate clay, and silt size gypsum (2); carbonate and silicate clay aggregates (5) are shown in black.

Table 3. Some properties of a Typic Gypsiorthid, mixed family (S74NM-35-7) from the Tularosa Basin, Otero County, N.M.

Depth	Horizon	Consistence†	EC MMHOS /CM	$CaCO_3$ Equiv.	Gypsum
cm				%	%
0– 4	A11	So	4	22	0
20– 41	B21	Sh	44	22	2
61– 82	C1	Sh	32	21	2
82–114	C2cs	Sh	28	21	24
114–145	C3	Sh	36	24	6

† Dry consistence: So = soft, Sh = slightly hard.

Table 4. Some properties of a Typic Gypsiorthid (S74NM-35-2) from the Tularosa Basin, Otero County, N.M.

Depth	Horizon	Consistence†	EC MMHOS /CM	$CaCO_3$ Equiv.	Gypsum
cm				%	%
0– 1	A11	So	3	25	1
4– 7	A13	So	3	28	7
7– 29	C1cs	Sh	3	13	75
62– 89	C3cs	Sh	21	31	49
115–123	C5cs	Sh	23	35	46
123–140	C6	Sh	22	21	11

† Dry consistence: So = soft, Sh = slightly hard.

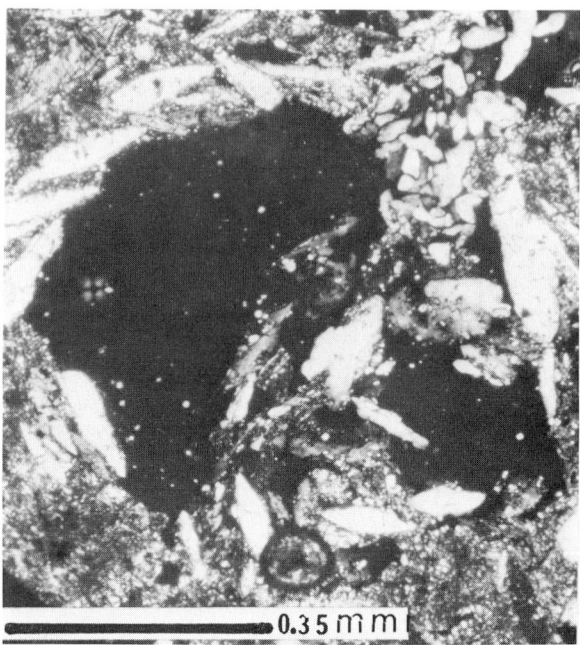

Fig. 9. Photomicrograph of the gypsic horizon of a Typic Gypsiorthid (S74NM-35-7) from the Tularosa Basin under crossed polarizers.

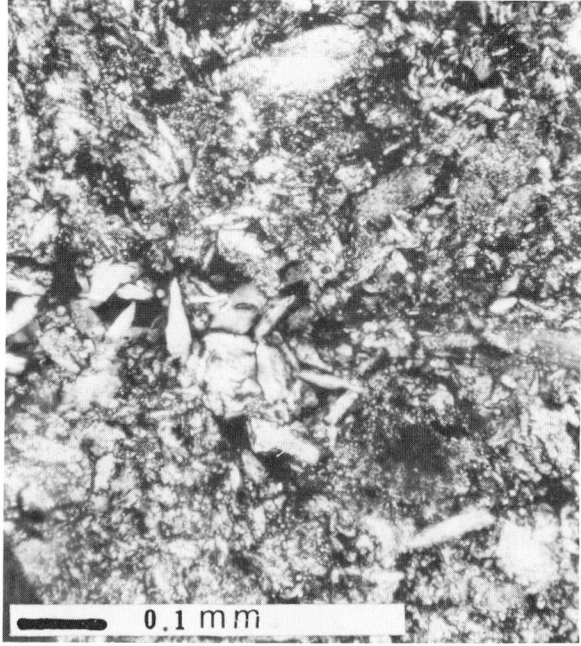

Fig. 10. Photomicrograph of the gypsic horizon of a Typic Gypsiorthid (S74NM-35-2) from the Tularosa Basin under crossed polarizers.

silasepic plasmic fabric (Fig. 11). Crystals of the gypsum are mostly euhedral spindles of very fine sand size. There are few cracks with interlocking subhedral gypsum crystals.

A Petrogypsic Gypsiorthid (S74NM-35-9) (study area 1 of Fig. 1), has about the same amount of gypsum but is indurated rather than being soft (Table 6). The upper part of the gypsic horizon is slightly hard, but like the petrogypsic horizon below (Fig. 12), it has an intertextic basic fabric. The upper part of the gypsic has a silasepic plasmic fabric, however, the petrogypsic horizon below has a crystic plasmic fabric (Fig. 12). Gypsum is the dominant skeleton grain mineral. Most of the gypsum crystals are subhedral, spindle-shaped rhombs. There are few cracks with equigranular crystals.

The petrogypsic horizon contains nearly continuous interlocking crystals of gypsum (Fig. 12). They range in size from coarse silt to medium sand. Whenever interlocking is prominent, the crystals are equigranular, subhedral rhombs. Cutans line many of the voids. The cutans are homogenized mixtures of fine silt, clay, and silt-size crystals of gypsum (Fig. 13 and 14).

Table 5. Some properties of an Ustollic Paleorthid (S74NM-35-1) from the Tularosa Basin, Otero County, N.M.

Depth	Horizon	Consistence†	EC MMHOS /CM	$CaCO_3$ Equiv.	Gypsum
cm				%	%
0– 2	A11	So	3	24	1
5– 23	C1cs	So	3	11	92
30– 38	C3cs	So	9	35	65
52– 66	C5cam	Ind	11	82	6
86– 94	C7ca	H	11	78	1
117–132	C10	Sh	11	1	22

† Dry consistence: So = soft, Ind = indurated, H = hard, Sh = slightly hard.

Table 6. Some properties of a Petrogypsic Gypsiorthid (S74NM-35-9) from the Tularosa Basin, Otero County, N.M.

Depth	Horizon	Consistence†	EC MMHOS /CM	$CaCO_3$ Equiv.	Gypsum
cm				%	%
0– 7	A1	So	68	23	1
7– 26	B2	So	61	28	1
26– 48	C1cs	Sh	19	12	75
48– 66	C2csM	Ind	12	10	89
66–127	C3cs	So	12	13	66
127–165	C4	So	12	24	40

† Dry consistence: So = soft, Sh = slightly hard, Ind = indurated.

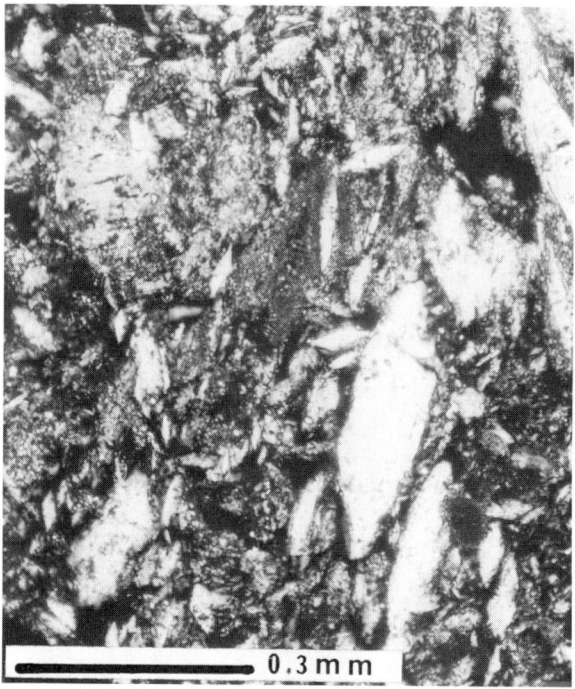

Fig. 11. Photomicrograph of the gypsic horizon of an Ustollic Paleorthid (S74NM-35-1) from the Tularosa Basin under crossed polarizers.

DISCUSSION

Genesis of Gypsum in Soils of the Western United States
Coastal Areas

Drainage of coastal wetlands causes any sulfides present to oxidize to acid sulfates. The acid sulfates, in turn, are neutralized by calcium carbonate, which either is present in the former wetlands or is added during reclamation. The gypsum persists in the soils if not leached out.

Dryland Areas

Gypsum, in well-drained dryland soils, evidently precipitates out of solution rapidly since crystals are sand size or finer (Fig. 9, 10, 11, 12, 13, and 14). It precipitates within the s-matrix in most soft gypsic horizons. In the early stages, few of the large voids or channels are lined with gypsans because the water is moving by unsaturated flow in the smaller voids and channels. The horizon may be recognized by its white color if carbonate is absent or less than a few percent. If the horizon is strongly calcareous, color alone is not enough to identify it as a gypsic horizon. Qualitative or semiquantitative tests like the acetone gypsum test (Richards, 1954) aid

Fig. 12. Photomicrograph of the petrogypsic horizon of a Petrogypsic Gypsiorthid (S74NM-35-9) from the Tularosa Basin under crossed polarizers.

recognition of gypsic horizons. The gypsum for the gypsic horizon comes from overlying horizons so that in gypsiferous materials absence of gypsum in A and B horizons is additional evidence of a gypsic horizon.

Potential gypsic horizons remain soft as they accumulate gypsum. Eventually the s-matrix becomes sufficiently filled with gypsum or other plasma to lower its permeability to water. With this somewhat reduced rate of water movement, crystal growth proceeds for a longer period of time, thus allowing larger crystals to form. Also the gypsum solutions are forced to move through the soil in successively larger voids. Crystal growth, thus, begins in the smaller voids and channels and then proceeds to larger and still larger voids until space for crystal growth becomes limiting causing crystals to interlock and lose some of their euhedral form (Fig. 12). Gypsic horizons then, like argillic horizons, calcic horizons, and duripans, form by feedback processes (Torrent and Nettleton, 1978).

Gypsum in Minesoils

Gypsum can also be a reaction product of the acid sulfate released by oxidation of pyrite and the carbonate present in the rock. Its identifica-

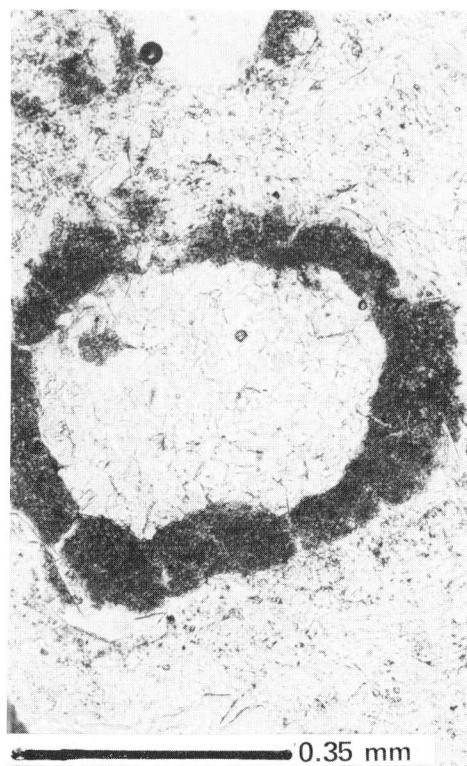

Fig. 13. Photomicrograph of a cutan in a void in the petrogypsic horizon of the Petrogypsic Gypsiorthid (S74NM-35-9) from the Tularosa Basin, New Mexico, in plain polarized light.

tion in a mine soil at the Rosebud Mine in southeastern Montana (study area 3 of Fig. 1) is direct evidence that the process has occurred because the original rock did not contain gypsum. Since there is an excess of carbonate remaining following oxidation of the pyrite, this mine soil will not become more acid through acid sulfate reactions.

Effects of Gypsum on Soils

SUBSIDENCE

Solution of gypsum from geologic formations containing relatively large amounts can cause subsidence (Fig. 15) (study area 2, Fig. 1). In extreme cases, anticlinal valleys like Gypsum Valley in southwest Colorado are formed (Stokes, 1948). Subsidence can be relatively small and yet make releveling necessary in soils used for irrigation farming. If not releveled, the subsiding areas can develop into large sink holes. Cracked building foundations, interrupted irrigation lines, and uneven roadbeds can be other victims of subsidence of soils through solution and removal of gypsum.

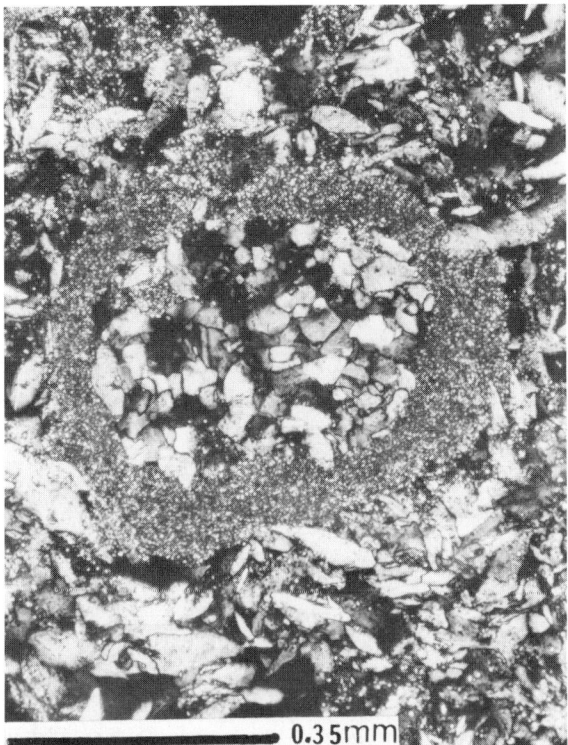

Fig. 14. Photomicrograph of a cutan in a void in the petrogypsic horizon of the Petrogypsic Gypsiorthid (S74NM-35-9) from the Tularosa Basin, New Mexico, under crossed nicols.

Fig. 15. Photograph of a depression about 35 m deep and 35 m across resulting from solution of gypsum, Clark County, Nev. (Muddy Creek fm., study area number 2, Fig. 1).

There is some increase in bulk density of soils as they subside upon loss of gypsum. This is partly a result of leachate residue with primary minerals having higher particle densities. Bulk density appears to be negatively correlated with gypsum content in nonindurated horizons (r = -0.65) when calculations are based on ovendry weights (see method 4A1b, Soil Conservation Service, 1972). When both gypsum and bulk density are corrected for the loss of the two water molecules of gypsum, the correlation is much lower (r = -0.25).

Loss of gypsum is just one of many factors controlling subsidence. Soils containing as little as 1.5% of gypsum have repeatedly failed when used in dry compacted structures such as irrigation canals and earthen dams (unpublished work, Soil Conservation Service Laboratory, Albuquerque, N.M.). Maximum packing densities are achieved in moist, but not wet, samples. Silty sediments which remain dry after deposition also undergo some subsidence when they become wet for the first time following deposition.

CONCRETE CORROSION

There are a number of ways in which gypsum deteriorates concrete. Most commonly, gypsum deteriorates concrete by releasing sulfate which then combines with sodium in the concrete to form such compounds as mirabilite ($Na_2SO_4 \cdot 10\ H_2O$) and thenardite (Na_2SO_4) and with calcium and aluminum to form ettringite [$Ca_6Al_2(SO_4)_3(OH)_{12} \cdot 26H_2O$] (Palache et al., 1951).

Concrete drain tiles in the Imperial Valley in California remain in good condition after 13 years while concrete linings of irrigation ditches deteriorated (Hobson, 1967). At the Imperial Valley Field Station, concrete drain tile has been in use continuously for 25 years without noticeable deterioration (Pillsbury, 1966). This suggests that drying under near surface conditions in addition to high sulfate is part of the corrosion process.

In the Las Vegas Area, Nevada, (study area 2, Fig. 1) winter temperatures frequently drop low enough to cause Na_2SO_4 to be converted to mirabilite (32.384 C). This nearly doubles its original volume and causes breakup of concrete slab structures (Downs, 1973).

Deterioration of concrete also occurs through the formation of crystals of ettringite and gypsum within the concrete. These crystals increase the volume of the solids within the concrete resulting in its eventual disintegration (Durand, 1956; Llamas Madurga, 1962; Biczok, 1964; Beutelspacher and van der Marel, 1966). Gypsum contents of 1.25% are high enough to provide enough sulfate to place the soil in a high corrosion class (greater than 7,000 ppm water soluble sulfate). Even detectable amounts are enough to show that the soil will have at least a moderate corrosion class (1,000 to 7,000 ppm sulfate) (Soil Conservation Service, 1971). Sodium and magnesium sulfates are believed to be more important than calcium sulfates in corroding concrete, but the presence of gypsum ensures that sufficient sulfate is present to make use of dense type 5 cement a good practice except for structures where high load-bearing capacity is critical (Double and Hellawell, 1977).

CONCLUSIONS

In 25 years of study at the National Soil Survey Laboratory, we have found that gypsum is the only calcium sulfate mineral found in soils. It is a common mineral in arid lands which have undergone little, if any leaching, and which contain sulfate as a major anion. Pedogenic gypsum may be recognized indirectly by its distribution with depth in the soil profile, that is, low amounts in the surface horizons and accumulation in subsurface horizons relative to underlying horizons and directly by its accumulation in pores and veins. A gypsic horizon can have as little as 6% gypsum if it is thick enough, and if the horizon below contains no more than 1% gypsum. One gypsic horizon we examined had 92% gypsum and still was soft. As the s-matrix in a developing gypsic horizon becomes plugged, progressively larger voids become filled with gypsum crystals. This feedback process decreases pore volume, restricts the hydraulic conductivity, and keeps the soil moist longer allowing the growing gypsum crystals to interlock and indurate the horizon. Gypsiferous soils have relatively low bulk densities and high hydraulic conductivities. Corrosion of concrete is more common on gypsiferous soils because they have sulfate as a major anion. Failure can be a problem in soils with as little as 1.5% gypsum.

LITERATURE CITED

1. Adams, J. E. 1944. Upper Permian Ochoa Series of Delaware Basin, West Texas and Southeastern New Mexico. Am. Assoc. of Petroleum Geol. 28:1596–1625.
2. Beutelspacher, H., and H. W. Van Der Marel. 1966. Toepassing van elektronen-microscopie in de grondmechanica. Meded. Lab. voor Grondmechanica 11. No. 2.
3. Biczok, I. 1964. Concrete corrosion and concrete protection. Akademiai Kiado, Budapest, Hungary.
4. Brewer, R. 1964. Fabric and mineral analysis of soils. Wiley, New York. 470 p.
5. Brown, G. 1961. The X-ray identification and crystal structure of clay minerals. Mineralogical Society, London. 544 p.
6. Burchett, R. R. 1970. Occurrence of gypsum in the Johnson shale (Permian) in Nemaha County, Nebraska. Resource Report No. 3. Univ. of Nebraska Conserv. and Survey Div. 23 p.
7. Byers, H., J. Sievers, and B. Tuffs. 1955. Distribution in the atmosphere of certain particles capable of serving as condensation nuclei. p. 47–70. In H. Weickmann and W. Smith (ed.) Artificial stimulation of rain, Proc. 1st Conf. Physics, Cloud and Precipitation Particles. Pergamon Press, New York (published 1957).
8. Cox, E. J. 1964. Gypsum and anhydrite. p. 133–137. In Mineral and water resources of South Dakota. U.S. Geological Survey and U.S. Bureau of Reclamation. 295 p.
9. Double, D. D., and A. Hellawell. 1977. The solidification of cement. Sci. Am. 237:82–91.
10. Downs, J. M. 1973. Interpretation of volume variations in soils due to sodium sulfate. p. 112. Agronomy Abstracts, 1973 Annual Meetings, ASA, SSSA, Las Vegas, Nevada.
11. Durand, J. H. 1958. Les Sols Irrigables.
12. Gifford, R. O., G. L. Ashcroft, and M. D. Magnuson. 1967. Probability of selected precipitation amounts in the western region of the United States. Agriculture Experiment Station, Univ. of Nevada.

13. Heart, E. W. 1966. Economic mineral deposits of the Great Valley. p. 249–252. In E. H. Bailey (ed.) USGS, Geology of Northern California. Bull. 190, California Div. of Mines and Geology, Ferry Building, San Francisco, 1966. 508 p.
14. Hedges, L. S. 1968. Geology and Water Resources of Beadle County, South Dakota, Part I, Geology. Bull. 18, South Dakota Geological Survey. 66 p.
15. Hobson, S. 1967. Soil surveys for predicting sulfate hazards to concrete irrigation structures. J. Trans. Am. Soc. Agric. Eng. 11:206–207.
16. Innes, R. P., and D. J. Pluth. 1970. Thin section preparation using an epoxy impregnation for petrographic and electron microprobe analysis. Soil Sci. Soc. Am. Proc. 34:483–485.
17. Junge, C. E. 1954. The chemical composition of atmospheric aerosols. 1. Measurements at Round Hill Field Station, June–July, 1953. J. Meteorol. II:323–333.
18. Keroher, G. C., and others. 1966. Lexicon of geologic names of the United States for 1936–1960. Geological Survey Bull. 1200. U.S. Government Printing Office, Washington, DC.
19. Llamas Madurga, M. R. 1962. Estudio geologico-tecnico de los terrenos yesiferos de la cuenca des Ebro y de los problemas que plantean en los canales. Serv. Geologico, Bol. 12, Informaciones y Estudios, 1962.
20. Lodge, J. P., Jr., K. C. Hill, J. B. Pate, E. Lorange, W. Basbergill, A. L. Lazrus, and G. S. Swanson. 1968. Chemistry of United States Precipitation, Final Report on the National Precipitation Sampling Network. National Center for Atmospheric Research, Boulder, Colorado. 66 p.
21. Martin, P. S., and P. J. Mehringer, Jr. 1965. Pleistocene pollen analysis and biography of the southwest. p. 433–451. In H. E. Wright and D. G. Frey (ed.) The quaternary of the United States. Princeton Univ. Press, Princeton, New Jersey, 1965. 922 p.
22. Murray, R. C. 1964. Origin and diagenesis of gypsum and anhydrite. J. Sediment. Petrol. 34:513–523.
23. Nelson, R. E., L. C. Klameth, and W. D. Nettleton. 1978. Determining soil gypsum content and expressing properties of gypsiferous soils. Soil Sci. Soc. Am. J. 42:659–661.
24. Palache, C., H. Berman, and C. Frondel. 1951. Dana's system of mineralogy II. Halides, nitrates, borates, carbonates, sulfates, phosphates, arsenates, tungstates, molydates, etc. John Wiley and Sons, New York. 1124 p.
25. Pillsbury, A. F. 1966. Durability of concrete irrigation and drainage pipe in alkali areas. p. 1–13. In Concrete irrigation and drainage pipe in sulfate soils. Am. Concrete Pipe Assoc. 1968. 30 p.
26. Richards, L. A. (ed.). 1954. Diagnosis and improvement of saline and alkali soils. Agric. Handb. No. 60. USDA. 160 p.
27. Soil Conservation Service. 1971. Guide for interpreting engineering uses of soils. USDA, Soil Conservation Service. 87 p.
28. ————. 1972. Soil Survey Investigations Report No. 1. Soil Survey Laboratory Methods and Procedures for Collecting Soil Samples. USDA. 63 p.
29. Soil Survey Staff. 1979. Classification of soil series of the United States, Part I, and Soil families of the United States and their included series, Part II. U.S. Government Printing Office, Washington, DC.
30. Stokes, W. L. 1948. Geology of the Utah-Colorado salt dome region with emphasis upon Gypsum Valley, Colorado, Utah Geol. Soc. Guidebook to the Geology of Utah, 3. p. 3–40.
31. Torrent, J., and W. D. Nettleton. 1978. Feedback processes in soil genesis. Geoderma. 20:281–287.
32. U.S. Geological Survey. 1932. Geologic map of the United States. Reprinted 1960.
33. VerPlanck, W. E. 1954. Salines in Southern California. p. 5–14. In Geology of southern California, Bull. 170, Chapter VIII, Mineral deposits and mineral industry, Div. of Mines, State of California, Div. of Natural Resources, San Francisco, Calif. 74 p.
34. Weber, R. H., and F. E. Kottlowski. 1959. Gypsum resources of New Mexico, Bull. 68, State Bureau of Mines and Mineral Resources, New Mexico Institute of Mining and Technology, Campus Station, Socorro, New Mexico. 68 p.

Chapter 10

Mineralogical Properties of Lignite Overburden as they Relate to Mine Spoil Reclamation[1]

J. B. DIXON, L. R. HOSSNER, A. L. SENKAYI, AND K. EGASHIRA[2]

ABSTRACT

The mineralogical properties of lignite overburdens differ significantly between the oxidized zone and the reduced zone. Reclamation of lignite mine spoil generally involves earthy materials that overlie the lignite. Hard rock types are exceptional. These overburden materials are sometimes capped by good soils that need to be reinstalled as illustrated by the loessial soils of West Germany. On the other hand reclamation may utilize mixed overburdens as is being done in the coastal plain deposits of eastern Texas where the soils may have poor physical and chemical properties and topsoiling is not practiced. In East Texas fresh overburden materials are left on the surface thus exposing carbonates, sulfides, and labile silicates to immediate weathering. The thickness of the weathered zone is about 5 to 20 m depending on the texture of the material and the depth to impermeable strata. Clayey or lignite strata impede penetration of air and water and thus the weathering front.

Iron sulfides and Fe-chlorite are generally absent from the oxidized zone. Labile iron compounds have weathered to iron oxides. Chlorite usually occurs in the unoxidized overburdens analyzed in eastern Texas. Pyrite is present in these deposits but occurs in local concentrations in varying amounts. Feldspars appear to have partially weathered out of the upper oxidized zone based on preliminary data. Feldspars occur mostly in the silt and sand fractions. Mica of the muscovite type is present in modest amounts throughout the overburden section. It is relatively resistant to weathering and persists to some extent to the soil surface. Smectite and kaolinite are abundant in the clay fraction throughout the overburden section except near the soil surface where much of the smectite has been removed. These

---

[1] Contribution from Texas Agric. Exp. Stn., Dep. of Soil and Crop Sci., Texas A&M Univ., College Station, TX 77843.
[2] Professors, research associate, and former research associate, Texas A&M Univ., College Station, TX 77843, respectively. The latter author is currently associate professor of soils, Kyushu Univ., Fukuoka, Japan.

Copyright © 1982 Soil Science Society of America, 677 S. Segoe Rd., Madison, WI 53711. *Acid Sulfate Weathering.*

conclusions are based on data from only a few locations in East Texas and require reevaluation throughout the lignite belt to account for possible variation along the strike of the beds and under different climatic conditions.

Carbonates occur infrequently in materials that overlie Wilcox lignite of East Texas. Thus they are not a reliable indicator of weathering. Calcite has been identified in a few layers. Siderite ($FeCO_3$) is important in local concentrations and weathers to goethite coated rocks that persist in exposed land surfaces.

Jarosite and gypsum are the most common sulfate minerals associated with lignite overburden of East Texas. Both minerals form on the exterior of overburden cores when they are allowed to oxidize and dry in storage. These localized fresh efflorescences suggest the rapid weathering of pyrite. Also, gypsum forms as a widespread white powder on dry shale surfaces suggesting that it formed as the interstitial solution evaporated in the absence of localized pyrite. Melanterite and szmolnokite also form on samples of lignite overburden from East Texas.

The discussions of mineralogical properties of lignite overburden presented here and the experience with reclaiming lignite overburdens in the United States and abroad reflect a favorable outlook for the continuation of such practices. However, it must be recognized that the extent of these experiences has been limited to a few locations and a short interval of time. In the East Texas lignite mining area many of the native upland soils, presumably formed from similar parent material to lignite overburdens, are infertile and difficult to manage. Thus continued investigations are needed of the overburden composition at different locations. Appropriate revegetation practices and long term management practices are needed to assure successful reclamation and return of the land to agricultural use.

INTRODUCTION

Lignite occurs in Alabama, Arkansas, Louisiana, Mississippi, and Texas (Pennington, 1978). It is being mined in Texas at several locations and a total of 21 mines is projected for operation by 1984 (Texas Railroad Commission, personal communication). Lignite is also extensive in North Dakota, Montana, South Dakota, and Wyoming. Lignite occurs in Alaska in several locations but the extent of the deposits is yet to be determined (United States Geological Survey, 1974). Low rank coal which includes subbituminous, lignite, and brown coal occurs in Canada adjacent to the North Dakota lignite, in the USSR, in widely disseminated deposits in western Europe, and southward to Turkey. A few areas of low rank coal are mapped in southeastern Australia, in Argentina and in Chile (Map of World Coal Resources and Major Trade Routes, 1975). Inquiry into factors relating to reclamation is merited because of the large and diverse areas of lignite and other low rank coals. The potential for rapid development of acidity and consequent acid weathering in mine spoils which contain sulfides make reclamation of many mined areas difficult.

The major lignites of Texas occur in the Wilcox Group of lower Eocene age (Kailer, 1974). The environments of lignite deposition were fluvial, deltaic, and lagoonal. The highest grade and most abundant lignites occur in eastern or central Texas in the deltaic and fluvial deposits, respectively. Lignite deposits of secondary grade and quantity occur in the Yegua Formation and Jackson Group of upper Eocene age.

Return of lignite mine spoil to agricultural production by dry and wet (sluicing) methods has been accomplished in West Germany (Petzold, 1978). Loessial soil material is returned to the surface. In East Texas, where lignite surface mining is currently expanding, the native soils tend to be low in fertility, very slowly permeable, low in water available to plants and difficult to till. Research results indicate that spoil which has been properly fertilized, planted, and maintained can effectively support crop production. The new soils that are forming from spoil have more desirable physical properties than the natural soils in much of the area (Hons et al., 1978). The properties of these new soils are largely determined by their mineralogical composition and particle size distribution. Oxidizable carbon in these overburdens is usually less than 1.5% but ranges from 0.0 to 3.9% (Arora et al., In press).

The layer silicate minerals largely control the supply of potassium and other cations available to plants from reclaimed spoil or soil. Pyrite and other sulfides are the major sources of acidity that is produced when they are oxidized to sulfates in exposed lignite overburden. Unfortunately, the amount of calcium carbonate present in overburden in eastern Texas is too limited to contribute much to neutralizing the acidity that will form in some overburden layers when they are oxidized. Weathering of feldspars will produce clay and as yet undetermined amounts of Ca, K, and Na, but the process is expected to be too slow to have an immediate impact on reclamation.

The lignite overburdens of eastern Texas include a full range of particle sizes from clays to sands; they are largely loamy materials heavily endowed with silt and clay. In certain areas where the Carrizo formation occurs the overburdens are extremely sandy. Except in the latter case the overburdens have more favorable particle size distribution than most of the native soils which have clayey subsoils. The overburden as a whole is composed of an upper oxidized zone and a lower reduced zone. The oxidized zone is free of acid forming sulfides. The reduced zone extends to the lignite and often contains sulfides which are frequently most concentrated near the lignite. Neither the oxidized or reduced overburden zones in eastern Texas contain adequate phosphorus or nitrogen for reclamation and crop production. Potassium immediately available is adequate but long-term supplies are modest in the medium and heavy textured material and low in the sands. Thus reclaiming spoil will require sustained fertilization to maintain vegetation.

The objectives of this paper are to discuss the mineralogical composition and properties of lignite overburden and how they influence reclamation of lignite mine spoil. The discussions apply to Wilcox Group lignite and related overburdens of eastern Texas except where noted otherwise.

DISCUSSION

Sulfide Minerals

Pyrite, marcasite, and other sulfide minerals are the major inorganic sources of acidity in lignite mine spoil. The abundance, distribution, and properties of these iron sulfides influence the potential acidity problem during reclamation. The lignites of the Wilcox Group and associated strata of Texas contain individual crystallites (Fig. 1A, 1C) and massive

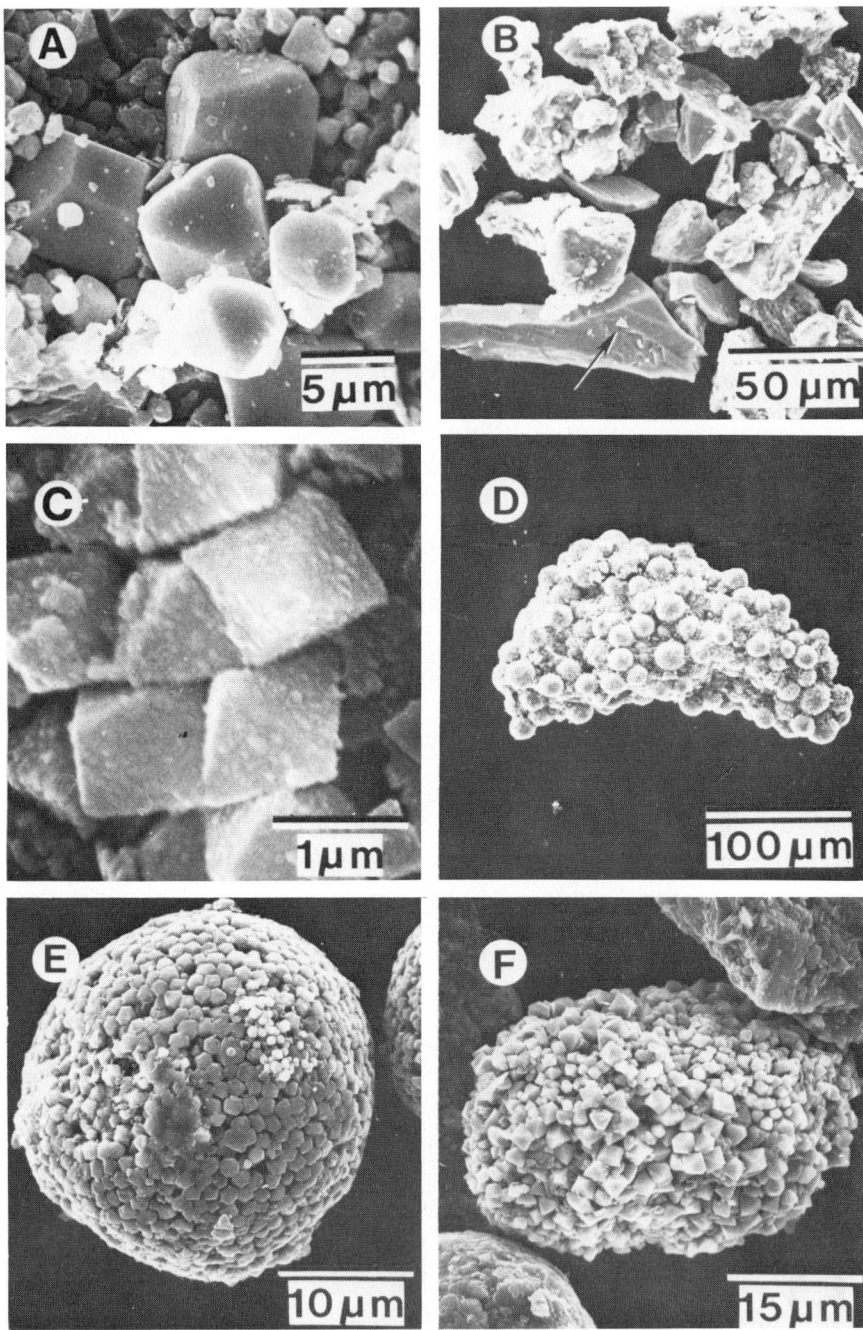

Fig. 1. Morphological forms of pyrite in lignite overburden of Texas (From Pugh, C. E. 1978. Influence of surface area and morphology on the oxidation of pyrite from Texas lignite. Doctoral dissertation. Texas A&M Univ., College Station, Tex. p. 50–53).

pyrite in particles of mostly sand and silt sizes (Fig. 1B). Pyrite also occurs in these deposits as framboids (Fig. 1E, 1F) and polyframboids (Fig. 1D). The surface area of framboidal pyrite is greater than that of the massive pyrite of the same particle size (Fig. 2). Large surface area is correlated with a rapid rate of acid formation (Smith and Shumate, 1970). The pyritic sulfur content of lignite and lignite plus overburden shale strata in the Wilcox Formation of Texas is 0.1 to 1.52% (Arora et al., In press). The amount of pyrite varies greatly in a given lignite bed; usually it is highest near the top and bottom of the bed. Spheroidal marcasite has been identified in the lignite overburden of Texas (Fig. 3) in rare instances.

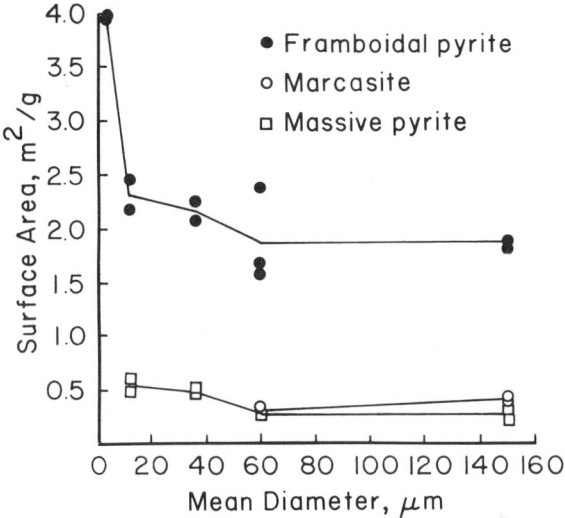

Fig. 2. Surface area of iron sulfides versus mean particle diameter (From Pugh et al., In press).

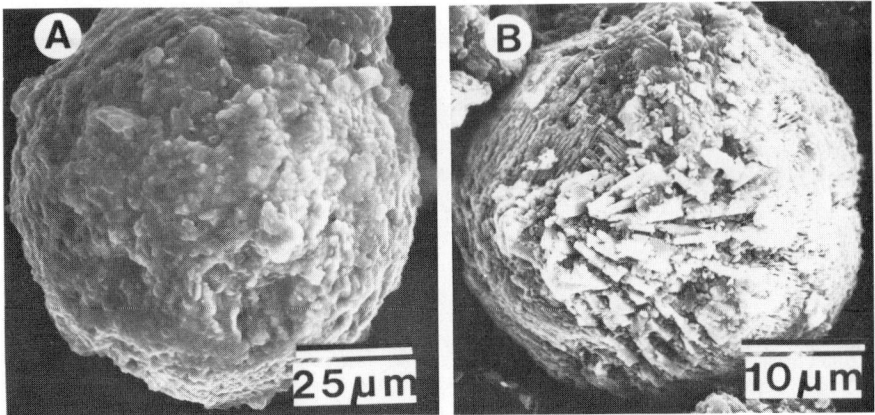

Fig. 3. Marcasite from lignite overburden of Texas (From Pugh, 1978. p. 53). Right particle identified by characteristic cockscomb morphology and left by XRD.

Marcasite is more abundant than pyrite in the overburden of the Frechen open cast lignite mine in the Rhenish soft coal field of West Germany (Brinkmann, 1977). The shallower (younger) overburdens contained a higher marcasite to pyrite ratio than the deeper layers. Samples from Zukunft-West and Fortuna-Garsdorf opencast mines in West Germany were found to have more pyrite than those from Frechen. The

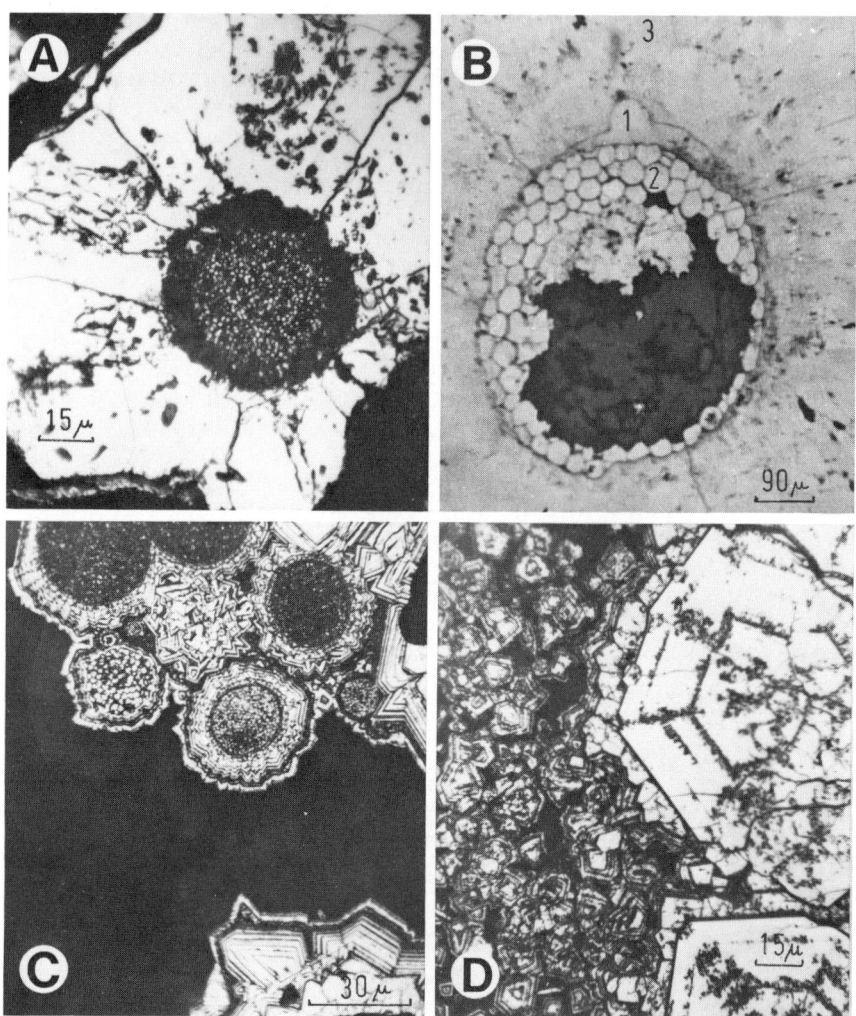

Fig. 4. A. Marcasite surrounding a pyrite framboid that contains iron oxides within the framboid and coating it. Oil immersion. B. Marcasite coated plant cross section. A gel-like seam enriched in Co and Ni (1) partially surrounds the plant cells (2). The plant cells and the marcasite are free of Co and Ni. C. Framboids with varying contents of pyrite and iron hydroxides. The framboids are surrounded by zoned pyrite, which also surrounds the non-zoned marcasite granules. The sulfides are coated by iron hydroxide gel (gray) and coal (black). Oil immersion. D. Large zoned marcasite crystals surrounded by smaller also zonally-built pyrites associated with limonite (dark). This figure elucidates the frequently observed granule size fluctuation of the two sulfide phases. Oil immersion. (Fig. 4A, 4B, 4C and 4D. Specimens were polished sections photographed by employing reflected plane polarized light. From Brinkman, 1977).

Fe to S atomic ratios of pyrite and marcasite of the lignite of West Germany was < 1:2. Samples of these minerals from Texas lignite had less than the ideal amount of S but Fe was not determined and impurities present in some samples contributed to the disparity (Pugh et al., In press).

Marcasite occurred as coatings around pyrite framboids and iron oxides (Fig. 4A). It commonly grew outward from the core material in a radial pattern. Plant parts were surrounded by marcasite and partially surrounded by a gel-like layer (Fig. 4B). Marcasite also crystallized in the form of a plant leaf (Brinkmann, 1977). Pyrite framboids contained iron oxides and were in turn coated by iron oxides or hydroxides (Fig. 4C). Both marcasite and pyrite occurred in banded (zoned) configurations (Fig. 4D). Other sulfide minerals (e.g., mackinawite, griegite, chalcopyrite, and pyrrhotite) though not reported in association with lignites or related spoil in the literature reviewed, occur in other sedimentary deposits (e.g. mackinawite and griegite in Rickard, 1973) and may be present; therefore they are included in Table 1 for the sake of completeness.

The occurrence of heavy metals in associations with iron sulfides is of interest because of geochemical implications concerning mineral formation and possible deleterious environmental influences in spoil or ash from lignite. The trace elements Co and Ni were most concentrated in gel-like seams associated with plant parts (Fig. 4B). The plant parts and the massive marcasite that surrounded the assemblage were free of Co and Ni. The two trace metals were almost without exception bound to gel-like material, "melnikovite-marcasite" (Brinkmann, 1977) which bordered plant remains now composed of marcasite. Although the data are limited, no evidence indicates toxic levels of trace metals in the sulfides associated with lignites investigated thus far.

Sulfate Minerals

Sulfate minerals are among the first reaction products of the sulfuric acid that is formed by sulfide oxidation in lignite and lignite overburden. Jarosite, gypsum, barite, and several other sulfate minerals have been found in such materials (Fig. 5A, B, and C).

Table 1. Sulfide minerals related or potentially related to lignite mine spoil reclamation.

Mineral	Formula	Three strongest XRD lines, Å†	PDF Card No.†
Chalcopyrite	$CuFeS_2$	3.03_x 1.85_8 1.59_6	9-423
Greigite	Fe_3S_4	2.98_x 1.75_8 2.47_6	16-713
Mackinawite	FeS	5.03_x 2.97_8 2.31_8	15-37
Marcasite	FeS_2	2.71_x 1.76_6 3.44_4	3-799
Pyrite, synthetic	FeS_2	1.63_x 2.71_9 2.42_7	6-710
Pyrrhotite	$Fe_{1-y}S$	2.06_x 2.64_5 2.98_4	22-1120

† Spacing and intensity subscript from Berry (1973). Intensity subscripts: x = 100% others are on the same relative scale and 1 = 10%, 2 = 20%, etc. y = 0.12.

Fig. 5. (A) Jarosite particles from fracture surface of weathered shale of the Claiborne group, College Station, Tex. The particles have a cubic appearance. (B) Barite tabular particles from weathered shale of the Claiborne group, College Station, Tex. (C) Gypsum needles protruding from surface of dry lignite overburden. Note apparent tubular particle at the upper left.

Distinct yellow (5Y 7/6) jarosite efflorescences often form on lignite overburden cores in storage. The mineral also commonly occurs in partings in shale in lignite overburdens (see color Plate 4A). Also, jarosite has been observed by scanning electron microscopy as cubes and octahedra that are believed to be in pseudomorphic forms of pyrite in soil (C. D. Carson and J. B. Dixon, unpublished). Jarosite is reported to form only under very acid (pH 3 to 4) conditions (van Breemen, 1973). Once formed, the low solubility of jarosite makes it a useful indicator of the weathering history of the deposit. Jarosite has been found in a calcareous Lufkin soil at College Station, Tex. (Carson et al., 1982) indicating an apparently marked increase in soil pH during the weathering history of this soil. Natrojarosite was experimentally formed from albite and jarosite from glauconite, illite, and microcline in 4 months. The Fe(II) in solution was oxidized by *Thiobacillus ferrooxidans* (Ivarson et al., 1978).

Gypsum occurs as macroscopic crystals (see color Plate 4B) associated with the lignite spoil at the San Miguel mine in southwestern Texas in upper Eocene deposits (D. S. Fanning, personal communication). Blades of gypsum have been found in Wilcox Group lignite overburden cores of northeastern Texas (Fig. 5B). The amount of native sulfate sulfur in Wilcox Group lignites of Texas is small (e.g., 0.01 to 0.18%; Arora et al., 1980) and gypsum is often present in local concentrations. Gypsum produces little or no acidity when it hydrolyzes and it is a beneficial source of Ca and SO_4. Although gypsum probably will not have an unfavorable influence on reclamation of lignite overburden for agricultural purposes it may cause engineering problems. Gypsum is produced by the oxidation of pyrite in the presence of Ca which may come from $CaCO_3$. Reaction of concrete with the acidity produced during pyrite oxidation could produce gypsum. In Canada gypsum crystals heaved concrete floors built on pyritic shales that contained $CaCO_3$ veins (Grattan-Bellew and Eden, 1975).

Melanterite has been observed in an apparently oxidized, moist core of overburden from Wilcox Group lignite. This and other sulfate minerals are seen as surface deposits on overburden materials. For example, szmolnokite was observed on samples of lignite mine spoil from Freestone and Milam Counties in Texas. Rozenite was identified on lignite mine spoil of Texas (see color Plate 4F) and on coal (see color Plate 4D) near Cannonburg, Ky. (D. S. Fanning, personal communication). Copiapite (aluminocopiapite) has formed on coal underclay in sheltered exposures near Cannonburg, Ky. (see color Plate 4C) (Fanning, personal communication). Copiapite may form on lignite as does gypsum even though both compounds are relatively soluble. Barite (see color Plate 3D) occurs in overburden associated with thin lignitic clay of the Claiborne Group at College Station, Tex. The barite may be associated with volcanic deposits as it is in the Catahoula Formation of Walker County, Tex. (McKee and Brown, 1977). Anhydrite forms in evaporites and conceivably could form from gypsum in association with lignites of dry regions (Table 2).

There are potentially many more sulfate minerals that can form due to the weathering of pyrite depending on the associated minerals, the climate, and other factors. Bandy (1938) described 76 minerals in the sulfate deposits of Atacama Desert of northern Chile which received less than 5 mm of rainfall annually. Pyrite was a major sulfide originally present in those deposits. Some of these relatively soluble minerals will not persist in moist climates but they may form temporarily during oxidation and drying of pyritic lignite or overburden. As lignite mining expands in dryer regions the probability of finding the more soluble phases will increase. As mentioned earlier copiapite occurs on exposed coal seams in areas protected from rainfall, in a humid climate. Gypsum, melanterite, szmolnokite, and jarosite efflorescences form during oxidation and drying of lignite overburden in Texas. Several other sulfate minerals are listed in Table 2 because of their potential importance in lignite spoil.

Table 2. Sulfate minerals related or potentially related to lignite mine spoil reclamation.

Mineral	Formula	Three strongest XRD lines, Å			PDF card no.†
Alunite	$(K,Na)Al_3(SO_4)_2(OH)_6$	2.99_x	2.89_x	2.29_8	14-136
Amarantite	$FeSO_4(OH) \cdot 3H_2O$	11.3_x	8.69_x	3.57_8	17-158
Anhydrite, synthetic	$CaSO_4$	3.49_x	2.85_3	2.33_2	6-226
Barite, synthetic	$BaSO_4$	3.44_x	3.10_x	2.12_8	5-448
Botryogen	$MgFe^{III}(SO_4)_2(OH) \cdot 7H_2O$	8.87_x	3.00_8	6.29_6	17-157
Copiapite (Aluminocopiapite)	$MgAlFe(SO_4)(OH)H_2O$	9.20_x	18.1_8	5.58_8	20-659
Coquimbite	$Fe_2(SO_4)_3 \cdot 9H_2O$	8.26_x	2.76_8	5.45_7	6.40
Epsomite	$MgSO_4 \cdot 7H_2O$	4.21_x	5.35_3	2.68_3	8-467
Ferrinatrite	$Na_3Fe(SO_4)_3 \cdot 3H_2O$	7.80_x	2.91_8	4.38_6	16-937
Fibroferrite	$Fe(SO_4)(OH) \cdot 5H_2O$	12.1_x	2.98_8	6.96_6	16-935
Gypsum	$CaSO_4 \cdot 2H_2O$	2.87_x	4.28_9	2.68_5	21-816
Halotrichite	$FeAl_2(SO_4)_4 \cdot 22H_2O$	4.77_x	3.48_x	4.29_6	11-506
Hohmannite	$Fe_2(SO_4)_2(OH)_2 7H_2O$	7.92_x	8.69_8	10.4_6	17-155
Jarosite	$KFe_3(SO_4)_2(OH)_6$	3.08_x	3.11_6	2.29_5	10-443
Kalinite	$KAl(SO_4)_2 \cdot 11H_2O$	4.83_x	4.32_x	4.11_x	17-133
Melanterite, synthetic	$FeSO_4 \cdot 7H_2O$	4.90_x	3.78_8	4.87_5	22-633
Mendozite	$NaAl(SO_4)_2 \cdot 11H_2O$	3.50_x	4.76_9	4.58_7	22-475
Metasideronatrite	$Na_4Fe_2(SO_4)_4(OH)_2 \cdot 3H_2O$	3.66_x	2.73_8	7.93_6	16-936
Natrojarosite	$NaFe_3(SO_4)_2(OH)_6$	5.06_x	3.06_8	3.12_7	11-302
Pickeringite	$MgAl_2(SO_4)_4 \cdot 22H_2O$	4.82_x	3.51_9	4.32_4	12-299
Quenstedtite	$Fe_2(SO_4)_3 \cdot 10H_2O$	4.08_x	5.78_8	4.19_8	17-160
Roemerite	$Fe_2^{III}Fe^{II}(SO_4)_4 \cdot 14H_2O$	4.79_x	4.03_9	5.05_5	13-530
Rozenite	$FeSO_4 \cdot 4H_2O$	4.47_x	5.46_9	3.97_7	16-699
Sideronatrite	$Na_2Fe(SO_4)_2(OH) \cdot 3H_2O$	10.2_x	3.01_8	3.38_6	17-156
Szomolnokite, synthetic	$FeSO_4 \cdot H_2O$	3.44_x	3.12_4	2.52_4	21-925
Tamarugite	$NaAl(SO_4)_2 \cdot 6H_2O$	4.22_x	4.21_8	3.65_6	19-1186

† Data were assembled from the powder diffraction file edited by Berry (1973).

Carbonate Minerals

The carbonate minerals listed in Table 3 occur or may occur in lignite overburden. Calcite nodules have been detected in overburdens of the Wilcox Group lignites of East Texas. Aragonite occurs in fossils of biological organisms and is therefore a potential component of lignite overburdens. Siderite occurs in lignite overburden of Texas usually as lenticular masses or veins up to about 10 cm thick (Arora et al., In press). It is transformed to a goethite encrusted mass by gradual weathering inward that produces a dark brown rind of goethite with a pale interior band around the pale yellow (2.5Y 7/4, dry) or light olive brown (2.5Y 5/4, moist) core of siderite (see color Plate 4E). Three siderite samples from Texas, contained layer silicates, quartz, and feldspar as accessory minerals (Hugh Durham, personal communication). These goethite coated rocks accumulate on eroded soil surfaces and also become a prominent feature of reclaimed spoil. Large lenticular masses of siderite up to 1 m in length have been observed on the Claiborne Group sediments at College Station, Tex. and in Wilcox Group lignite overburden in Freestone County, Tex. Magnesite, dolomite, and rhodochrosite are components of sedimentary rocks and are potential constituents of lignite overburden.

Table 3. Carbonate minerals related to lignite mine spoil reclamation.

Mineral	Formula	Three strongest XRD lines, Å †	PDF card no. †
Aragonite, synthetic	$CaCO_3$	3.40_x 1.98_7 3.27_5	5-483
Calcite, synthetic	$CaCO_3$	3.04_x 2.29_2 2.10_2	5-586
Dolomite	$CaMg(CO_3)_2$	2.89_x 2.19_3 1.79_3	11-78
Magnesite, synthetic	$MgCO_3$	2.74_x 2.10_5 1.70_4	8-479
Rhodochrosite, synthetic	$MnCO_3$	2.84_x 3.66_4 1.76_4	7-268
Siderite	$FeCO_3$	2.79_x 1.73_8 3.59_6	8-133

† Data were assembled from the power diffraction file edited by Berry (1973).

Carbonates are of particular interest in lignite and coal overburdens because of their potential role in neutralizing acidity. Since the weathering of Fe and Mn carbonates may involve oxidation and hydrolysis of the cations and consumption of hydroxyl during precipitation as Fe or Mn oxides or oxyhydroxides, they may not aid in neutralizing acidity; their influence on mine spoil reclamation merits further investigation where they occur.

Silicate Minerals

Any silicate may occur in lignite overburden and the more common ones are discussed here because of their relative importance. When lignite mine spoil weathers under acid oxidizing conditions, ferromagnesian (mafic) chlorite becomes labile as indicated by overburden studies reported later and by laboratory investigations of Ross (1969). On exposure to acidity and oxidation, chlorite alters to interstratified vermiculite and chlorite (Ross and Kodama, 1976). Senkayi et al. (1981) separated chlorite from lignite mine spoil of Texas and transformed it to vermiculite and eventually to smectite by acid oxidation with bromine water. As a component of sediments above Wilcox Group lignite of Texas, chlorite is a potential source of smectite (Table 4). Formation of smectite in these medium-textured deposits could lead to development of a clayey horizon with poor physical properties. The amount of smectite that can form in this way needs to be determined. In principle, prevention of acid formation would improve soil-forming reactions by reducing the rate of smectite formation in medium to fine textured materials that contain chlorite.

In laboratory experiments, chlorite alteration yielded Mg and Fe in solution from the hydroxide sheet while the 2:1 structural layer remained intact (Senkayi et al., 1981). During the first 3.6 days of treatment, Mg and Fe were lost at a much greater rate than Al or Si. The rate of loss of all four elements was much lower during the final 7.9 days than in the first 3.6 days of the experiment. Magnesium and Fe were preferentially removed from the trioctahedral hydroxide sheet of chlorite thus leaving the dioctahedral 2:1 part of the structure which is smectite. It is assumed that the 2:1 layer of the chlorite was dioctahedral prior to the oxidation treatment. Potassium was released at a lower rate than the other four elements

Table 4. Minerals present in clay fractions of overburden from Wilcox Group lignite of East Texas (From Arora et al., In press).

Sample	Depth	Clay	Sm†	Vr	Mi	Ch‡	Kl	Qr
2–0.2 μm	m		%					
FRA 1	25.9–27.4	29	9	6	16	X	XX	XX
FRA 6	22.0–23.5	32	10	6	21	X	XX	X
FRA 11	36.8–38.2	20	24	10	20	T	XX	X
MIR 2S1	24.5–24.6	21	9	6	26	T	XX	XX
MIR 3S3	60.8–61.0	50	5	4	30	X	XXX	XX
FRF 2S6	14.5–14.6	34	11	6	23	X	XX	XX
FRF 2S25	16.2–16.3	38	5	4	20	X	XXX	XX
FRF 3S6	18.3–18.4	57	14	6	19	T	XX	X
<0.2 μm								
FRA 1	25.9–27.4	17	30	11	9	T	X	0
FRA 6	22.0–23.5	21	46	1	10	T	X	0
FRA 11	36.8–38.2	15	ND	ND	14	T	XX	0
MIR 2S1	24.5–24.6	18	33	9	14	0	X	T
MIR 3S3	60.8–61.0	41	ND	ND	19	0	XX	0
FRF 2S6	14.5–14.6	17	34	10	18	0	X	T
FRF 2S25	16.2–16.3	31	21	9	13	T	XX	T
FRF 3S6	18.3–18.4	36	27	8	12	0	XX	T

† Sm = smectite; Vr = vermiculite; Mi = mica; Ch = chlorite; Kl = kaolinite; Qr = quartz; ND — not determined.
‡ Relative mineral content estimated from XRD data; T = trace amount.

throughout the experiment indicating the relative stability of the mica contaminant in the chlorite specimen. As noted below, mica separated from the overburden was a muscovite type.

Mica is present in small to moderate amounts in the clay fractions of strata overlying the Wilcox Group in Texas (Table 4). Mica obtained from the 2 to 0.2 μm fraction of lignite overburden clay by high gradient magnetic separation is dioctahedral. Since this mineral is a muscovite-type mica it is expected to release K very slowly. Although full assessment of the K supplying power of lignite overburden in Texas is not complete, the limited data available indicate a low K supplying power to plants from the mica present in the overburden. Exchangeable K will provide an immediate short term supply of K for plants.

Vermiculite is a small component of lignite overburdens studied thus far in Texas. Based on K-fixation (Alexiades and Jackson, 1965) vermiculite ranges from 1 to 11% in coarse and fine clay fractions (Table 4). As discussed earlier, some vermiculite may form by weathering of chlorite. Vermiculite from the shale and that formed by weathering is likely to remain a small percentage of the clay fraction. Yet it may be important in the K exchange of these materials. Recent findings indicate that K is fixed by the high charge smectite in lignite overburden. Thus the amount of vermiculite may be overestimated when based on K-fixation.

Smectite is one of the most important minerals in the clay fractions of Wilcox Group lignite overburdens from Freestone and Milam Counties in Texas. It comprises 5 to 24% of the eight 2 to 0.2 μm clays and 21 to 46% of six <0.2 μm clays (Table 4). The coarse clay fractions were more abundant than the fine clays in these samples. Its high cation exchange capacity, high surface area, and high shrink-swell potential make the influence of smectite greater than most other clay constituents. The potential of chlorite weathering to smectite suggests that more smectite may be added to these spoils if acid oxidative weathering occurs. The role of smectite in holding plant nutrients is a positive factor for the reclamation of these materials when they become spoils. Where the material is clayey and smectite is the dominant layer silicate, the potential for developing an acid soil that is extremely difficult to reclaim is a decisively negative factor. Smectitic clayey acid soils of the southeastern USA contain significant amounts of exchangeable acidity and have slowly permeable subsoil horizons that contribute to their unproductivity and resistance to improvement (Dixon and Nash, 1968).

Kaolinite is often the most abundant clay mineral in the 2 to 0.2 μm clay fraction and second (after smectite) in abundance in < 0.2 μm clay fractions of Wilcox Group lignite overburdens (Table 4). Kaolinite adds to the anion holding capacity and reduces the cohesive and adhesive properties of clay as compared to smectite. Since iron oxides and humus, which also reduce cohesiveness, are deficient in many lignite overburden materials, moderate kaolinite content is a positive factor for reclamation. The kaolinite in Wilcox Group lignite overburdens of Texas is generally well crystallized and primarily coarse clay in size. Presumably it has a low cation exchange capacity like other coarse kaolinite clays, but the smectite will contribute to this property of soils that may develop from spoil.

Quartz is present in small to moderate amounts in 2 to 0.2 μm clays of lignite overburdens (Table 4). Quartz does not contribute any significant chemical activity to soils, but as a component of coarser fractions it is a contributor to the physical properties of lignite overburden. Quartz particles are less cohesive than layer silicate particles (such as smectite) and they improve the friability of soils. Silica cement was observed to bond quartz sand grains together in lignite overburden in West Germany (Brinkmann, 1977).

Feldspars are absent in most lignite overburden clay fractions of the Wilcox Group. Trace amounts of feldspar have been suggested, but not confirmed, in a few 2 to 0.2 μm clays of lignite overburden. The potential importance of feldspars in silt and sand fractions of lignite overburdens requires further investigations of the amounts present, their composition, and their weathering properties.

Mineralogical Changes During Weathering

Investigations of several cores from the major lignite bearing sediments (Wilcox Group) in northeast Texas indicated that these sediments have been appreciably affected by surface weathering processes. The depth of weathering ranges between 8 and 20 m depending on the texture of the materials. Mineralogical alterations associated with weathering of these sediments are represented by Cores Ro-2 and Ha-6 and are summarized in Fig. 6 and 7. The two cores reported were taken some 300 km apart and the rainfall averaged 900 and 1,200 mm per year for Ro-2 and Ha-6, respectively.

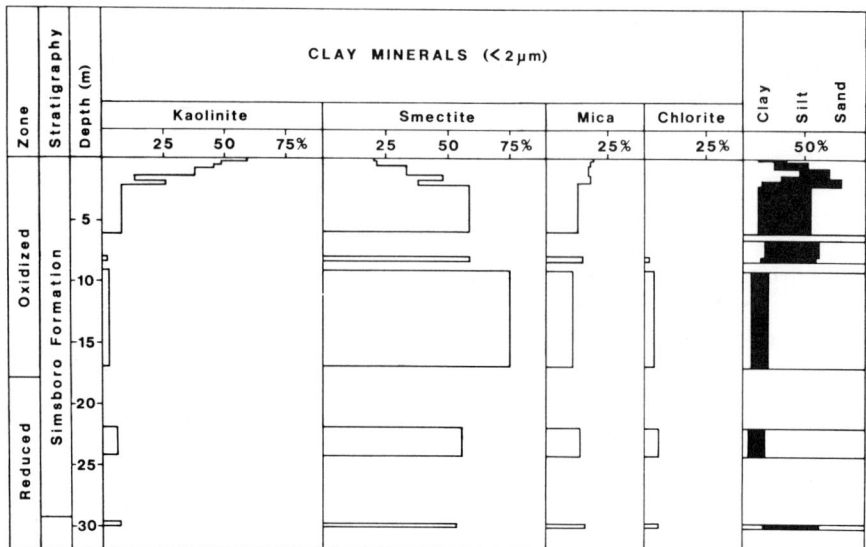

Fig. 6. Clay mineral distribution in core Ro-2.

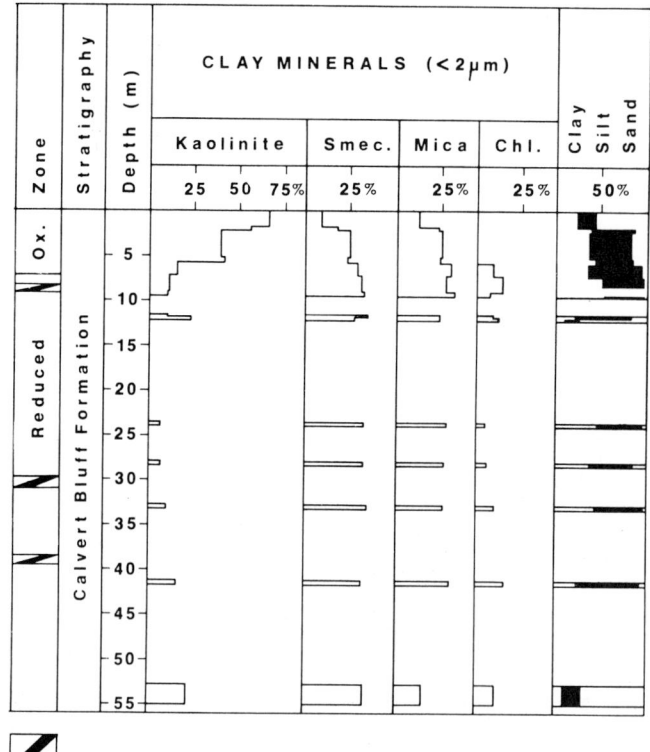

Fig. 7. Clay mineral distribution in core Ha-6.

Due to the relatively coarse texture of the materials a fairly deep oxidized zone developed in the section represented by core Ro-2 (Fig. 6). Smectite is the most abundant clay mineral throughout this core (Fig. 6). Smectite, however, decreases and kaolinite increases towards the top of the core. The absence of chlorite in the upper half of the oxidized zone is attributed to weathering; it is present in the lower oxidized zone and in the unoxidized part of the core. Alteration of smectite to kaolinite during pedogenic weathering (Altschuler et al., 1963), weathering of feldspar or differential removal of the smectite by leaching may explain the lower quantities of smectite and the relatively high concentration of kaolinite in the soil at the top of the core. Below the soil horizons the content of kaolinite remains fairly low throughout the core. Except for a slight increase toward the surface, the mica content also remains fairly constant throughout the core. The possibility of coarser mica weathering into the clay fraction precludes a final conclusion on its resistance to weathering relative to other minerals. The clay mineralogy of these sediments is largely inherited from pre-existing Eocene soils (Fisher and McGowen, 1967; Burst, 1959). Some chlorite may have been inherited from pre-existing sources but a diagenetic origin seems more likely.

Due to the generally fine texture of the materials in core Ha-6, the oxidized zone is only about 8 m deep. Mineralogical weathering trends similar to those in core Ro-2 were also observed in core Ha-6 (Fig. 7). Kaolinite is the most abundant clay mineral in the most weathered upper 5 m depth of the core. The content of kaolinite increases toward the top of the core and reaches a maximum in the soil horizons (Fig. 7). This distribution indicates that kaolinite forms during weathering of these sediments. Both smectite and mica gradually decrease towards the top of the core. Leaching may be partly responsible for the differential removal of smectite from the upper portion of the core. However, leaching of smectite alone cannot entirely account for the concentration of kaolinite throughout the upper 5 m depth of the core. It appears reasonable to assume that weathering processes in the upper portion of the core favored formation of kaolinite and degradation of smectite. Weathering may also explain the decrease in the content of mica and absence of chlorite in the upper portion of the core.

The most important changes in mineralogical properties associated with weathering of the lignite overburden sediments in East Texas are best illustrated in core Ha-6 (Fig. 8). Kaolinite (7.2 and 3.56 Å peaks) increases dramatically toward the top of the core. Secondly, smectite (18.0 Å) decreases toward the surface. Mica is highest in the 5.2 to 9.0 m zone and then decreases toward the surface. The 18 Å peaks gradually become broader and less defined toward the top of the core. This may be due to loss in the regularity of the crystalline structure or to decreasing particle size. Chlorite (14.2, 7.2, 4.71, and 3.53 Å) is only present below 6 m. The disappearance of chlorite coincides with the boundary of the oxidized and reduced zones (Fig. 7). Quartz (4.25 and 3.34 Å) appears to be present throughout the core, indicating that the energy of the sedimentation environment was relatively uniform.

X-ray diffraction analyses of Wilcox Group lignite overburden from Troup, Texas revealed a weathering trend of layer silicates in the silt frac-

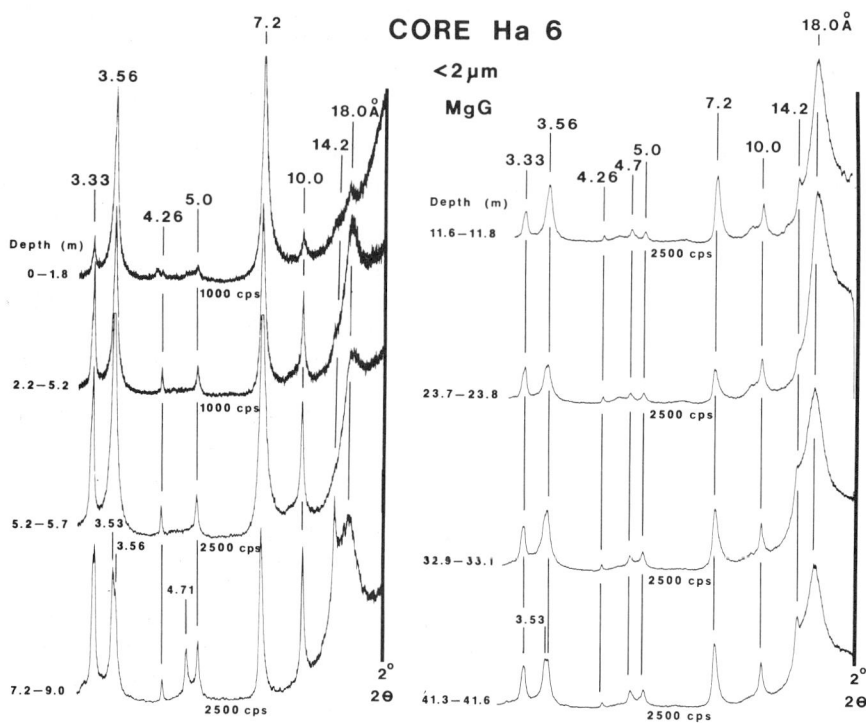

Fig. 8. X-ray diffraction patterns of representative clay samples from core Ha-6.

tions (Table 5). Weathering in the upper layers is shown by the absence of Fe-Mg chlorite and in the color profile observed in the field and in cores indicating iron oxide formation (see color Plate 2B). The chlorite reported in the silts is of the ferromagnesian type. The coarse clay fraction showed some evidence of intergradient chlorite-smectite development in the upper two layers of overburden. Quartz was the major mineral in addition to the layer silicates in the silt fractions.

Whole soil X-ray diffraction (XRD) analyses revealed feldspar in the 4th, 7th, 9th, and 11th layers suggesting a similar trend as observed for layer silicates in the silts discussed above. Feldspars occurred in trace amounts in the 1st and 13th layers but were absent in the 3rd layer indicating variability in the sediments. The sand content ranged from 0 to 79% among the eight layers analyzed which reflects diversity in sediment properties and weatherability. Gypsum was present in the 4th, 7th, and 11th layers but absent in the 9th and 13th layers possibly due to localized occurrence of pyrite in the parent material.

Mineralogy of Two Lignite Overburden Soils in East Texas

Two soils have been chosen for comparison with Wilcox Group overburdens. These soils were formed in an area of Claiborne Group sediments that overlie the Wilcox materials in Rusk County, Tex. These two soils represent the range of particle sizes of soils in the region where Wilcox Group sediments are mapped although they probably formed from the younger (Claiborne) material. The first of these soils is Darco (formerly Troup) series. Darco soils are classified as loamy, siliceous, thermic Grossarenic Paleudults. Troup soils were mapped in neighboring Panola County and are recognized as forming from Carrizo sand of the Wilcox Group in the field soil survey (Dolezel, 1975).

Deep sandy soils of East Texas, such as the Darco or Troup series, have thick sand to loamy sand textures in the upper 2 to 3 m. The clay content increases gradually with depth and the texture becomes sandy loam at about 3 m (Table 6). The clay fraction of these soils contains kaolinite, mica, smectite, vermiculite, and aluminum-rich chlorite, in order of decreasing abundance (Table 7). The fine clay fraction contains very thin particles of vermiculite and smectite indicated by extreme peak broadening (Hons et al., 1976). Interstratification of vermiculite and smectite in the fine clay was also suggested by XRD peak position. The mica content of the <5 μm fraction correlates with exchangeable potassium, suggesting that it is a source of available potassium for plants. Even though potassium feldspar is a minor constituent of the sand and two coarser silt fractions, it apparently does not contribute much potassium to plants because of its relatively large particle size and low surface area. Quartz is the most abundant mineral in the sand and silt fraction of the Darco soil and is present in smaller quantities in the clay fraction.

Table 5. Particle size distribution and layer silicate mineralogy of silt fractions of lignite overburden from Troup in northeast Texas.

Layer	Depth m	Sand 2,000–50 μm	Coarse silt, 50–20 μm	Medium silt 20–5 μm	Fine silt, 5–2 μm	Clay	Relative mineral abundance†		
							50–20 μm	20–5 μm	5–2 μm
				%					
1	0 –0.2	44.0	14.2	21.6	4.9	15.3	None‡	Mi	Mi>Kl
2	0.2–1.1	62.5	13.8	7.6	1.7	14.3	None	Mi>Kl	Mi>Kl
3	1.1–2.3	79.1	7.3	3.5	0.6	9.6	None	Mi>Kl	Mi>Kl
4	2.3–4.2	0.2	13.0	38.5	6.8	41.5	Mi>Ch	Mi>Ch>Kl	Mi>Ch>Kl>Sm
7	4.2§–10.7	0.7	13.6	35.8	10.4	39.5	Mi>Ch	Mi>Ch>Kl	Mi>Ch>Kl>Sm

† Ch = chlorite; Kl = kaolinite; Mi = mica; Sm = smectite.
‡ No layer silicates detected.
§ Zones of lignite were excluded from this interval in sampling the silicates.

Table 6. Particle size distribution of Darco soil as percentages of inorganic material.†

Depth	Particle size, μm						Free iron oxides as Fe_2O_3	Textural class‡
	2,000–50	50–20	20–5	5–2	2–0.2	<0.2		
cm			%					
0–17	89.9	4.8	3.5	0.6	0.8	0.4	0.1	sand
62–87	87.2	5.3	5.0	0.9	1.2	0.4	0.1	sand
122–160	86.5	5.8	5.0	0.8	1.3	0.5	0.2	sand
205–235	81.2	5.4	5.5	1.1	2.7	3.7	0.5	loamy sand
272–300	88.5	1.7	1.9	0.8	2.2	4.3	0.5	loamy sand
320–335	83.9	1.5	1.7	0.7	5.5	6.0	0.8	sandy loam

† Source: Hons et al., 1976.
‡ Free iron oxides included with clay in textural classification.

Table 7. Mineral composition of selected Darco soil horizons.†

Depth cm	Size fraction	Minerals on whole soil basis§							
		Fd‡	Qr	Mi	Vr	Sm	Kl	Ch	Sum
	μm	%							
0–17	2,000–50	0.79	88.05	0.62	ND	ND	ND	ND	89.5
	50–20	0.29	4.31	0.46	ND	ND	ND	ND	5.1
	20–5	0.24	2.83	0.27	ND	ND	ND	ND	3.3
	5–2	0.02	0.42	0.11	<0.01	ND	0.06	0.01	0.6
	<2	0.01	0.37	0.16	0.02	0.06	0.33	0.02	1.0
	Total	1.35	95.98	1.62	0.02	0.06	0.39	0.03	99.5
122–160	2,000–50	0.76	84.81	1.38	ND	ND	ND	ND	87.0
	50–20	0.36	5.12	0.60	ND	ND	ND	ND	6.1
	20–5	0.31	3.97	0.50	ND	ND	ND	ND	4.8
	5–2	0.03	0.56	0.15	<0.01	ND	0.06	0.01	0.8
	<2	0.02	0.57	0.30	0.03	0.08	0.62	0.04	1.7
	Total	1.48	95.03	2.93	0.03	0.08	0.68	0.05	100.4
205–235	2,000–50	0.28	78.52	1.78	ND	ND	ND	ND	80.6
	50–20	0.31	4.79	0.51	ND	ND	ND	ND	5.6
	20–5	0.30	4.41	0.58	ND	ND	ND	ND	5.3
	5–2	0.02	0.59	0.21	<0.01	ND	0.24	0.01	1.1
	<2	0.04	0.91	0.94	0.16	0.80	1.91	0.11	4.9
	Total	0.95	89.22	4.02	0.16	0.80	2.15	0.12	97.5

† Source: Hons et al., 1976.
‡ Fd = K-feldspar, Qr = quartz, Mi = K-mica, Vr = vermiculite, Sm = smectite, Kl = kaolinite, Ch = chlorite.
§ Mineral percentages are given to hundredths to preserve evidence for small amounts determined on fractions and expressed on a whole soil basis. ND = not detected by XRD and not determined.

Soils with sandy surface horizons and clayey subsoils are illustrated by the Sacul series (clayey, mixed, thermic Aquic Hapludults) formerly classified as the Boswell series. The Sacul soil has a sandy loam A horizon that overlies a clay textured B horizon that extends from 22 cm to the C horizon at 138 cm. The clay textured C horizon extends to 2 m, the full depth of sampling. Kaolinite, smectite and mica are the most abundant clay minerals in the soil. Vermiculite is present in trace amounts in the clay fraction. Quartz, mica, and feldspar occur in the silts and sands. No aluminous chlorite was present in Sacul as it was in the Darco soil. There was less evidence for mixed layering in the clays of Sacul than in Darco soil. The smectite, kaolinite, and mica in the Sacul soil occur in larger particle sizes than in the Darco soil which indicates less physical dispersion and less intensive weathering than in the latter, more sandy soil.

Free iron oxides in the Darco soil horizons range from 0.1 to 0.8% and in the Sacul soil horizons from 0.4 to 4.7%. Iron oxides provide these soils with their yellowish brown to dark red colors. Iron oxides correlate with clay content and degree of weathering in the soils discussed here. In the more clayey, Sacul soil, the highest iron oxide content is in the B2t horizon, from 22 to 88 cm (Table 8). The colors of free iron oxides indicate the accumulation of weathering products in a given soil profile. The decline in free iron contents in the 78 to 200 cm horizons of the Sacul soil attests to the lower degree of weathering as did the higher quantities of coarser clay particles mentioned earlier.

Table 8. Particle size distribution of Sacul soil as percentages of the inorganic material.[†]

Depth	Particle size, μm						Free iron oxides as Fe_2O_3	Textural class[‡]
	2,000–50	50–20	20–5	5–2	2–0.2	<0.2		
cm	%							
0–15	66.2	20.1	6.7	1.2	2.9	2.5	0.4	sandy loam
22–58	31.2	9.7	4.7	2.0	10.8	36.8	4.7	clay
78–110	44.0	12.7	4.7	2.8	8.7	24.4	2.9	clay loam
138–200	17.0	13.2	9.4	5.4	20.3	33.3	1.5	clay

† Source: Hons, F. M. 1974. Potassium sources and availability in three east Texas soils. Master's Thesis. Texas A&M Univ. College Station. 77 p.
‡ Free iron oxides included with clay in textural classification.

The influence of clay mineral suite on reclamation of lignite mine spoil has been investigated in the north Bohemian lignite district of Czechoslovakia (Jonas, 1972). Four suites of major clay minerals were ranked as follows:

Preference for reclamation	Color	Clay mineral composition
1st	Grey to brownish grey	Montmorillonite > kaolinite > illite
2nd	Yellow to yellowish brown	Kaolinite > illite > montmorillonite
3rd	Grey	Kaolinite > illite
4th	Yellow and yellowish brown	Kaolinite > illite

The above ranking was chosen because of the poorer physical properties of the clays with less montmorillonite according to Jonas (1972). The preferred clays had better structure, better water regime, and were less subject to dispersion in laboratory tests. This classification does not include the influence that sand and silt fractions may have in some spoils. Also, the role of iron oxides, carbonates, and exchange ions should be considered. Thus the ranking may be considered as only part of the soil-plant-water system and it even may be misleading under some conditions. The rating of Jonas (1972) seems especially risky for fine textured soils. Future research on spoil may find a place for this clay mineral rating system while also including other factors that control soil behavior under specified conditions.

The mineralogy in the oxidized zone of lignite overburdens of East Texas is much like the overlying soils. Thus weathering has proceeded well beyond the 2 to 3 m normally included in soil sampling. Chlorite and pyrite are absent from most oxidized zone material. Kaolinite is more abundant in the oxidized zone than in the unoxidized zone. The clay fraction of the sandy soil (Darco or Troup) contained aluminous-chlorite which usually was not present below the soil zone. Ferromagnesian chlorite was not present in either the soils or the upper oxidized zone; it was present in the reduced zone and in one core extended into the lower oxidized zone. The weathering of silt and sand size feldspars to kaolinite needs assessment; both particle size reduction and mineral transformation may occur. Although only two cores and two soils were compared, considerable similarity was evident. The most difficult material to reclaim, of those studied, would probably be the 138 to 200-cm layer beneath the Sacul soil profile. It contained 55% clay, 28% silt, and 17% sand. Further reduction in particle size, which is likely to occur during weathering, would only make the material more difficult to cultivate and less permeable to water.

Lignite was formerly mined in Miocene deposits in Fasterholt, Denmark, and the sedimentary deposits associated with the lignite strata were analyzed to determine the reclamation requirements of spoil produced from these strata (Christensson and Jensen, 1970). Three sediment samples associated with the three consecutive lignite beds in Fasterholt were taken for analysis. Their pyrite contents were 1, 3, and 6%. The three sediment samples were mixed together into one composite sample and particle size and mineralogical analyses were made after controlled oxidation (pH 5 to 8) of the pyrite with hydrogen peroxide to minimize alteration of clay minerals. The composite sample was composed of 22% 1 mm to 20 μm "sand", 34% 2 to 20 μm silt, and 44% < 2 μm clay.

The sand was largely quartz. Microcline and oligoclase feldspars were also present. Other light minerals present were chalcedony, flint, glauconite, and muscovite. Hornblende, garnet, and magnetic grains were detected in the heavy mineral fraction. In the silt fraction illite, kaolinite, muscovite, and quartz were identified. Illite, kaolinite, and quartz were identified in the clay fraction. In all three fractions illite was abundant and quartz was abundant in the silt and sand fractions. There was a considerable fraction of amorphous or unidentified material in the clay. In view of the illite and mica components the question arises as to smectite possibly being present in a poorly crystalline or complex interstratified form. A second lignite-related bed sampled in Soby, Denmark was analyzed and it contained no glauconite in the sand fraction. The Soby clay fraction (20% < 2 μm clay) contained about the same composition of crystalline components as the Fasterholt clay but less non-crystalline material.

Glauconite in the Fasterholt sample was mainly concentrated in the silt and fine-grained sand fractions. The grains were composed of extremely thin crystallites estimated by XRD line broadening at 54 Å (Tovborg Jensen, 1943 according to Christensson and Jenson, 1970). The glauconite grains were soft and easily broken down in a mortar but did not disperse in water by shaking or ultrasonic treatments. Results of these tests were interpreted to mean that the glauconite grains were cemented or grown together rather than having been formed by flocculation. The presence of glauconite was considered an asset to agricultural soil development because of the potential source of potassium.

Alteration of silicate minerals in spoil-bank clay subsoil layer has been investigated in Yugoslavia (Pavel and Jonas, 1978). The increase in extractable Al indicated accelerated mineral weathering as the pH declined from 7.1 to 5.2 after 13 years exposure of the spoil. Careful XRD techniques revealed small changes that appear to be due to acid weathering. A smectite valley to peak ratio declined very slightly which suggested greater dispersion of the smectite layers after 15 years of acid weathering in the field. The 10 Å mica (illite) peak was broader in weathered than unweathered spoil which also suggested reduction in the particle size of the mica. Lowered intensity of the second vs. the first basal order of the illite was interpreted as a decrease in octahedrally coordinated Fe due to changes in oxidation-reduction conditions and acidity in the weathering spoil. Suggestive evidence for organization of a chlorite-like interlayer was presented in the form of resistance to collapse in potassium saturated samples after extraction with citrate. Although these apparent mineralogical changes are small in relation to total mineral composition of the clays, they imply considerable influence of 15 years of acid weathering under oxidizing conditions on the chemical behavior of the soil mass. They support the view that fresh spoil requires continued investigation over a period of several years to assure a steady, relatively uniform set of chemical and mineralogical conditions that can be matched by fertilization, liming, and cultural practices for plant growth and production.

It is generally recognized that the presence of sulfides and their acidity forming potential are major deterrents to reclamation of lignite mine spoil. The addition of lime to neutralize that acidity has been investigated for brown coal spoil in the Moscow Coal Field of the USSR (Izhevskaya et al., 1974). The acidity from the spoil changes the soil properties and interferes with plant growth and yield when soil is layered over it during reclamation. Counteracting the acidity from sulfide oxidation by applying high doses of lime and layering of lime in between the soil and the spoil were not economic. Therefore, burial of the material that contained sulfides at the bottom of the spoil banks was recommended for reclamation of the Moscow Coal Field.

The foregoing data and discussions indicate that lignite overburden especially those in eastern Texas are composed of an acid oxidized zone and an alkaline reduced zone (Fig. 9). Any sulfides originally present and

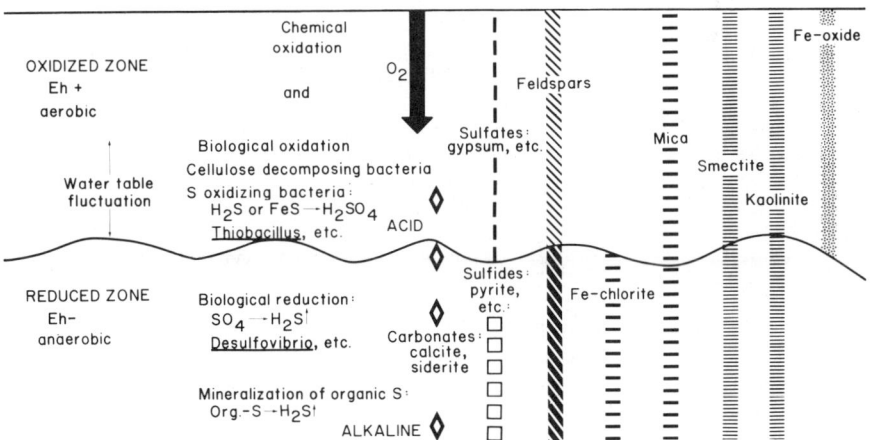

Fig. 9. Schematic diagram of lignite overburden showing zones of oxidation and reduction and mineral components that may occur in each (Parts of this diagram were taken from Brinkmann, 1976).

most of the Fe-chlorite have been altered thus yielding red, brown, and yellow colors due to the formation of iron oxides and localized jarosite in the oxidized zone. Gypsum occurs in the oxidized zone. The few carbonates originally present have been removed or, in the case of siderite, occluded within a shell of goethite in the oxidized zone. Feldspars have been depleted at least slightly in the oxidized zone. Mica composition is small and it has changed very little due to weathering. Kaolinite and smectite are the major clay minerals. Kaolinite has been increased in the clay fraction in the oxidized zone and smectite has been depleted near the surface.

ACKNOWLEDGMENT

The authors owe a debt of gratitude to Dr. Hans Brawand for translating selected materials from German to English.

LITERATURE CITED

1. Alexiades, C. A., and M. L. Jackson. 1965. Quantitative determination of vermiculite in soils. Soil Sci. Soc. Am. Proc. 29:522–527.
2. Altschuler, Z. S., E. J. Dwornik, and H. Kramer. 1963. Transformation of montmorillinite to kaolinite during weathering. Sci. 141:148–152.
3. Arora, H. S., J. B. Dixon, and L. R. Hossner. Mineralogy of lignitic coal overburdens of the Wilcox Group of east Texas. Soil Sci. (In press).
4. ————, C. Pugh, L. R. Hossner, and J. B. Dixon. 1980. Forms of sulfur in east Texas lignitic coal. J. Environ. Qual. 9:383–386.
5. Bandy, M. C. 1938. Mineralogy of three sulphate deposits in northern Chile. Am. Mineral. 23:669–760.
6. Berry, L. G. (ed.). 1973. Powder diffraction file. Joint committee on Powder Diffraction Standards. 1601 Park Lane, Swarthmore, Penn.
7. Brinkmann, K. 1976. Genese und oxydation der Eisensulfide im Deckgebirge der rheinischen Braunkohle. Braunkohle 28:448–457. (In German).
8. ————. 1977. Mineralogy and geochemistry of iron sulfides in the overburden of the Frechen open cast mine (Rhenish soft coal field). Neues Jahrb. Mineral. Abh. 129(3): 333–352. (In German with English abstract).
9. Burst, T. F. 1959. Postdiagenetic clay mineral environmental relationships in the Gulf Coast Eocene. Clays and Clay Miner. 6:327–341.
10. Carson, C. D., D. S. Fanning, and J. B. Dixon. 1982. Alfisols and ultisols with acid sulfate weathering features in Texas. p. 129–149. In J. A. Kittrick, D. S. Fanning, and L. R. Hossner (ed.) Acid sulfate weathering. SSSA Spec. Pub. no. 10. Madison, Wis.
11. Christensson, F., and E. Jensen. 1970. The mineralogical composition of the pyrite-bearing lignite clay from southwest Jutland. Denmark Vet-landbohojsk Arsskr. p. 123–139.
12. Dixon, J. B., and V. E. Nash. 1968. Chemical, mineralogical and engineering properties of Alabama and Mississippi Black Belt soils. Southern Coop. Ser. No. 130. Ala. Agric. Exp. Stn., Auburn, Ala. 69 p.
13. Dolezel, R. 1975. Soil survey of Panola County, Texas. USDA-SCS. p. 1–55.
14. Fisher, W. L., and J. H. McGowen. 1967. Depositional systems in the Wilcox Group of Texas and their relationship to the occurrence of oil and gas. Gulf Coast Assoc. of Geol. Soc. Trans. 17:105–125.
15. Gratten-Bellew, P. E., and W. J. Eden. 1975. Concrete deterioration and floor heave due to biogeochemical weathering of underlying shale. Canadian Geotech. J. 12:372–378.
16. Hons, F. M., J. B. Dixon, and J. E. Matocha. 1976. Potassium sources and availability in a deep, sandy soil of East Texas. Soil Sci. Soc. Am. J. 40:370–373.

17. ————, P. E. Askenasy, L. R. Hossner, and E. L. Whiteley. 1978. Physical and chemical properties of lignite spoil material as it influences revegetation. p. 209–217. In W. R. Kaiser (ed.) Proc. of Gulf Coast Lignite Conf.: geology, utilization and environmental aspects. Rep. of Investigations No. 90. Bureau of Economic Geology, Univ. of Texas at Austin.
18. Ivarson, K. C., G. J. Ross, and N. M. Miles. 1978. Alterations of micas and feldspars during microbial formation of basic ferric sulfates in the laboratory. Soil Sci. Soc. Am. J. 42:518–524.
19. Izhevskaya, T. I., A. I. Savich, and V. N. Cheklina. 1974. Sulphide-containing rocks and their significance in recultivation of spoil banks of the brown coal open pits. 10th Int. Congress of Soil Science. Fertility of soils. Moscow IV:427–432. Moscow, USSR.
20. Jonas, F. 1972. Soil formation on the reclaimed spoil banks of the north Bohemian lignite districts. Research Institute for Land Reclamation and Improvement. Zbraslaw N. Vlt: p. 18–31.
21. Kaiser, W. R. 1974. Texas lignite: near surface and deep-basin resources. Rep. of Investigations No. 79. Bureau of Economic Geology. Austin. p. 1–79.
22. Map of world coal resources and major trade routes. 1975. Miller Freeman Publications, Inc., San Francisco, Calif.
23. McKee, T. R., and J. L. Brown. 1977. Preparation of specimens for electron microscopic examination. p. 809–846. In J. B. Dixon and S. B. Weed (ed.) Minerals in soil environments. Soil Sci. Soc. Am., Madison, Wis.
24. United States Geological Survey. 1974. National atlas, coal resources and distribution. Sheet No. 6-426.
25. Pavel, L., and F. Jonas. 1978. Changes occurring on clay minerals during initial soil formation on a recultivated Miocene spoil bank clay. p. 142–154. In Int. Symp.: Soil fertility improvement and clay minerals. Prague.
26. Pennington, D. 1978. Hydrogeologic factors which influence the reclamation of lignite mines. p. 218–227. In W. R. Kaiser (ed.) Proc. of Gulf Coast Lignite Conf.: geology, utilization and environmental aspects. Rep. of Investigations No. 90. Bureau of Economic Geology, Univ. of Texas at Austin.
27. Petzold, V. E. 1978. Agricultural reclamation and problems of the first cultivation in the Rhenish brown coal mining. Z. Kulturtech. Flurbereinig. 19:2–9.
28. Pugh, C. E., L. R. Hossner, and J. B. Dixon. Pyrite and marcasite surface area versus morphology and particle diameter. Soil Sci. Soc. Am. J. (In press).
29. Rickard, D. T. 1973. Sedimentary iron sulfide formation. p. 28–65. In H. Dost (ed.) Acid sulphate soils: Proc. of the Int. Symp. on Acid Sulphate Soils. Wageningen, The Netherlands. International Institute for Land Reclam. and Improve. Wageningen.
30. Ross, G. J. 1969. Acid dissolution of chlorites: Release of magnesium, iron and aluminum and mode of acid attack. Clays and Clay Miner. 17:347–354.
31. ————, and H. Kodama. 1976. Experimental alteration of a chlorite into a regularly interstratified chlorite-vermiculite by chemical oxidation. Clays and Clay Miner. 24:183–190.
32. Senkayi, A. L., J. B. Dixon, and L. R. Hossner. 1981. Transformation of chlorite to smectite through regularly interstratified intermediates. Soil Sci. Soc. Am. J. 45:650–656.
33. Smith, E. E., and K. S. Shumate. 1970. Sulfide to sulfate reaction mechanism. Grant 14010FPS. Prepared for Federal Water Qual. Admin., Dep. of the Interior. 115 p.
34. van Breemen, N. 1973. Soil forming processes in acid sulphate soils. p. 66–130. In H. Dost (ed.) Acid sulphate soils ILRI Publ. 18, Vol. 1. Int. Inst. for Land Reclam. and Improv. Wageningen.

Chapter 11

Relation of Pyritic Sandstone Weathering to Soil and Minesoil Properties[1]

R. N. SINGH, W. E. GRUBE, JR., R. M. SMITH, AND R. F. KEEFER[2]

ABSTRACT

Mahoning sandstone rock strata above surface mineable coal was compared chemically and mineralogically with minesoils developed there—from and with adjacent soils to evaluate weathering and soil development. The weathered zone of rock was found to be about 6 m deep and was acidic, high in Al and free Fe, but low in S and in exchangeable bases. This weathered (high chroma) sandstone contained quartz, kaolinite, and minor amounts of mica and vermiculite. An unweathered zone of rock below 6 m was basic to slightly acid, low in exchangeable Al and free Fe, but higher in S and exchangeable bases. This unweathered (low chroma) sandstone contained quartz, kaolinite, and mica along with authigenic pyrite and carbonates. Weathering resulted in vermiculitization of the highly crystalline mica of the Mahoning sandstone. Soil development was directly related to the underlying rock material as determined by chemical and mineralogical analyses. The data show that proper placement of overburden rock resulted in edaphologically desirable minesoils, whereas haphazard overburden replacement enhanced pyritic oxidation and acidity which inhibited successful revegetation.

INTRODUCTION

Coal has been reported in 53 of the 55 counties in West Virginia, but 15 counties account for 90% of total production (Barlow, 1974). In 1974, 35 counties of West Virginia produced 124 million tons. About 78% of this came from deep mining and 22% from surface mining. West Virginia University studies (1971, 1974) indicated that large reserves of coal lie close to the surface and can be surface-mined with modern equipment.

---

[1] Contribution from the Div. of Plant and Soil Sciences, West Virginia Univ., Morgantown, WV 26506. Approved for publication by the Director of the West Virginia Univ. Agric. and For. Exp. Station as Scientific Paper No. 1637. Paper presented before the Div. S-9, Soil Sci. Soc. Am., 6 Aug. 1979, Ft. Collins, CO.

[2] Professor, chemist (EPA), professor and professor, respectively.

Copyright © 1982 Soil Science Society of America, 677 S. Segoe Rd., Madison, WI 53711.
  Acid Sulfate Weathering.

The production of coal by surface mining leaves large areas of rock materials exposed to soil forming factors and thus provides an ideal situation in which to study the relationship between bedrock, soil, and minesoil properties. The objectives of this study were: (a) to describe and characterize quantitatively certain physical and chemical properties of coal overburden strata and adjacent soils and (b) to determine chemical and mineralogical characteristics of minesoils and relate them to those of nearby native[3] soils.

In order to examine and compare the bedrock, native soil and minesoil relationships, sampling localities were selected so that differences in climate, slope, and vegetation would be minimized. The sites selected for this study were located near the top of gently sloping hills where soil erosion would be slight and accumulation of material from above would be minimal. Preston County, in northern West Virginia, was ideally located because of its topography, geological stability, and active surface mining operations since 1950.

The soils selected for this study were in a northern hardwood forest (Strausbaugh and Core, 1970). These soils developed from lower Mahoning sandstone which is considered to have been deposited up to 20 m thick (West Virginia Univ., 1971) in a complex river channel system. These soils are members of the Inceptisol and Ultisol orders, and include the Dekalb series (Dystrochrepts) (sites X and H) and Cookport series (Fragiudults) (sites E and L).

MATERIALS AND METHODS

Description of Field Study Area

The area for study was in north central Preston County, W. V., in the unglaciated Allegheny plateau. Slope grade ranged from 5 to 10%. Rock samples were collected through cooperation with surface mining operators. Rotary drilling of bedrock before blasting provided columns of crushed rock samples in depth increments of 32 cm to a total depth of 18 m. Sampling from the drilling bench up to the adjacent soil surface was done by hand to reveal soil morphological horizons. At each rock sampling location, (Fig. 1), the soil was described and samples were taken from each genetic horizon. Since properties of minesoils are more closely related to their parent material than those of other native soils, two minesoils were sampled by morphological horizons. Minesoils 1 and 2 were located near the site where sandstones X and E were sampled, respectively. This area was mined in 1971 and reclaimed in 1972 according to the guidelines established by the West Virginia Univ. (1971) that recommend covering toxic or potentially toxic unweathered rock with weathered or other favorable overburden. At the second site where minesoils 3 and 4 were located, Mahoning sandstone was prevalent but rock cores were not

---

[3] Native soils include all soils that are neither drastically disturbed nor made by man.

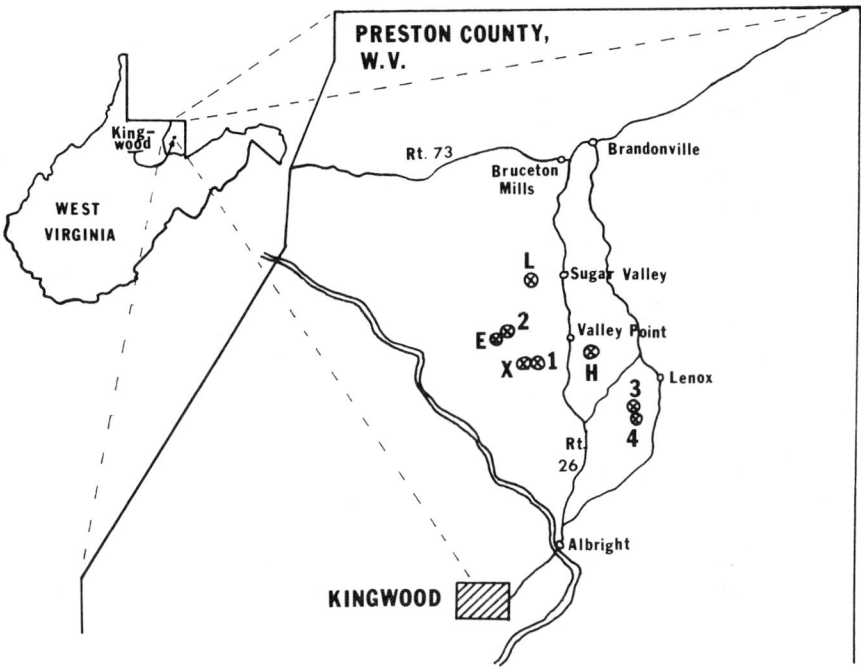

Fig. 1. Sampling location of native soils, minesoils, and rocks in Preston County, West Virginia.

sampled for this study. This area was mined and reclaimed in 1970 before the establishment of regulations which require covering of potentially toxic mine spoil material with nontoxic overburden. Due to lack of separation of weathered and unweathered rock material during mining, high chroma (weathered) sandstone, low chroma (unweathered) sandstone, and carbolithic material were present throughout the minesoil profiles at this site.

Laboratory Procedures

After air drying, native soil and minesoil samples were passed through a 2 mm sieve to remove coarse fragments. Particle size distribution was determined by the pipette method described by Kilmer and Alexander (1949). The pH of soils and pulverized sandstone was measured in saturated paste with a Corning Model 12 research pH meter. Total organic matter (OM) was determined by wet combustion (Allison, 1965). Cation exchange capacity (CEC) was determined by the method of Rich (1961). Exchangeable Ca, Mg, and K were extracted by the neutral NH_4OAc procedure of Jackson (1958). Potassium was determined by flame emission and Ca and Mg by atomic absorption spectrophotometry. Exchangeable Al was extracted with $1\ N$ KCl (Coleman et al., 1959) and determined by atomic absorption spectroscopy. Total Fe was determined by the procedure of Jackson (1958). Free Fe was determined by atomic absorption (Raad et al., 1969) following sodium dithionite-citrate-bicarbonate extraction (Jackson, 1956). Total S in overburden material and minesoils was determined using a Leco induction furnace with automatic titrator (Grube et al., 1973).

The clay fractions of each soil and pulverized rock sample were separated by decantation and super-centrifugation after removal of organic matter by H_2O_2 digestion as described by Jackson (1956). Mineralogical analyses were carried out on 0.2 to 2 µm and < 0.2 µm clay fractions. Parallel oriented specimens on glass slides for X-ray examination were prepared by slowly drying glycerol-solvated Mg-saturated clays. The effect of heat treatment on K-saturated clays was also studied by X-ray diffraction (XRD). The clays were investigated using a Siemens crystalloflex IV, X-ray diffractometer, using nickel filtered copper radiation and a scintillation detector.

For differential thermal analyses (DTA), the clay samples were saturated with Mg and kept at 55% relative humidity for 48 hours. The samples were heated from 25 to 100 C at a rate of 15 C per min on a Robert Stone DTA Unit. Percent of mica was estimated by total K analysis and percent smectite and vermiculite by the procedure of Alexiades and Jackson (1965). The quantitative determination of quartz was carried out by sodium pyrosulfate fusion method of Kiely and Jackson (1965). The remaining minerals in the soils were not determined and are listed as "other".

RESULTS AND DISCUSSION

Physical, chemical, and mineralogical analyses were carried out on all rock, native soil, and minesoil profiles. Because of the similarity in properties of soils X and H (Dystrochrepts) and of soils E and L (Fragiudults) results of only one soil from each of these great groups are discussed.

Bedrock

Bedrock Description and Properties: The soils described in this study are underlain by lower Mahoning sandstone which extends from near the top of the Upper Freeport Coal seam to the C horizon of the soils. The probable environment during deposition of the Mahoning sandstone is described by Grube et al. (1982). The sandstone in the present landscape consisted of two distinct zones: the lower, dull gray portion (low chroma); and the upper weathered zone colored by free iron "oxides" (high chroma). As shown in Fig. 2, the weathered zone of sandstone, which reached a thickness of 6 m, was very low in total S and Ca at all four locations. On the other hand, the unweathered, low chroma sandstone below 6 m contained from 0.4 to 1.0% S and from 1 to 3.5% Ca (Fig. 2). Petrographic examination of the low chroma sandstone revealed the presence of small and scattered grains of pyrite, but no pyrite grains were observed in the weathered high chroma rock. Apparently aeration and groundwater percolation oxidized the pyrite and produced sulfuric acid. The

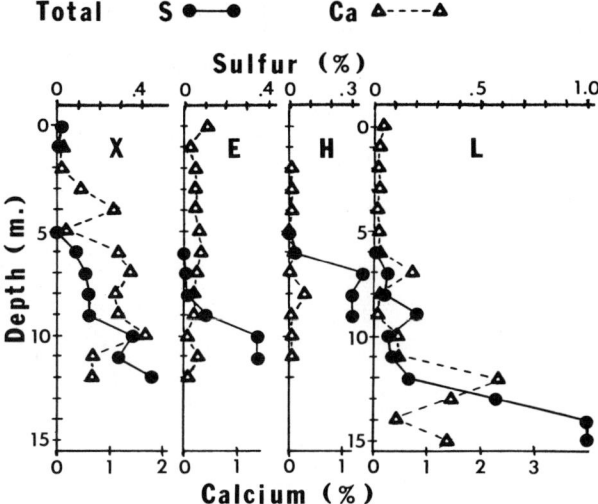

Fig. 2. Total sulfur and total calcium in overburden rock material from four sites X, E, H, and L.

acidic percolating solution presumably reacted with the sandstone rock and became less acid as it continued moving downward and laterally. As this process continued, each successive solution remained acidic somewhat longer and carried sulfates a little farther. Because of this process, negligible amounts of pyrite and sulfur were found in the weathered zone (Fig. 2).

The hypothesis of sulfur weathering is supported by exchangeable cation data given in Tables 1 to 4. In all cases, in the weathered zone the concentration of exchangeable bases (Ca, Mg, and K) was low and the exchange complex appeared to be saturated with H and Al as reflected by low pH and high levels of exchangeable Al. On the other hand, at a depth of about 7 m and below, the concentration of exchangeable bases and pH increased abruptly with exchangeable Al level reaching zero in most cases. This zone of high pH and high exchangeable bases was also higher in S (Fig. 2) as previously mentioned. Apparently at this greater depth, i.e., below 6 m, sulfur weathering of the Mahoning sandstone had not occurred and thus chemical composition of these two zones differed. This contention is further supported by total and free Fe data given in Fig. 3. Although the distribution of total and free Fe in rock profiles was somewhat erratic with depth, generally free Fe closely followed the amount of total Fe in the weathered zone. The free Fe maximum was generally found in the weathered zone of Mahoning sandstone at all locations. Even though in several instances (sites X and L, Fig. 3) the concentration of total Fe in the unweathered zone exceeded the amount found in the weathered zone, the amount of free Fe was extremely low (<0.14%). A study by Blume and Schwertmann (1969) of several soil profiles of Entisols and Dystrochrepts indicated that the free Fe maximum occurred in the A horizons which were also more susceptible to weathering. This increase in free Fe was attributed to a loss of easily soluble minerals (car-

Table 1. Some chemical properties of overburden material at site X.

Depth	Horizon	pH	Exchangeable cations			
			Ca	Mg	K	Al
m			meq/100 g			
0		--	--	--	--	--
1	Weathered	5.4	0.13	0.05	0.11	0.54
2	Sandstone	5.4	0.57	0.15	0.18	0.44
3	(High chroma)	5.7	1.04	0.33	0.17	0.33
4		5.7	0.77	0.29	0.19	1.00
5		5.4	0.42	0.12	0.14	0.56
6		5.8	1.37	0.40	0.16	0
7	Unweathered	7.6	3.00	0.97	0.16	0
8	Sandstone	7.9	3.20	1.11	0.16	0
9	(Low chroma)	7.1	3.30	1.25	0.17	0
10		7.5	2.90	1.15	0.18	0
11		7.0	1.93	0.80	0.17	0
12		6.3	1.30	0.42	0.22	0

Table 2. Some chemical properties of overburden material at site E.

Depth	Horizon	pH	Exchangeable cations			
			Ca	Mg	K	Al
m			meq/100 g			
0	Weathered	4.8	0.45	0.50	0.41	4.20
1	Sandstone	5.1	0.16	0.14	0.32	2.78
2	(High chroma)	5.2	0.18	0.14	0.41	0.69
3		5.6	1.03	0.90	0.36	0.46
4		5.4	0.90	0.81	0.35	0.25
5		5.7	0.90	0.62	0.21	0.05
6	Unweathered	6.3	1.03	1.00	0.24	0
7	Sandstone	6.2	0.97	0.78	0.19	0
8	(Low chroma)	6.3	1.35	1.21	0.26	0
9		5.4	0.65	0.99	0.25	0
10		5.8	1.88	1.61	0.30	0
11		4.7	1.45	0.81	0.18	0
12		4.9	2.50	1.45	0.43	0

Table 3. Some chemical properties of overburden material at site H.

Depth	Horizon	pH	Exchangeable cations			
			Ca	Mg	K	Al
m			meq/100 g			
0		--	--	--	--	--
1		--	--	--	--	--
2	Weathered	4.4	0.70	0.12	1.12	1.30
3	Sandstone	4.4	0.53	0.12	1.32	1.50
4	(High chroma)	4.6	0.33	0.13	0.68	1.00
5		4.8	0.29	0.10	0.70	1.40
6		4.5	0.45	0.16	0.77	1.40
7	Unweathered	5.2	1.95	1.40	0.60	0.70
8	Sandstone	5.1	2.07	1.53	0.66	0.10
9	(Low chroma)	4.4	1.42	1.00	0.61	0.70
10		3.4	1.30	0.85	0.88	0.22
11		4.0	1.35	1.51	0.71	0.12

Table 4. Some chemical properties of overburden material at site L.

Depth	Horizon	pH	Exchangeable cations			
			Ca	Mg	K	Al
m			meq/100 g			
0	Weathered	5.3	0.85	0.18	0.18	4.85
1	Sandstone	5.2	0.36	0.15	0.18	1.93
2	(High chroma)	5.4	0.35	0.14	0.12	0.66
3		5.6	0.24	0.21	0.38	0.14
4		5.4	0.88	0.54	0.24	0.47
5		5.7	0.67	0.33	0.20	0
6	Unweathered	6.0	2.65	0.88	0.24	0
7	Sandstone	7.0	5.00	2.12	0.16	0
8	(Low chroma)	7.0	3.29	0.99	0.17	0
9		6.9	2.30	0.71	0.23	0
10		7.2	3.69	1.38	0.22	0
11		7.0	3.60	1.56	0.21	0
12		6.8	2.38	0.76	0.22	0
13		7.0	4.42	1.65	0.17	0

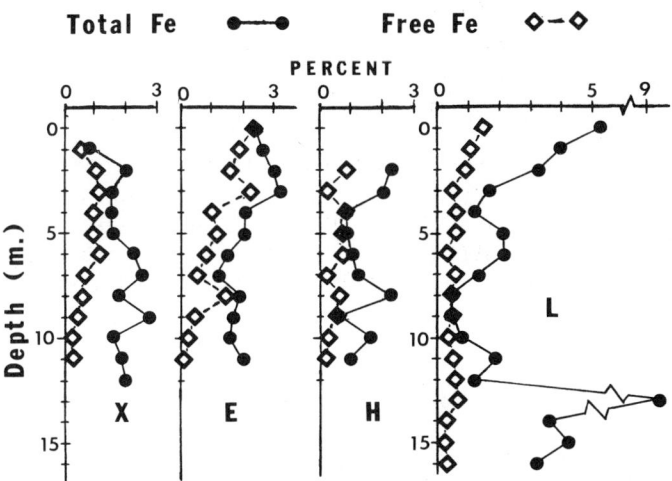

Fig. 3. Total iron and free iron in overburden rock material from four sites X, E, H, and L.

bonates) and release of Fe from silicates and carbonates. In the case of the Mahoning sandstone, the increase in free Fe in the weathered zone can be expected from pyrite oxidation and weathering of carbonates (siderite) and silicates (e.g., biotite).

Mineralogical data of the sandstone rocks (Fig. 4) show that the clay minerals present in the weathered zone were kaolinite, mica and vermiculite in the < 2 µm clay fractions. In the unweathered zone (low chroma) the minerals present were kaolinite, mica, and a small amount of smectite (site E). As shown in Fig. 4, the distribution of these clay minerals at two rock depths (5 to 6 m and 9 to 10 m at site X, and 4 to 5 m and 8 to 10 m at site E) was a good indicator of the weathered and unweathered conditions of the rock strata. These rock data show that more vermiculite and less

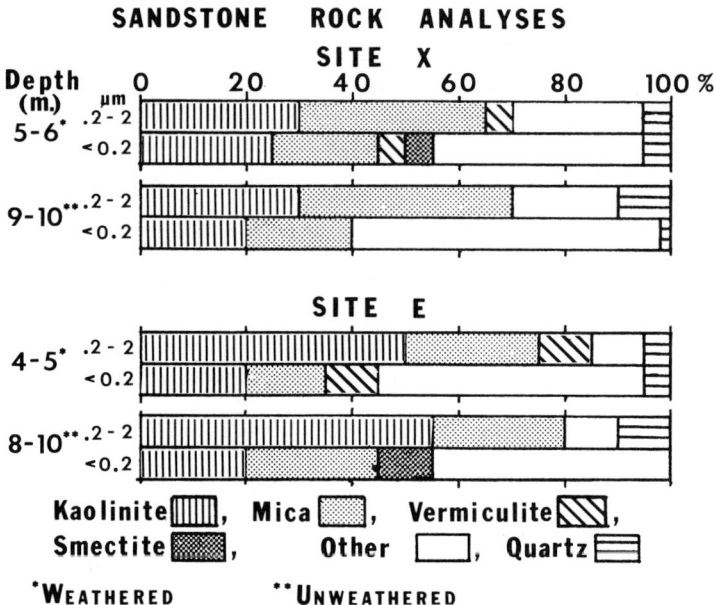

Fig. 4. Distribution of clay minerals, in clay fractions 2 to 0.2 μm and <0.2 μm for weathered and unweathered rock strata from two sites X and E.

mica were present in the weathered zone than in the unweathered zone. Several researchers (Douglas, 1965; Rich and Obenshain, 1955; Sawhney, 1960) reported an increase in the proportion of vermiculite to mica near the soil surface as a direct result of the removal of weathered products in solution along with continuous replenishing of vermiculite by weathering of mica.

Native Soils and Minesoils

Description and Properties of Native Soils: The soil reaction of the lower B horizons of the soils X and E was between 4.5 and 4.7 (Table 5) and was quite similar to the underlying weathered bedrock. The higher pH of the Ap horizons and some of the upper B horizons was obviously due to liming and vegetative recycling as reflected by high contents of exchangeable Ca. Generally, the exchangeable Mg in the B horizons was low (0.1 to 0.3 meq/100 g) compared to Ca (1 to 3.8 meq/100 g) at both soil locations. This was expected because Mg compounds such as sulfates are more soluble than $CaSO_4$ and may have leached out during acid weathering of the parent rock. This was reflected in the rock profile analysis (Tables 1 and 2). At site X, the amount of exchangeable Ca varied from 0.13 meq/100 g at the top of the rock strata to 1.37 meq/100 g at a depth of 6 m, whereas the amount of exchangeable Mg varied from 0.05 to 0.4 meq at these two depths. Below 7 m where the low chroma (unweathered) sandstone was encountered, the amount of exchangeable Ca and Mg increased to about 3 and 1 meq/100 g, respectively. Rich and

Table 5. Some chemical properties of two typical soils developed from Mahoning sandstone.

Site	Horizon	Depth	pH	Organic matter	CEC	Exchangeable cations				Base saturation
						Ca	Mg	K	Al	
		cm		%	——— meq/100 g ———					%
X	Ap	0–15	4.5	5.2	10.0	2.2	0.1	0.2	0.3	25
	B1	15–25	4.5	3.6	6.6	1.7	0.2	0.2	1.6	32
	B2	30–45	4.5	2.0	6.2	1.0	0.1	0.2	2.0	21
	C	50–60	4.4	0.4	6.2	1.0	0.1	0.1	1.9	19
E	Ap	0–20	5.7	4.0	14.5	11.2	0.3	0.1	0.1	80
	B1	20–30	5.5	3.4	15.2	10.2	0.2	0.1	0.7	69
	B2	30–48	5.2	2.9	15.5	3.8	0.1	0.1	3.1	26
	Bx1	48–54	4.7	2.8	17.5	2.3	0.2	0.1	4.5	15
	Bx2	54–68	4.7	0.1	18.5	2.4	0.3	0.1	5.9	15

Obenshain (1955), in their study of some Virginia soils, attributed the accumulation of Mg compared to Ca at lower profile depths to the greater mobility of Mg than Ca.

These soils likewise resembled the underlying rock in the amount of exchangeable Al and total and free Fe in the soil profiles. The amount of exchangeable Al at sites X and E were as much as 2 and 5.9 meq/100 g, respectively (Table 5). These exchangeable Al values corresponded very closely to the amounts of exchangeable Al found in the weathered sandstone at location X and the amount of exchangeable Al in the weathered zone of the rock strata was as much as 1 meq/100 g for site X (Table 1) and as much as 4.2 meq/100 g (Table 2) at site E. The lower values of exchangeable Al in soil X compared to soil E, possibly reflected the differences in the texture of the rocks observed at these locations. Field observation indicated more coarse textured sandstone at site X than at site E. This may explain why at site X the amount of sand in the soil profile varied between 38 to 44% and clay between 19 to 26% (Table 6). The corresponding values for sand and clay at site E were 6.7 to 24.1% and 21 to 34%, respectively. The sandstone at site E was fine textured due to the large amount of < 2 µm clay. The higher CEC in the soil profile and sandstone at site E than site X probably resulted from greater amount of clay at site E (Table 5).

Distribution of total and free Fe in the soil profiles (Table 6) and in the sandstone rocks (Fig. 3) also suggests that sandstones at sites X and E were indeed the parent rocks of soils X and E, respectively. At site X, the total and free Fe values in the weathered rock (0 to 6 m) were < 2.1 and 1.4%, respectively (Fig. 3). Total and free Fe values in soil X were about 2.1 and 1.4%, respectively (Table 6). On the other hand, at site E, the amount of total and free Fe were 3.5 and 2.1% in the rock (Fig. 3) and approximately 5 and 1.7% in the soil profile, respectively (Table 6).

The soil mineralogical data in Fig. 5 and 6 closely resembled the mineralogical data of the sandstone rock (Fig. 4) except that there were higher amounts of vermiculite in the upper horizons of these soils. Kaolinite was the dominant clay mineral in soil E as well as in the sandstone rock at location E. Mica and kaolinite were the principal clay minerals in soil X as well as in the rock at location X. Presence of smectite,

Table 6. Particle size distribution, total and free iron in two typical soils developed from Mahoning sandstone.

Site	Horizon	Depth	Particle size distribution			Texture‡	Total Fe	Free Fe
		cm	Sand	Silt	Clay		%	%
			——— % ———					
X	Ap	0–15	38.0	40.6	22.0	l	1.7	1.5
	B1	15–25	44.0	36.6	19.2	l	2.0	1.4
	B2	30–45	40.3	33.0	26.2	Ch.†l.	2.0	1.6
	C	50–60	38.6	41.6	19.8	v.Ch.l.	2.0	0.9
E	Ap	0–20	20.3	58.6	21.1	sil	3.4	1.6
	B1	20–30	24.1	49.0	26.9	l	4.2	1.6
	B2t	30–48	11.9	57.1	30.9	sicl	3.8	1.7
	Bx1	48–54	6.7	58.8	34.4	sicl	5.0	1.8
	Bx2	54–68	13.3	54.2	32.5	sicl	4.8	1.8

† Ch. = channery.
‡ Abbreviations follow Soil Survey Manual (Soil Survey Staff, 1951), p. 139.

as revealed by XRD of coarse (2 to 0.2 μm) and fine clay (< 2.0 μm) fractions of soil and sandstone rock, further indicated inherited mineralogical characteristics of the soils from the parent rock.

The gradual increase in vermiculite with corresponding decrease in mica from weathered rock (Fig. 4) in the upper horizons of the soils (Fig. 5 and 6) and absence of vermiculite and dominance of mica in the unweathered zone (low chroma rock) indicated that vermiculite may have formed from mica during pyrite oxidation. The sulfuric acid produced could dissolve the easily-soluble bases, such as Ca, Mg, and K, and facilitate breakdown of mica and other minerals. Thus the upper horizons would have been enriched with more resistant clay minerals.

Description and Properties of Minesoils: Minesoils 1 and 2 were located near sites X and E (Fig. 1). Weathered and unweathered sandstone materials were stockpiled separately during excavation of coal by surface mining. During reclamation of these sites, the toxic or potentially toxic unweathered sandstone was buried beneath the weathered sandstone and the area was seeded to legume-grass mixtures including Kentucky 31 tall fescue (*Festuca arundinacea* L.). At locations 3 and 4, which were also on Mahoning sandstone, the weathered and unweathered rock materials were not sorted. During grading at this location, overburden materials were haphazardly mixed prior to seeding. At locations 1 and 2, the ground was completely covered with grass and vegetation seemed to be thriving. At locations 3 and 4, the vegetation was sparse and "wet spots", due to the hygroscopic nature of free acid produced by pyrite oxidation, were quite common. Chemical data (Table 7) show that, at sites 1 and 2, the acidity in the minesoil profiles was low (relatively high pH) with accompanying low levels of exchangeable Al in the top layers of these minesoil profiles, probably due to recent liming during reclamation. The upper part of the control horizon (24 to 60 m depth) shows chemical characteristics similar to those found in weathered parent rock and native soils at these sites except that at site X the amount of exchangeable Mg was 1.6 meq/100 g, probably reflecting a high magnesium zone in the parent rock. From the limited information available about the liming and ferti-

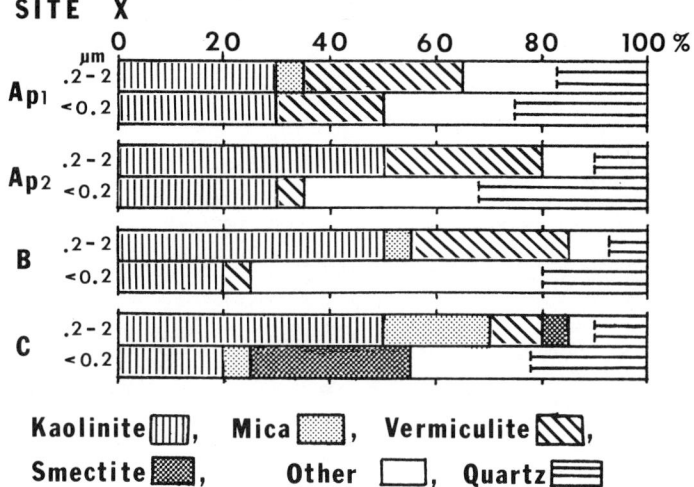

Fig. 5. Distribution of clay minerals, and quartz in clay fractions 2 to 0.2 μm and <0.2 μm for horizons Ap1, Ap2, and B and C from a natural soil developed at site X.

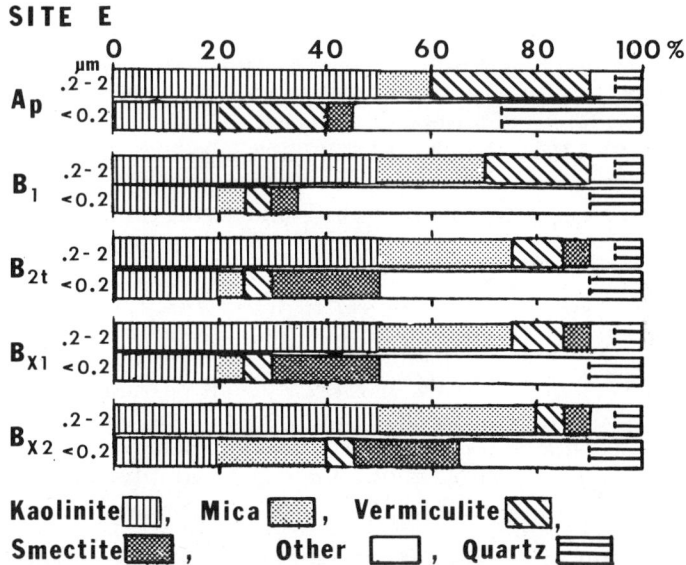

Fig. 6. Distribution of clay minerals, and quartz in clay fractions 2 to 0.2 μm and <0.2 μm for horizons Ap, B1, B2t, Bx1, and Bx2 from a natural soil developed at site E.

lizing material used, it appears unlikely that much magnesium was applied as an amendment. If Mg found at this site came from weathering of Mg bearing minerals or from surface applied limestone which might have been dolomitic, it may have moved downward and accumulated in the lower horizons where concentration of Ca was low, in accordance with high mobility of Mg reported by Rich and Obenshain (1955).

Table 7. Some chemical properties of four mine soil profiles.

Site	Depth	pH	Organic matter	CEC	Exchangeable cations				Base saturation	Total S
					Ca	Mg	K	Al		
	cm		%		meq/100 g				%	%
1	0–3	6.8	4.9	13.8	12.8	0.7	0.3	0.0	100	0.03
	3–24	5.6	3.6	9.5	6.5	1.3	0.2	0.0	84	0.01
	24–60	4.9	2.5	8.0	3.9	1.6	0.2	0.5	72	0.01
2	0–3	4.7	5.9	10.3	4.8	0.6	0.5	2.0	57	0.03
	3–24	5.0	4.6	7.5	4.5	0.4	0.2	1.0	68	0.02
	24–60	3.6	3.6	4.3	0.9	0.3	0.2	1.8	32	0.10
3	0–10	5.0	4.6	11.3	12.3	0.8	0.3	0.5	118	0.04
	10–20	3.8	4.4	8.0	1.4	0.3	0.2	3.7	24	0.04
	20–30	3.7	4.4	5.5	1.4	0.4	0.2	3.1	36	0.07
	30–40	3.8	4.1	4.5	0.9	0.5	0.2	3.4	35	0.05
4	0–10	3.2	5.1	5.0	0.6	1.3	0.1	4.2	40	0.16
	10–20	3.0	4.9	11.3	0.8	1.9	0.1	4.7	25	0.28
	20–30	3.0	4.6	9.0	1.1	3.0	0.2	5.6	48	0.28
	30–40	3.0	4.4	8.5	1.5	3.5	0.2	5.3	61	0.35
	40–50	3.2	4.1	8.5	3.8	3.3	0.2	4.1	85	0.73
	50–60	3.4	3.6	11.0	1.6	2.6	0.2	3.5	40	0.46

The CEC of minesoils 1 and 2 decreased with depth as did OM values (Table 7). This trend in values of CEC and OM was also found in the native soils around this area. The amount of total S was very low and was very close to the amount of S found in the weathered Mahoning sandstone. Thus, it appears that these two minesoils were free of pyritic S. This may explain why even several years after lime application, these minesoils had relatively high pH and quite high percentage base saturation (Table 7).

On the other hand, at locations 3 and 4, the minesoil reaction was generally between 3 and 4 which indicates the presence of free acid in these soils. At these locations, the presence of "wet spots", even during dry periods demonstrated the production of sulfuric acid from pyrite exposed to the atmosphere. This is supported by total S which varied between 0.04 and 0.73% and was considerably higher than found in the weathered rock material, particularly at site 4. Field observations of these soils (3 and 4) showed considerable amount of dull grey sandstone (low chroma) and carbolithic material resembling dark shale. Relative uniformity in the distribution of OM in profiles of minesoils 3 and 4 further confirms the mixing of carbolithic material throughout the soil profile (Table 7). As shown in Table 7, these soils were also quite high in exchangeable Al (3.1 to 5.6 meq/100 g). Apparently mixing of low chroma sandstone, high chroma sandstone, and carbolithic material in the top portion of these soils resulted in the oxidation of pyrite which produced sulfuric acid. The acid attacked the primary and secondary minerals and released considerable amounts of Al and Fe and also other cations, such as Mg. The amount of exchangeable Mg in these soils varied between 0.3 and 3.5 meq/100 g, but in most cases was between 2 and 3 meq/100 g. The minesoil at site 4 contained the highest amount of Mg and appeared never to have been limed as indicated by low pH and low exchangeable Ca. Thus, the source of Mg could not have been dolomitic limestone.

Table 8. Particle size distribution, total and free iron in four mine soils profiles.

Site	Depth	>2 mm	Particle size distribution			Texture†	Total Fe	Free Fe
			Sand	Silt	Clay			
	cm	%	%				%	
1	0-3	47	46.7	35.0	18.3	l	3.6	1.6
	3-24	51	44.1	37.9	18.0	l	4.5	1.0
	24-60	37	49.8	36.3	13.9	l	7.1	1.3
2	0-3	32	25.2	51.3	23.5	sil	5.0	1.3
	3-24	40	26.4	50.4	23.2	sil	11.7	1.5
	24-60	67	39.0	43.2	17.8	l	3.9	1.4
3	0-10	39	36.1	31.4	32.5	cl	3.9	1.6
	10-20	34	34.2	32.3	33.5	cl	3.1	1.5
	20-30	25	39.3	31.0	29.7	cl	10.9	1.5
	30-40	44	36.9	31.6	31.5	cl	7.5	1.0
4	0-10	33	31.6	33.1	35.3	cl	7.8	2.0
	10-20	43	31.2	33.5	35.3	cl	4.8	2.2
	20-30	33	30.1	33.9	36.0	cl	10.3	2.1
	30-40	50	31.9	34.4	33.7	cl	10.9	2.0
	40-50	52	30.3	34.2	35.5	cl	8.9	2.0
	50-60	38	25.9	35.0	39.1	cl	10.6	1.5

† Abbreviations follow Soil Survey Manual (Soil Survey Staff, 1951), p. 139.

Probably minerals, e.g., biotite (which is quite stable under basic conditions as found when carbolithic and low chroma sandstone are present), became unstable on exposure to free acid and low pH (Wilson, 1967; Jackson, 1965; Barnhisel and Rotromel, 1974). The presence of high amounts of total Fe in these minesoils (Table 8) indicates that pyrite may not be the source of all of the Fe found in these soils. Some of this Fe may have come from other sources such as biotite and siderite. The mineralogy of these minesoils (to be reported elsewhere), which has not been completed, will provide more insight as to the source of Mg and iron in these soils.

Particle size distribution, total and free Fe data (Table 8) of the minesoil profiles at sites 1 and 2 show that percent sand, clay, total and free Fe were very similar to those found in native soils at sites X and E (Table 6). A slightly lower amount of clay and a higher amount of sand and coarse fragments of these minesoils compared to the native soils indicate that soil forming factors, described by Jenny (1941) have not acted over enough time to cause significant clay production and movement.

The higher percent clay and lower percent sand in minesoils 3 and 4 than in soils 1 and 2 indicated that fine textured material was mixed throughout these soils. Uniformity in the distribution of clay at all depths in minesoil 3 (30 to 33% clay) and minesoil 4 (33 to 35% clay) reflects texture of the parent material placed by mining. Field observations during sampling of these profiles also indicated that pyrite in dull grey sandstone (low chroma) was undergoing oxidation and weathering as revealed by color change to bright colors (high chroma) and easy crushability of sandstone by hand. The oxidation of pyrite in deeper layers was expected since oxygen could have been dissolved in the water percolating through the unconsolidated rock material. Also, large rock fragments leave voids in the minesoil that favor oxygen diffusion into the minesoil profiles.

Thus, it appears that minesoils 3 and 4 will be finer textured than minesoils 1 and 2 with time. The future properties of these developing soils can be assessed more easily following mineralogical analyses (to be reported elsewhere).

SUMMARY AND CONCLUSIONS

Changes in chemical and physical properties of rock strata due to natural weathering have long been recognized (Krumbein and Pettijohn, 1966). The depth from the land surface to which these forces penetrate ranged from tens of cm in the Arctic to hundreds of meters in the Tropics (Ollier, 1965). The weathered zone found in the present study in northern West Virginia was between 6 and 7 m, however, along local fractures, the oxidized minerals may be found to depths of 16 m or deeper. This limited weathering to 6 m produced low levels of bases and high levels of Al in the soils developed from Mahoning sandstone which probably was at least partially related to leaching of the parent rock with sulfuric acid produced by pyrite oxidation. This relationship is supported by the absence of pyrite, low S, and exchangeable bases content of the weathered high chroma sandstone immediately beneath the soils. However, this weathered rock was high in total Fe, free Fe, and exchangeable Al. The loss of bases from this weathered rock presumably initiated the formation of vermiculite from the indigenous mica. In the unweathered (low chroma) sandstone rock below 6 m, sulfur was abundant, some as pyrite. Exchangeable Al was generally absent in this unweathered rock, but exchangeable bases were quite high.

The distribution of total Fe, free Fe, and clay minerals in nearby soils was similar to that in the rock material underneath which indicates that the soils have developed from rock similar to underlying strata. This relationship is further supported by the presence of smectite in both the sandstone and in the C horizons of the soils studied. The most noteworthy clay mineral contribution of parent material to the soils was highly crystalline mica. The Mahoning sandstone was conspicuously micaceous. The weathering trend observed in the soils was similar to that noted in the sandstone. The main effect of weathering of the clay minerals was the vermiculitization of mica in the high chroma rock and in the soil horizons.

Proper placement of the overburden rock material (high chroma materials at the surface, low chroma materials buried) resulted in minesoils which could be made edaphologically desirable by treatment with lime and fertilizer. The pH, OM, CEC, exchangeable bases, exchangeable Al, base saturation, and total S were all within a range adaptable to plant growth. The vegetation reflected these data. On the other hand, those minesoils where the overburden rock material was haphazardly replaced following coal mining resulted in very sparse and poor plant growth. The presence of pyrite, along with its oxidation product, i.e., sulfuric acid, produced such a low pH, high exchangeable Al content, low base saturation, and deficiency of exchangeable bases that revegetation was largely unsuccessful. Several attempts to correct this by liming, fertilization, and seeding have given poor results.

ACKNOWLEDGMENTS

This work has been financed in part with the Federal funds from the U.S. Environmental Protection Agency under grants 14010EJE, S800745, R802603, through CSRS under grant 684-15-15, and from Hatch 217. The contents do not necessarily reflect the views and policies of the U.S. Environmental Protection Agency, nor does mention of trade names or commercial products constitute endorsement or recommendation for use.

Some funding for analyses was also provided by the W.V.U. Energy Research Center on grant no. MI79R02.

LITERATURE CITED

1. Alexiades, C. A., and M. L. Jackson. 1965. Quantitative determination of vermiculite in soils. Soil Sci. Soc. Am. Proc. 29:522–527.
2. Allison, L. E. 1965. Organic carbon. p. 1367–1378. In C. A. Black (ed.) Methods of soil analysis. Part 2. Agronomy Series No. 9. Am. Soc. Agron., Madison, Wis.
3. Barlow, J. A. 1974. Coal and coal mining in West Virginia. W. Va. Geological Bull. No. 2.
4. Barnhisel, R. I., and A. L. Rotromel. 1974. Weathering of clay minerals by simulated acid coal spoil-bank solutions. Soil Sci. 118:22–27.
5. Blume, H. P., and U. Schwertmann. 1969. Genetic evaluation of profile distribution of aluminum, iron, and manganese oxides. Soil Sci. Soc. Am. Proc. 33:438–444.
6. Coleman, N. T., S. B. Weed, and R. J. McCracken. 1959. Cation-exchange capacity and exchangeable cations in Piedmont soils of North Carolina. Soil Sci. Soc. Am. Proc. 23:146–149.
7. Douglas, L. A. 1965. Clay mineralogy of a Sassafras soil in New Jersey. Soil Sci. Soc. Am. Proc. 29:163–167.
8. Grube, W. E., Jr., R. M. Smith, R. N. Singh, and A. A. Sobek. 1973. Characterization of coal overburden materials and minesoils in advance of surface mining. p. 134–152. In Proc. Res. Applied Tech. Symp. Mined-Land Reclamation, Pittsburgh, Pa.
9. ————, ————, and J. T. Ammons. 1982. Mineralogical alterations that affect pedogenesis in minesoils from bituminous coal overburdens. p. 209–223. In J. A. Kittrick, D. S. Fanning, and L. R. Hossner (ed.) Acid sulfate weathering. SSSA Spec. Pub. no. 10. Madison, Wis.
10. Jackson, M. L. 1956. Soil chemical analysis. Advanced course publ. by the author. Dep. Soils, Univ. Wisconsin, Madison, Wis.
11. ————. 1958. Soil chemical analysis, Prentice-Hall, Englewood Cliffs, N.J.
12. ————. 1965. Clay transformation in soil genesis during the Quaternary. Soil Sci. 99:15–22.
13. Jenny, H. 1941. Factors of soil formation. McGraw Hill Book Co., New York, NY. 182 p.
14. Kiely, P. V., and M. L. Jackson. 1965. Quartz, feldspar and mica determination for soils by sodium pyrosulfate fusion. Soil Sci. Soc. Am. Proc. 29:159–163.
15. Kilmer, V. J., and L. T. Alexander. 1949. Methods of making mechanical analysis of soils. Soil Sci. 68:15–24.
16. Krumbein, W. C., and F. J. Pettijohn. 1966. Manual of sedimentary petrography. Appleton Century-Crofts, New York. p. 21.
17. Ollier, C. D. 1965. Some features of granite weathering in Australia. Zeit. f. Geomorph. 9:285–304.
18. Rich, C. I. 1961. Calcium determination for cation exchange capacity measurements. Soil Sci. 9:226–231.
19. ————, and S. S. Obenshain. 1955. Chemical and clay mineral properties of a Red-Yellow Podzolic soil derived from muscovite schist. Soil Sci. Soc. Am. Proc. 19:334–339.

20. Raad, A. T., R. Protz, and R. L. Thomas. 1969. Determination of Na-dithionite and NH_4-oxalate extractable Fe, Al, and Mn in soils by atomic absorption spectroscopy. Can. J. Soil Sci. 49:89–94.
21. Sawhney, B. L. 1960. Weathering and aluminum interlayers in a soil catena: Hollis-Charlton-Sutton-Leicester. Soil Sci. Soc. Am. Proc. 24:221–226.
22. Soil Survey Staff. 1951. Soil Survey Manual. U.S. Dep. Agriculture, Agric. Res. Admin., Washington, DC p. 503 (See p. 139).
23. Strausbaugh, P. D., and E. L. Core. 1970. Flora of West Virginia, part 1, 2nd ed. W. Va. Univ. Bull. Ser. 70, No. 7-2.
24. West Virginia University. 1971. Mine spoil potentials for water quality and controlled erosion. Water Pollution Control Res. Ser. 14010EJE Environmental Protection Agency, Washington, DC.
25. ―――. 1974. Minesoil potentials for soil and water quality. EPA-600/2-74-070. NERC, ORD, USEPA, Cincinnati, Ohio.
26. Wilson, M. J. 1967. The clay mineralogy of some soils derived from a biotite-rich quartz-gabbro in the Strathdon area Aberdeenshire. Clay Miner. 7:91–100.

Chapter 12

Mineralogical Alterations that Affect Pedogenesis in Minesoils from Bituminous Coal Overburdens[1]

W. E. GRUBE, JR., R. M. SMITH, AND J. T. AMMONS[2]

ABSTRACT

Studies were initiated to provide sound geologic and pedologic bases for improved mined land reclamation and pollution abatement. Geologic features of multi-county regions of coal surface mining activity, examined from the viewpoint of sediment depositional and compositional trends, demonstrated variations in overburden composition associated with the local geographic position within the coal depositional swamp. Data, which are exemplary of more extensive arrays cited and published elsewhere, are presented showing the trends in composition of rock strata of the upper Pennsylvanian System in northern West Virginia and southwestern Pennsylvania. Chemical analyses for pyritic sulfur and soluble carbonates in these coal overburden materials verify the relative base-rich quality of the younger part of the section (Conemaugh Formation and above) and the both base-poor and low-pyrite composition of the older rocks in the region for which data are presented. These data also quantitate the penetration of the zone of oxidative weathering to depths ranging from 6 to 12 m, varying with rock type and whether pyrite or carbonate minerals are considered. The balance, or Acid-Base Account, of net acid-producing potential calculated from pyritic sulfur content and intrinsic calcium carbonate equivalent permits evaluation of the rock with respect to the ultimate acidity or basicity that might be expected in a new soil developing in excavated rock materials. Comparison of recognized properties of native soils on an area with projections from analyses of properties of rocks from which they were believed to have been formed shows agreement sufficient to demonstrate the value of studying potential soil parent materials prior to their placement into a pedologic setting. This aggregation of geologic, pedologic, and chemical information has re-

---

[1] Contribution from U.S. Environmental Protection Agency, Cincinnati, Ohio; West Virginia Univ., Morgantown, W. V.; and Tennessee Technological Univ., Cookeville, Tenn.
[2] Chemist, U.S. Environmental Protection Agency, Cincinnati; professor emeritus, Div. of Plant and Soil Sciences, West Virginia Univ., Morgantown; and assistant professor of plant and soil science, Plant and Soil Science Dep., Tennessee Technological Univ., Cookeville.

Copyright © 1982 Soil Science Society of America, 677 S. Segoe Rd., Madison, WI 53711. *Acid Sulfate Weathering.*

sulted in development of a basis for predicting the nature of soils that form in specified rock environments. Knowledge of minesoil properties is being used by land reclamation and pollution abatement planners, and soil scientists who are involved in both agronomic studies of minesoil capabilities and genetic studies of new soil development.

INTRODUCTION

Extensive surface mining of coal in the central Appalachian region of the USA has resulted in the exposure of vast quantities of disrupted geologic materials to soil forming processes. Sedimentary rocks which have been previously buried at depths not ordinarily affected by near-surface pedogenic processes suddenly become soil parent material. Investigations were initiated to discover remedial practices that would reduce water pollution associated with geochemical alterations within the disrupted rock strata in this region. These studies have provided geological, chemical, and pedological insights regarding both short-term alterations which affect noticeable pollution episodes and long-term mineralogical transformations which affect the solum forming in the disturbed earth materials.

Embarkation into somewhat unknown study areas is probably best approached by first obtaining a perspective overview of the position of specific sites within a broad setting, followed by discrete studies directed toward individual points of interest. In the studies reported herein, we included geological expertise among our first field site visits. We include in this discussion a brief overview of sedimentary bed depositional patterns found within the central Appalachian study area, followed by some examples of results of analytical studies of coal overburden composition. Interpretations of these data have led to conclusions regarding the significance of the composition of these imminent soil parent materials relative to both the expected "minesoil"[3] properties and the observable properties of contiguous native soils.

GEOLOGICAL OVERVIEW

Bituminous coals in the United States (characterized by high carbonaceous content, between 15 and 50% volatile matter, and fuel properties between anthracite and lignite) provide the vast majority of the coal that is readily obtainable to meet the nation's energy needs. Approximately half of this energy reserve is considered recoverable by surface mining methods (National Coal Association, 1975). These coal deposits were laid down during the Carboniferous Period (350 to 270 million years ago) of the Palezoic Era (Holmes, 1965). Because of the age of these deposits, and

---

[3] The term "minesoil" is a popular term that has not been rigidly defined in a technical sense. In accordance with *Soil Taxonomy's* definition of soil (Soil Survey Staff, 1975), which specifies that soils may be modified or even made by man, minesoil is soil made by mining or mining related activity (Smith et al., 1974, p. 254). Usage of this term by researchers has increased considerably since the early 1970's, with a growing acceptance that minesoils may originate from a wide variety of earth manipulation activities. In contrast, a "native soil" is considered to be one developed in place over past geologic time.

geological processes occurring subsequent to this Period, the organic masses that became coal, and overlying sediments, were compacted to such an extent that durable rocks were formed. Thus, today the overburdens that need to be removed to recover bituminous coals by surface-mining methods in the eastern and central USA are comprised of sedimentary rocks of various types: sandstone, mudstones, limestone, and an endless variety of combinations of these. In some localities in the Central States where bituminous coal is surface-mined, the once-exposed rocks have been covered during the Pleistocene or Holocene with mantles of unconsolidated glacial deposits and silt or loess to depths of up to several tens of meters.

Figure 1 presents the generally recognized geologic column as it occurs in the study area. Figure 2 illustrates a schematic geologic section that might be found in bituminous coal fields. A wide variety of characteristics is illustrated by this pictorial. The contrasting rock types contribute to the wide variety of undisturbed native soils of the region, even before the effects of colluvial activity, vegetation, etc. are considered. An ever present feature, still perhaps not appreciated for its pedogenic significance, especially in humid regions, is the zone of oxidative weathering alteration. This zone, whose depth of penetration varies with the parent rock type, land slope, and other factors, is consistently observable wherever the geologic section from the surface is exposed. In humid regions (primarily the eastern and midwestern USA) this weathered zone invariably has most of the pyritic and some of the silicate iron converted to oxides and is highly leached of many plant nutrients. In regions of the Midwest, where there has been significant loess and till deposition, a weathered rock zone is recognizable underneath these recent deposits.

Arkle (1973) has discussed the deposition of upper Pennsylvanian sediments (the dominant coal-bearing strata of the Carboniferous Period in the USA) throughout the central Appalachians. Although he describes the formation of these strata only in this region, the concepts of regionality of depositional environmental of coal and superjacent beds may be applied to understanding the occurrence of minesoil parent materials throughout the entire eastern and midwestern bituminous coal fields.

As pointed out by Arkle, the Pennsylvanian System is probably more completely developed within the confines of the central Appalachians of southwestern Pennsylvania and West Virginia than in any other area. Thus, intensive study of pedogenic process upon the rocks in this region is remarkably productive. Surface mining of coal and extensive road construction have made a significant number of exposures of fresh rock strata available for study. Detailed characterization (Smith et al., 1974, 1976) of coal-bearing rock sections has strengthened Arkle's thesis of major depositional trends which resulted in more base-rich deposits as younger Pennsylvanian rocks were formed in northern West Virginia and southwestern Pennsylvania. An arbitrary geologic/pedologic division line, between relatively acid forming and predominantly base-influenced rock strata has been drawn (about 30 m above the base of the Conemaugh Formation in a region of the Central Appalachians that has been intensively studied, Smith et al., 1974) in an attempt to indicate the transition time between these sedimentary environments. Younger strata,

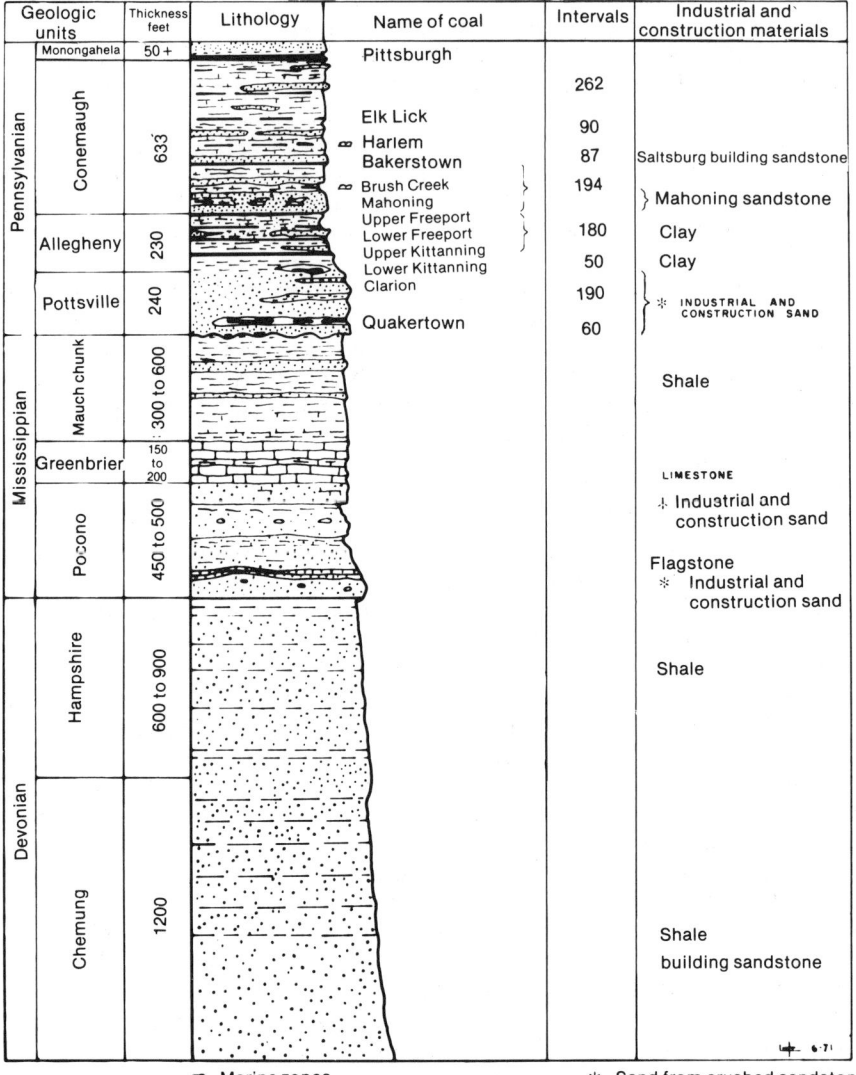

Fig. 1. Geologic Section in Preston County, West Virginia, showing the generally recognized formations within the Pennsylvanian of the Central Appalachians.

found stratigraphically above this line tend to weather into less acid soils. Strata below the division line generally lead to formation of residual soils of generally low fertility, greater acidity, and low buffering capacity.

Concurrently with stratigraphic (time of deposition) changes in rock character, the variations in overburden composition associated with local geographic position within the coal depositional swamp need to be considered. Basin peripheries appear to be dominated by coarse-grained sedi-

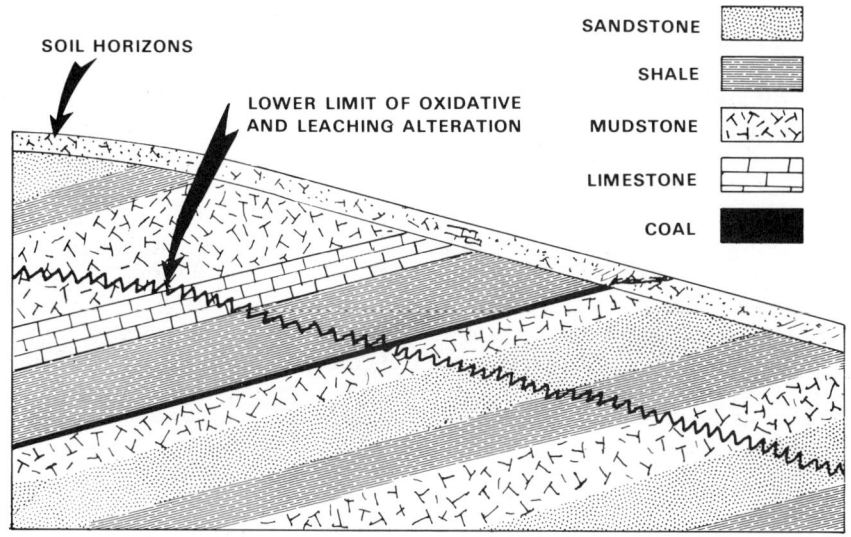

Fig. 2. Idealized section showing variety and gradations of sedimentary beds within a small section of geologic column.

ments, and usually have the least carbonate content. With progression toward the apparent center of the coal depositional swamp, overburden rock types grade toward the finer particle sizes and any calcareous influences become more pronounced. Edges of the extent of the coal seam in a geographic region commonly contain the channel and deltaic sands associated with encroachment of stream systems into lakes, bays, or other large and quiet waterbodies. The sands have proven to be a common source of difficulty in reclaiming overburden materials. Where mudstones dominate between thinly separated coal horizons, they too tend to be quite pyritic, and present a land reclamation challenge (Smith et al., 1976). These apparently are fine sediments associated with immediate pre and post-swamp environmental conditions. Investigations of overburden properties and minesoil development have added further support to the significance of the postulated compositional trends.

This presents a background of the general types and origins of materials that need to be moved during the surface-mining of bituminous coal, and generally describes the nature of the parent materials available to form new soils that are resulting from coal overburden displacement.

Overburden analysis data reported (West Virginia Univ., 1971; Smith et al., 1974, 1976) tend to reflect the environments of deposition of the coals and adjacent rocks. Preliminary evidence gathered to date shows that the nature of minesoils developing in specified materials can be predicted from knowledge of the chemical character of constituent rocks. Studies of the properties of native soils believed to have developed in similar, defined parent rock also supported this concept.[4]

---

[4] Grube, W. E., Jr. 1974. Pedologic potential of selected upper Pennsylvanian sedimentary rocks using chemical parameters. Ph.D. Diss. West Virginia Univ., Morgantown, WV 26506.

GEOCHEMICAL PROCESSES AFFECTING PEDOGENESIS

Soil development within specific rock materials of the nature described above appears to have received little attention. Perhaps the complexity of such systems has been a deterrent. Some simple relationships have been hypothesized: Joffe (1949) argued that the more complex the mineral composition of the rock, the more easily it weathers. Fitzpatrick (1963) concluded that igneous rocks weather more easily and deeper than metamorphic rocks, which weather more easily and deeper than sedimentary rocks.

The importance of the mineral acid potential (presence of pyritic minerals) and the neutralizing potential (readily soluble bases, including carbonates) of soil parent materials has been passively mentioned (Hunt, 1972), but has not been previously shown to be of great significance to genesis in upland soils.

Joffe (1949) recognized the conversion of pyrite to sulfuric acid, which "...increases the solvent effect of water." Hunt (1972) states, "sulfuric acid is widespread and is an important factor in the weathering of surface deposits in most coalfields and metal mining districts, for in these areas the sulfide minerals are abundant in the bedrock and become freed and oxidized in the course of mining or by circulation of groundwater."

The fundamental chemical reactions involved in weathering of the most labile minesoil minerals, $S^{2-} \rightarrow SO_4^{2-}$, and $H_2SO_4 + CaCO_3 \rightarrow$ are exothermic, suggesting the spontaneity of intense weathering of freshly exposed rock materials containing sulfide and carbonate minerals. The reactions result ultimately in lower energy soil parent material. Joffe (1949) also suggested the beneficial effect on some soil organisms in that *Thiobacillus thiooxidans* utilizes heat evolved in the oxidation of sulfur (S). These reactions go to completion only over a period of many years, so it is possible for soil management systems to obtain the beneficial effects of release of essential elements from weathering rock minerals.

The development of the solum within the zone of already highly weathered and leached geologic materials has (perhaps inadvertantly) been recognized in many profile descriptions found in soils literature. Where the color of the C horizon and/or underlying parent material is noted, a high Munsell Color Chroma (indicating oxidized iron compounds) is frequently noted. Grube et al. (1973a) discussed in detail soils developed in highly weathered and leached fine-grained sandstone that had once been pyritic. Several authors (Ruhe et al., 1965; Eswaran and Bin, 1978) have discussed pedogenesis in parent materials found in tropical regions, where oxidation and weathering alteration reach depths of a few tens of meters. However, these studies have described changes found mainly in granitic and other magmatic rock types.

Simonson (1970) pointed out some extremes in nutrient supply of native soils that could be expected as a result of depletion or accumulation of nutrient elements from parent material during soil formation. He stated that low levels of exchangeable bases, particularly in Ultisols, result from a long history of weathering and leaching, and that a major proportion of nutrient elements being lost downstream comes from the weathering of rock at the bottom of the regolith rather than from the soil. Ultisols and

Inceptisols (dominantly leached Dystrochrepts) are the soils which predominate in the regions included within this study, and it is these soils that appear to have incurred the largest net losses of leachable nutrient elements in humid temperate zones during soil formation.

We emphasize here the importance of simple mechanisms that profoundly influence acidity and alkalinity of soils developed from Pennsylvanian and, by implication, other rocks. Although many investigations and reports of chemical and mineralogical species in soil profiles proclaim the results of changes in the soil properties over the long time of soil genesis, few recognize the developmental influence of minerals in the native unweathered rocks which will develop acidity or alkalinity on a relatively grand scale when the rocks are finally exposed to weathering forces at the land surface. These minerals are mainly pyrites (FeS_2 and related sulfides) and carbonates ($CaCO_3$ and other alkaline earth carbonates) naturally present in the sedimentary rocks. The effects of the strong acid or base resulting from oxidation and dissolution of these minerals will be tempered by the presence or absence of buffering materials in the weathering rock/soil-development system. Data (Smith et al., 1974, 1976) show that fine-grained sedimentary rocks (shales and mudstones) generally contain somewhat higher percentages of pyrite than the coarser textured rock types (sandstones) in the areas studied. These textural observations probably are a direct reflection of the environment of deposition of the sediments. Likewise, the fine textured rocks appear to be generally richer in titratable bases, which counteract acidity. The increased buffering capacity afforded by the greater amount of clay minerals in the shales and mudstones further counteracts the strong acidity which develops as the reduced S compounds are oxidized. From this it is evident that sandstone-derived soils may be more acidic than those developed from finer-grained rock; given similar rock sulfide content and periods of weathering exposure. Some sandstones analyzed have been found to contain 2.5% calcium carbonate equivalent and higher in narrow zones, but frequently there are insufficient bases present to neutralize the acidity that is potentially present as pyrite.

The formation of strong acidity from the oxidation of reduced S minerals in soil has been investigated by many workers, primarily in studies of acid waters associated with coal mines and in studies of "acid sulfate soils" and "cat clays" (Brinkman and Pons, 1973). Reported work appears to deal mostly with immediate soil acidity problems and their alleviation, however, only when amounts of pyrite, up to 1% expressed as total S, are present. That very small proportions of pyrite in a medium can oxidize to form significant amounts of acidity may be shown by calculations. (For a material containing 0.1% S, all as pyrite, complete oxidation will yield hydrogen ions, from 1 g of material in 1 ml of water in quantity that if unneutralized would result in a pH of 1.20. It can be similarly concluded that even if only 1/10 of that pyrite was oxidized, the resulting sample would have a pH slightly above 2.0). Thus, the quantity of acid, corresponding to low pH solutions, which results from only a small amount of pyrite oxidation may greatly influence the acidity of soil or weathering rock. The profound effect of very low pH on soil mineral alteration has been documented (Keller, 1964).

STUDY OF SECTION EXPOSED BY COAL SURFACE-MINING

Segments of the coal mining industry have recently realized that the best advantage of reclamation possibilities can be made if thorough study is initiated before the first shovel is turned. Early studies of overburden cores generally included only a cursory geologic logging, with emphasis on the coal strata. Premining studies now include characterization of the overburden rock, primarily to seek clues to potentially toxic strata that may restrict reclamation success. Methods of identifying potentially toxic strata and also relatively favorable rock types were developed and put into extensive use by major coal mining companies and some regulatory bodies in the early 1970's (Grube et al., 1973b; Wiram, 1977). These overburden characterization methods were specifically designed to produce improved reclamation success in regions of the USA where serious difficulties were recognized, and it was suspected that the presence of pyritic S in the overburden strata was the major hindering factor.

Popular viewpoints and legal mandates in many areas now require preremoval and subsequent replacement of recognized native soil horizons on land that is disturbed for surface-mining of coal. Some of these bituminous coal deposits are overlain by mollic epipedons (Soil Survey Staff, 1975), particularly in the Midwest. These make excellent topsoil for replacement during reclamation. In large sections of the surface-minable bituminous coal regions, however, and especially in the eastern hilly and mountainous areas, the native soil cover is thin, highly weathered, and deficient in many agronomic qualities. It is in these latter regions where utilization of carefully selected deeper sections of the overburden strata for the new soil parent materials can lead to soils with qualities improved over those which had developed previously. The approach used can be widely applied outside the geographic area studied, and also to earth disturbance of materials for purposes other than coal recovery.

Study of coal strata properties in parallel with properties of the native soils of the same region has resulted in revelation of fundamental influences of the parent rock on soil development from Pennsylvanian rocks.[4] Recognition that rock strata may contain base-rich components (mainly as carbonates), as well as potentially acid-producing compounds, dictates an evaluation of this potentially rich source of acid-neutralizing material. The concentration of readily available bases in rock materials can be estimated by applying a modification of the procedure use to measure the neutralizing equivalence of agricultural limestone (Jackson, 1958). Where sulfur in rock is present exclusively as pyrite, the total sulfur content accurately quantifies the acid-producing potential. From the stoichiometry of the reaction of the oxidation of FeS_2 it can be calculated that for a material containing 0.1% S, all as pyrite, complete oxidation of both the S and the iron (Fe), plus hydrolysis of the ferric iron, will yield a quantity of acid that will require nearly 3 metric tons (1 metric ton = 1,000 kg) of calcium carbonate to neutralize 900 metric tons of material. The quantity of inherent readily soluble bases in a freshly exposed rock may be sufficient to neutralize acid at a rate equal to or exceeding the rate of acid production if pyrite is also present. Therefore, by comparing the total quantity of bases that would be required to neutralize potential acidity as calculated from S content measurements with analysis of the intrinsic neutralization potential of the sample, a balance can be drawn which indicates the ultimate acidity or alkalinity that can be expected in material.

An acid/base account, as illustrated by Fig. 3 and 4, demonstrates the ultimate acidity or basicity of different rock zones. These figures were obtained by plotting the values calculated from analyses for pyritic S and calcium carbonate equivalent in pulverized rock samples. Total S content (in the pyritic form) of the rock was determined by LECO Induction Furnace and Titration Analyzer (West Virginia Univ., 1971). From the stoichiometric equation of pyrite oxidation, the maximum acidity that can be generated from this quantity of reduced S is calculated. The neutralization potential, a measure of the amount of neutralizers present in overburden material, is established by treating a sample with a known

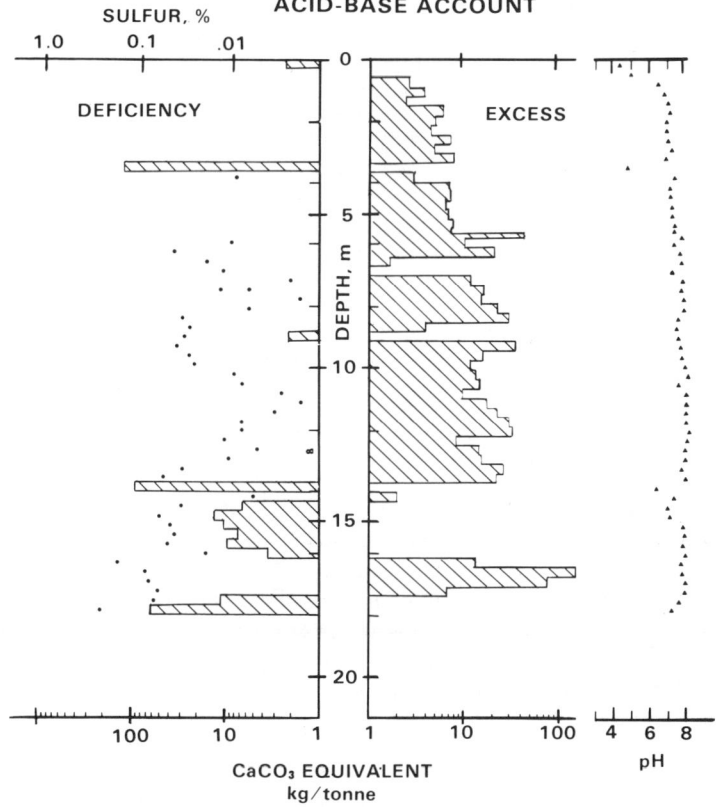

Fig. 3. Acid-Base Account, sulfur content (dots), and saturated paste pH data of rock section overlying the Elk Lick coal seam near Mt. Storm, W.V. Shaded bars to the left of the center of the figure indicate the degree to which the acidity or potential acidity of a material exceeds the neutralizing capacity of the material; shaded bars to the right indicate an excess of neutralizing potential.

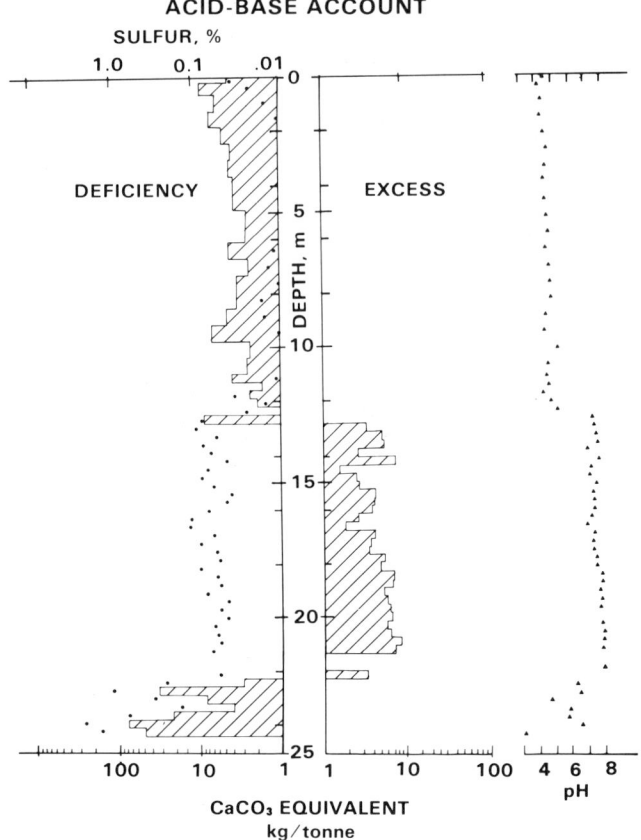

Fig. 4. Acid-Base Account, sulfur content (dots) and saturated paste pH data of rock section overlying the Lower Kittanning coal seam near Brandonville, W.V. Shaded bars to the left of the center of the figure indicate the degree to which the acidity or potential acidity of a material exceeds the neutralizing capacity of the material; shaded bars to the right indicate an excess of neutralizing potential.

amount of standardized hydrochloric acid, heating to assure complete reaction, and titration with standardized base. The result is expressed as calcium carbonate equivalents. Equivalents of acidity that could be generated are compared to the measured calcium carbonate equivalents and a net acid-base account is reached, and either tabulated or plotted as shown in Fig. 3 and 4 (units were converted to kg/metric ton for these figures) (Grube et al., 1973a).

Figure 3 includes results from characterization of a section of younger sediments, occurring stratigraphically above the geologic/pedologic division line recognized within the Pennsylvanian of the middle Appalachians (see earlier section of this paper). Figure 4 represents analyses of strata lying below the geologic/pedologic division, and underlying the strata described in Fig. 3 by about 76 m. These two sets of data show the distinctive change from base-influence in the upper section to very little influence from either pyrite or calcium carbonate in the lower strata.

Other data from many correlated sites published elsewhere (Grube et al., 1973a; Grube, 1974; Smith et al., 1974; Smith et al., 1976) demonstrate similar rock compositions. Soils derived from strata similar to those characterized in Fig. 3 are common throughout the counties of northern West Virginia, southwestern Pennsylvania and eastern Ohio, and nearly always contain high exchangeable bases in the lower soil horizons. Likewise, minesoils derived from the upper section of strata in this region are routinely recognized as agriculturally quite productive. Conversely, strata similar to those characterized in Fig. 4 are the parent materials for native soils in the highly dissected plateau region of southern West Virginia, southwestern Virginia, eastern Kentucky, and northern Tennessee. This latter geologic/pedologic region is characterized by relatively poor agricultural soils, along with sharply developed topography that inhibits extensive land development. Landscapes associated with strata found above the geologic/pedologic division generally have gentle slopes with wide valleys of agricultural development. Application of the acid-base balance concept in determining whether the ultimate soil chemical regime will be acidic or alkaline allows prediction of the probable nature of soils developing where bituminous coal has been or will be recovered by surface mining in this region.

These figures also contain data quantitating the extent of oxidative weathering from the surface downward, to a depth of about 6 m in Fig. 3 and to a depth of about 12.2 m in Fig. 4. The pH of pulverized rock samples with depth shows a dramatic increase where the unweathered material was encountered in Fig. 4, but only a slight increase in Fig. 3. Sulfur is nearly absent in the entire weathered zone of both sites. Figure 4 shows no available alkalinity in the weathered rock but the site illustrated in Fig. 3 contains significant quantities of bases in excess of any required to neutralize pyritic acidity throughout the entire geologic section represented. Similar data from other sites representing strata above the geologic/pedologic division (for example, see Fig. 1 in Smith and Sobek, 1978) suggest that carbonates are removed during the weathering process at a slower rate than pyrite is converted to acidity.

PREDICTION OF PEDOGENESIS IN MINESOILS

Since information concerning the makeup of the overburden to be excavated is readily obtainable, the final step is to utilize all available information to plan the placement of removed rock strata so that the best sequence results for development of the best new soil that is possible with the materials available. It should be noted that surface layers and amendments placed for the immediate purpose of rapidly establishing vegetative cover to meet legal reclamation requirements are not certain to have long lasting effects on soil development. However, Caruccio and Geidel (1978) have calculated that the increase in partial pressure of carbon dioxide in soils with good mulch cover (root systems) may increase the available alkalinity of infiltrating waters by a factor of eight. They also have discussed the reduction of pyrite oxidation in neutral to alkaline media, in addition to simple neutralization of acidity. The existence of the weathered zone in undisturbed strata certainly dispells any thought that pyrite will be stabilized over geologic time. However, even for terms of a few years, the kinetics of chemical interactions among basic cations released from calcareous strata and acidity developed from pyrite oxidation have not been well characterized in the field setting. Pedersen et al. (1978) compared several minesoil profiles along with several native soil profiles in a continguous undisturbed area. They concluded that, in general, the chemical constitution of the minesoils resembled that of the native soils. The mineralogical data from analyses of both areas suggested that the minesoils and the native soils were derived from the same materials, i.e., silt shale and thin bedded sandstone within the middle section of the Allegheny Group of the Pennsylvanian.

By implication, and proven by chemical analyses, the strata underlying the weathered zone contain a wealth and diversity of minerals that immediately begin alteration when excavated and placed on the land surface by mining activities. The most significant mineralogical changes found in Pennsylvanian sedimentary rocks are: (1) oxidation of pyrite, (2) solubilization of bases from carbonates and other minerals, and (3) formation of vermiculite and breakdown of other clay minerals with potassium release being the most notable result.

Environmental considerations of mined-land reclamation predominantly revolve around inhibiting generation of water pollutants, both soluble and particulate, and aesthetic viewpoints. The latter are accomplished by "good workmanship", physical stability, and chemical fertility plus judicious selection of vegetation species and their placement. Soluble water pollutants generated from Pennsylvanian rocks are dominated by those associated with acid mine drainage and include, predominantly, the most acid soluble metals in affected strata (Wilmoth et al., 1978). Overburden placement and mixing designed to result in the best minesoil properties nearly always also results in effective neutralization of generated acidity before the soil water has completely permeated the entire minesoil mass and entered the surface or groundwater transport system. Vegetative cover sufficient to satisfy current legal requirements is designed to control soil erosion. Overburden placement to effect the most productive soil in mined land is nearly always compatible with actions necessary to minimize pollution, but the converse—planning first for minimizing short-term pollution—may not result in the best minesoil, and may not even result in the minimum pollution over the long term.

SUMMARY

Data, which are exemplary of more extensive arrays cited and published elsewhere, are presented showing the trends in composition of rock strata of the upper Pennsylvanian System in northern West Virginia and southwestern Pennsylvania. Included are results of chemical analyses that show concentrations of elements that directly influence the acidic or basic status of developing minesoils. The discrete data presented in this paper have been integrated with regional depositional environmental trends exposed by geological studies of other investigators. Discussion of data presented here and elsewhere shows the relevance of rock analyses to understanding significant mineralogical alternations that can be expected in rock strata disrupted and excavated during coal surface mining operations. This aggregation of geologic, pedologic, and chemical information has resulted in development of a basis for predicting the nature of soils that form in specified rock environments.

The strong influence of pyritic S content of the rocks in contributing strong acid to pre-weather and leach soil parent material is an important consideration in this region. The highly weathered and leached character of the parent materials of the native soils is sharply contrasted with the nutrient and base rich status of underlying unweathered rock.

The lasting effect of several percent Ca carbonate (or equivalent) intrinsic to many of the younger rocks in the geologic section is evident in the recognition of high exchangeable bases in the lower horizons of present native soil profiles originating from calcareous shales and mudstones.

The balance, or Acid-Base Account, of net acid-producing potential calculated from pyritic S content and intrinsic Ca carbonate equivalent is a method of evaluation of the rock with respect to the ultimate acidity or basicity that might be expected in a new soil developing in excavated rock materials. The value of this methodology is also illustrated by its wide acceptance among coal mining companies as a valuable aid in planning efficient mined land reclamation.

Significant reserves of available plant nutrients exist in unweathered rocks, compared to native soils and rocks from the weathered zone of an undisturbed section. This source of "free" nutrient elements should play an important role in decisions about the future use potential of mined land areas for plant growth and production.

Comparison of recognized properties of native soils on an area with projections from analyses of properties of rocks from which they were believed to have been formed, shows agreement sufficient to demonstrate the value of studying potential soil parent materials prior to their placement into a pedologic setting. Such data can be used as a background upon which, in addition to use as a predictor of properties of soils developing in recently excavated land, creation of new minesoil can be soundly based.

ACKNOWLEDGMENTS

The work reported here has been financed in part with Federal funds from the Environmental Protection Agency under grants 14010EJE, S800745, and R802603. The contents do not necessarily reflect the views and policies of the Environmental Protection Agency, nor does mention of trade names or commercial products constitute endorsement or recommendation for use.

LITERATURE CITED

1. Arkle, T., Jr. 1973. The stratigraphy of the Pennsylvanian and Permian of the Central Appalachians. Geol. Soc. Am. Spec. Paper 148.
2. Brinkman, R., and L. J. Pons. 1973. Recognition and prediction of acid sulfate soil conditions. p. 169–203. In H. Dost (ed.) Acid sulphate soils. Publ. 18, Vol. 1. Int. Inst. Land Reclam. and Improv., Wageningen. Proc. Int. Symp. Acid Sulfate Soils, Wageningen, The Netherlands.
3. Caruccio, F. T., and G. Geidel. 1978. Geochemical factors affecting coal mine drainage quality. p. 129–148. In F. W. Schaller and P. Sutton (ed.) Reclamation of drastically distributed lands. Am. Soc. of Agron., Madison, Wis.
4. Eswaran, H., and W. C. Bin. 1978. A study of a deep weathering profile on granite in Peninsular Malaysia: I. Physico-chemical and micromorphological properties. Soil Sci. Soc. Am. J. 42:154–158.
5. Fitzpatrick, E. A. 1963. Deeply weathered rock in Scotland, its ocurrence, age, and contribution to the soils. J. Soil Sci. 14:33–43.
6. Grube, W. E. 1974. Pedologic potential of selected upper Pennsylvanian sedimentary rocks using chemical parameters. Ph.D. Diss. West Virginia Univ., Morgantown, W.V.
7. Grube, W. E., Jr., R. M. Smith, and R. N. Singh. 1973a. Interpretations of mottled profiles in surficial Ultisols and fine-grained Pennsylvanian age sandstones. In E. Schlichting and U. S. Schwertmann (ed.) Pseudogley and gley, transactions of commissions V and VI, ISSS, 255–262. Verlag Chemie, Weinheim/Bergst. West Germany.
8. ————, ————, ————, and A. A. Sobek. 1973b. Characterization of coal overburden materials and minesoils in advance of surface mining. p. 134–152. In Proc. Research and Applied Technology Symp. on Mined-Land Reclamation, Bituminous Coal Research, Inc., Monroeville, Pa.
9. Holmes, A. 1965. Principles of physical geology. Ronald Press, New York.
10. Hunt, C. B. 1972. Geology of soils. W. H. Freeman & Co., San Francisco.
11. Jackson, M. L. 1958. Soil chemical analysis. Prentice Hall, Inc. Englewood Cliffs, New Jersey.
12. Joffe, J. S. 1949. Pedology. 2nd Ed. Pedology Publications, New Brunswick, N.J.
13. Keller, W. D. 1964. Processes of origin and alteration of clay minerals in soil clay mineralogy. p. 3–76. In C. I. Rich and G. W. Kunze (ed.) Univ. of North Carolina Press, Chapel Hill.
14. National Coal Association. 1975. Coal facts. NCA, Coal Building, Washington, DC.
15. Pedersen, T. A., A. S. Rogowski, and R. Pennock, Jr. 1978. Comparison of morphological and chemical characteristics of some soil and minesoils. Reclamation Rev. 1:143–156 (Pergamon Press Ltd.) (Also discussed in Comparison of some properties of minesoils and contiguous natural soils. EPA-600/7-78-162.).
16. Ruhe, R. V., J. M. Williams, R. C. Shuman, and E. L. Hill. 1965. Nature of soil parent materials in Ewa-Eaipahu area, Oahu, Hawaii. Soil Sci. Soc. Am. Proc. 29:282–287.
17. Simonson, R. W. 1970. Loss of nutrient elements during soil formation. p. 21–45. In Nutrient mobility in soils: Accumulation & losses. O. P. Englestad (ed.). SSSA Special Pub. no. 4. Soil Sci. Soc. Am., Madison, Wis.
18. Smith, R. M., W. E. Grube, Jr., T. Arkle, Jr., and A. A. Sobek. 1974. Mine spoil potentials for soil and water quality. EPA-600/2-74-070. NERC, ORD, USEPA, Cincinnati, Ohio 45268.
19. ————, and A. A. Sobek. 1978. Physical and chemical properties of overburdens, spoils, wastes, and new soils. p. 149–172. In F. W. Schaller and P. Sutton (ed.) Reclamation of drastically disturbed lands. Am. Soc. of Agron., Madison, Wis.
20. ————, A. A. Sobek, T. Arkle, Jr., J. C. Sencindiver, and J. R. Freeman. 1976. Extensive overburden potentials for soil and water quality. EPA-600/2-76-184. IERL, ORD, USEPA, Cincinnati, Ohio 45268.

21. Soil Survey Staff. 1975. Soil taxonomy: A basic system of soil classification for making and interpreting soil surveys. Agric. Handb. No. 436. USDA-SCS. 754 p.
22. West Virginia University. 1971. Mine spoil potentials for water quality and controlled erosion. Water Pollution Control Research Series, 14010 EJE 12/71. USEPA, Cincinnati, Ohio 45268.
23. Wilmoth, R. C., T. L. Baugh, and D. W. Decker. 1978. Removal of selected trace elements from acid mine drainage using existing technology. p. 886–894. *In* Proc. of the 33rd Industrial Waste Conference, Purdue Univ. Ann Arbor Science Publishers, Ann Arbor, Mich.
24. Wiram, V. P. 1977. Keeping things clean with soap and water. Proc. 5th Symp. on Surface Mining and Reclamation. Bituminous Coal Research Inc., Monroeville, Penn.

Chapter 13

Characteristics and Reclamation of "Acid Sulfate" Mine Spoils[1]

R. I. BARNHISEL, J. L. POWELL, G. W. AKIN, AND M. W. EBELHAR[2]

ABSTRACT

Several factors have been identified, any one of which may seriously limit the establishment of vegetative cover on acid sulfate coal mine spoils. The addition of liming agents such as calcium carbonate as required by sample testing will reduce the acidity, but spoils in which large amounts of lime are required often have other factors limiting plant growth. Acid sulfate coal mine spoils frequently tend to have very low levels of available phosphorus, occasionally low levels of potassium, and these spoils are usually droughty.

Several experiments have been established in Kentucky to learn which successful combinations of soil fertility amendments and which soil test for lime requirement (i.e., pH, buffer, or total acidity) may be best used in reclamation practices on acid sulfate coal mine spoils. Success in revegetation of these spoils was achieved with the addition(s) of lime and plant nutrients in combination with appropriate steps to reduce runoff by providing a rough micro-relief and with the selection of adapted species and/or varieties.

INTRODUCTION

During the 1950s, surface mining for coal increased significantly in eastern USA. Several states initiated laws and associated regulations in attempts to promote reclamation of disturbed lands. These laws varied among states, but in general more stringent reclamation standards were adopted with each meeting of the respective general assemblies.

---

[1] Contribution from the Dep. of Agronomy, Kentucky Agric. Exp. Stn., Lexington, KY 40546. This paper (No. 79-3-170) is published with approval of the Director of the Kentucky Agric. Exp. Stn. Presented before Div. S-9, S-5, S-2, and S-6 of the Soil Sci. Soc. Am. as a portion of the symposium "Acid Sulfate Weathering II" during the ASA meetings, 6 Aug. 1979, in Fort Collins, CO.

[2] Professor of agronomy; reclamation supervisor, Peabody Coal Co., Greenville, Kentucky; Agronomist, TVA, Muscle Shoals, Alabama; Assist. Professor of Agronomy, Delta Branch Agric. Exp. Stn., Stoneville, Alabama, respectively.

Copyright © 1982 Soil Science Society of America, 677 S. Segoe Rd., Madison, WI 53711. *Acid Sulfate Weathering.*

Progress was slow primarily because of the lack of financial support and the general lack of interest by agronomists and others in conducting research in reclamation. Various environmental groups were instrumental in passage of federal legislation to control surface mining of coal, and eventually this resulted in Public Law 95-87, "The Surface Mining Control and Reclamation Act of 1977." In the opinion of many, this law was the most complex piece of legislation ever enacted by Congress. The law addresses essentially all aspects of surface and underground coal mining both past and present. The law was followed by a complex and detailed set of regulations that in many cases are vague or difficult to interpret.

Although the words "acid sulfate mine spoils" do not appear in the Act, portions of at least two sections are indirectly related to the effects of acidity: "Title IV—Abandoned Mine Reclamation" and "Title V—Control of the Environmental Impacts of Surface Coal Mining."

The primary purpose of Title IV is the reclamation and restoration of land and water resources adversely affected by past coal mining. One of the major tasks associated with this reclamation is the selection or planting of species to prevent erosion, since, in most cases, the spoil materials in the humid east are acidic. Part of this reclamation effort will be directed toward reclamation of rural lands through cost sharing projects.

Title V is directed toward the reclamation of land disturbed by current mining activities. This program includes the restoration of disturbed land to conditions capable of supporting at least the uses which it had prior to mining. For agricultural crops, this has been interpreted to mean that crop yields will be equal to or greater in the case of prime land or greater than 90% of the yield in the case of non-prime land to that of an adjacent reference area (or some other reference yield).

Mine operators must take appropriate measures that will avoid acid mine drainage that may affect the hydrologic balance. All acid-forming materials will be buried or otherwise treated during the reclamation process to provide for adequate rooting media.

LIME REQUIREMENT TESTS

Three approaches are frequently used in determining the lime requirement of mine spoils; a) measurement of the pH of samples suspended in water or salt; b) measurement of the pH of a buffer-spoil mixture such as Shoemaker et al., 1961; and c) determination of the total oxidizable sulfide mineral content or the potential total acidity (unpublished method by senior author modified from Smith et al., 1974). Characterization of some mine spoils as to their acidity or potential to produce acids may be found in several published bibliographies (Funk, 1962; NCA/BCR, 1975; Czapowskyj, 1976; Gleason and Russell, 1976; Weiss et al., 1977).

The Univ. of Kentucky Agric. Exp. Stn. initiated a spoil testing service in 1975. Since this program began, several thousand samples of surface mine coal spoils or potential overburden samples have been tested and subsequent fertilizer and lime recommendations made. A wide range in values has been observed from alkaline spoils (pH 7.0 to 8.3) to highly acidic spoils (pH 2.0), the latter of which may require as much as 200

Table 1. Chemical and mineralogical properties of typical Kentucky shale spoil materials.

Property	Black shale mod. acidic	Gray shale Slightly acidic	Neutral
pH	2.2	5.1	7.2
Sulfur	3.9%	1.0%	0.3%
Lime requirement			
SMP† (metric tons/ha)	21.1	1.1	0
P.A.‡ (metric tons/ha)	43.7	4.0	0
Mineralogy (Total)§			
Kaolinite	45%	26%	23%
Mica	35	40	52
Quartz	20	28	22
Chlorite	0	6	3

† Shoemaker et al., 1961.
‡ Potential acidity—A modified H_2O_2 method adapted from Smith et al., 1974.
§ X-ray diffraction data—based on peak height measurements and weighted averages from particle size analysis.

metric tons of agricultural lime per hectare to neutralize the active and potential acidity. The majority of coal spoils in Kentucky are derived from shale or siltstone overburden materials, and fall into three groups: a) neutral, b) slightly acidic, and c) moderately acidic. Chemical and mineralogical properties of typical Kentucky spoils are illustrated in Table 1.

Lime recommendations based on pH values measured in water or salt solutions are poor when the sample contains sulfide minerals. Freshly exposed samples may have pH's near neutrality yet contain enough sulfide minerals which upon oxidation would require lime applications in excess of 150 metric tons per hectare. Measurement of water or salt pH of a sample is useful information for samples taken from both unreclaimed or reclaimed spoils since it gives the current acidity level and allows judgements to be made as to the continued success or potential for success of any given reclamation project.

Lime recommendations based on buffer methods are also subject to error if the spoil sample contains sulfide minerals. It is a better method than water pH measurements to estimate current lime needs since a buffer method would reflect differences in cation exchange capacities and exchangeable Al.

Measurement of total or potential acidity (P.A.) frequently results in the best data for making lime recommendation as such methods account for the unoxidized sulfide minerals. These methods may occasionally give incorrect values as the result of the sample preparation technique used, because most methods require that the sample be finely ground. If the sample contains carbonates, either naturally occurring or as lime applied prior to sampling, underestimation of the lime requirement may occur. This is because native carbonates usually occur as large particles, which are inefficient in neutralizing acids generated by the oxidation of the sulfides. In the finely ground test sample, this is not the case. Lime applied prior to sampling, particularly when not uniformly incorporated, may become coated with a thin layer of iron oxide. This iron oxide coating may be broken upon sample preparation, thus exposing the lime to neutralization reactions in the reaction flask and causing low lime recommendations.

The practice used in Kentucky (Barnhisel, 1976a) for making lime recommendations is to review data from all three measurements: pH, SMP-buffer test (Shoemaker et al., 1961), and P.A. test. Whichever of these three tests results in the largest lime rate is the one recommended for the reclamation project. In other words, if both the SMP-buffer and P.A. tests indicate zero lime rate, but the pH measured in water suspension is less than 5.5, a rate of 2 to 4 metric tons per hectare is recommended. This example occurs with sandstone spoils, which have been limed. The lime has been coated with iron oxides and such spoils have very low cation exchange capacities, thus resulting in an inaccurate SMP buffer lime requirement test value.

CROP AND pH RESPONSE TO LIME APPLIED TO SURFACE MINED COAL SPOILS

A series of experiments have been established in Kentucky to evaluate crop response to applied lime. These observations are summarized in Tables 2, 3, and 4. The experimental designs and other supporting data are given elsewhere (Akin, 1976;[3] Ebelhar, 1977;[4] and Barnhisel, 1977).

The least acidic site (Table 2) had a SMP-lime requirement of 5.5 metric tons/ha. However, the pH did not maintain an acceptable pH level of 5.5 for 20 months unless twice that lime rate was applied or 11 metric tons/ha. The first increment of lime resulted in a significant increase in yield of forage harvested between 13 and 20 months after establishment of the experiment.

A similar trend with regard to pH response to applied lime was observed for acidic sandstone spoils (Table 3). The 18 metric tons/ha lime rate was equivalent to the SMP buffer lime requirement test (Shoemaker et al., 1961), whereas the 36 metric tons/ha rate corresponds to the total potential acidity test level. By 17 months, the pH level from all plots for both the 18 and 36 metric tons/ha lime rates had approached unacceptable levels, less than pH 5.5. For the treatment in which twice the lime rate required to neutralize the acid produced from sulfides (P.A. test) was applied, the pH exceeded values of 6.0 after 17 months.

Significant increases in forage yields as the result of applied lime were not obtained with the acidic sandstone spoils (Table 3); in fact, the 72 metric tons/ha lime rate resulted in a significant yield reduction when compared with the 36 metric tons/ha lime rate. The lack of positive yield responses may be attributed to the low test levels of available K. Prior to harvesting forage, some subplots had received as much as 80 kg K/ha with little change in soil test levels from initial very low levels (For details see Ebelhar, 1977[4]). Additions of K have subsequently been applied, the

---

[3] Akin, G. W. 1976. Evaluation of chemical changes and forage responses as a result of lime and phosphorus treatments on orphan surface-mined coal spoils. Unpublished M.S. Thesis, Univ. of Kentucky, Lexington. 177 p.

[4] Ebelhar, M. W. 1977. The response of common bermudagrass to nitrogen, lime, potassium and phosphorus treatments and their effect on the chemistry of aced surface-mined coal spoils in western Kentucky. Unpublished M.S. Thesis, Univ. of Kentucky, Lexington. 178 p.

Table 2. Effect of lime applied to acidic shale orphan spoils on pH measured over time and on forage yield (Akin, 1976[3]).

Lime rate	Time in months					Forage yield[‡]
	0[†]	3	7	13	20	
metric ton/ha			pH			kg/ha
0	4.5	5.3	4.7	4.9	4.7	1,253 a
5.5	4.5	5.2	5.3	5.5	5.2	1,377 b
11	4.5	5.6	5.8	6.2	5.6	1,395 b

[†] pH values prior to applying lime.
[‡] Total yield of Ky 31 tall fescue–red clover forage for the second growing season obtained between 13 and 20 months. Duncan Multiple Range Test at alpha of 0.05.

Table 3. Effect of lime applied to acidic sandstone spoils on pH measured over time and on forage yield (Ebelhar, 1977[4]).

Lime rate	Time in months					Yield[‡]
	0[†]	2	4	13	17	
metric ton/ha			pH			kg/ha
18	3.4	6.0	4.6	5.4	4.6	740 ab
36	3.4	6.4	5.6	6.0	5.7	829 b
72	3.4	7.1	6.4	6.8	6.3	659 a

[†] pH value prior to applying lime.
[‡] Total forage of common bermudagrass for the second growing season obtained between 13 and 17 months. Duncan Multiple Range Test at alpha of 0.05.

Table 4. Effect of lime applied to highly acidic sandstone spoils and the method of lime incorporation on the pH at three-depth increments and on forage yield (Barnhisel, 1977).

Lime rate	Incorporation method	Sample depth in cm			Yield[†]
		0–5	5–10	10–15	
metric tons/ha			pH		kg/ha
67	Disked	6.5	4.8	3.3	165 b
134	Disked	6.8	5.1	3.4	123 b
67	Chisel plowed	6.2	5.1	3.8	175 b
134	Chisel plowed	6.8	4.9	3.4	190 b
67	Not incorporated	5.4	3.8	3.0	0 a
134	Not incorporated	5.1	3.5	3.2	0 a

[†] Total yield of Ky 31 tall fescue for the second growing season. Duncan Multiple Range Test at alpha of 0.05.

largest of which was 320 kg K/ha with but a small increase in soil test levels. Although not verified, we suspect that K is being precipitated as the mineral jarosite. The very low soil test level of K may subsequently explain the lack of forage response to applied lime.

Data summarized in Table 4 are from an experiment established on sandstone spoils in which the potential total acidity equaled 134 metric tons/ha (For details, see Barnhisel, 1977). Very low forage yields were obtained and with little effect between the two methods used for lime incorporation or for the applied lime rate. Where lime was not incorporated, but allowed to remain on the surface, the pH's were low and the plots void of vegetation.

In this study, the lime was incorporated to a depth of 10 cm by two methods, disking and chisel plowing. Samples from which pH's are reported in Table 4 were collected 1 year after the lime was incorporated. There was a sharp contrast in color as a result of the lime application, and this color change corresponded with the pH values. The upper 10 cm had a red color (2.5 YR 4/6) whereas the acid zone below had a yellow color (10 YR 7/8).

Data presented in Tables 2, 3, and 4 for responses of pH to applied lime are similar to those of soil as given in ASA Monograph 12 (Pearson and Adams, ed., 1967).

FORAGE RESPONSE TO PHOSPHORUS

From analyses of thousands of samples by the Soil Testing Lab., Univ. of Kentucky, and numerous published reports (Berg, 1969; Berg and May, 1979; Plass and Vogel, 1973; Barnhisel et al., 1975; Bauer et al., 1978) it is apparent that if any one plant nutrient can be considered as being limiting for all spoils, it would be P. This is followed closely by N.

Several surface mine reclamation projects have been established in Kentucky that illustrate the response of forages to P (Powell, 1973;[5] Akin, 1976;[3] Ebelhar, 1977;[4] Barnhisel, 1977; Powell et al., 1980). Data presented in Table 5 were obtained over 3 years on neutral spoils in which the rates of 0, 84, and 186 kg P/ha were applied as a single application (for complete details see Powell, 1973[5] and Powell et al., 1980). A significant yield response was obtained at the 0.05 level of significance for each increment of P for the 1st harvest year for both the smooth and rough surface roughness treatments. In succeeding years differences in yields from the initial P application rates were less dramatic and in a number of instances not significant at the 0.05 level.

Soil test levels at the end of the 1st harvest year were in the medium range for plots which had received 186 kg P/ha 17 months earlier when the experiment was established. The P soil test levels for all other treatments were in the low range (less than 133 kg/ha). For plots to which phosphorus was not applied, the soil test level tended to increase with time although these various test levels were not significantly different (For soil test values, see Table 1 of Powell et al., 1980).

As can be seen in Table 5, there was an increase in yield at all levels of applied P as a result of surface roughness (See Powell et al., 1980 for more details). The vegetation on plots that were roughed by ripping or subsoiling were better able to utilize the added P since they apparently had more moisture as the result of less surface runoff. The effect of surface roughness on forage yield decreased with time but this was the result of lower available P levels, since the integrity of the depressions in the surface created by ripping were maintained throughout the experiment.

---

[5] Powell, J. L. 1973. Evaluation of tillage and phosphorus fertilizer on reclamation of surface-mined coal spoils. Unpublished M.S. Thesis. Univ. of Kentucky, Lexington. 94 p.

Table 5. Effect of P and surface roughness on yield of fescue–red clover forage over time (Powell et al., 1980).

Roughness treatment and harvest year†	Total P applied (kg/ha)		
	0	84	186
	——————— kg/ha ———————		
Smooth graded surface			
1st Harvest year	113 a	1,469 b	2,035 c
2nd Harvest year	525 a	798 b	887 b
3rd Harvest year	913 a	1,246 ab	1,271 b
Rough surface (ripped)			
1st Harvest year	423 a	2,638 b	3,508 c
2nd Harvest year	788 a	1,005 b	1,257 c
3rd Harvest year	1,280 a	1,023 a	1,169 a

† Yield for first harvest taken during the second growing season—Plots established in March. Duncan Multiple Range Test at an alpha of 0.05. These letters apply (across a row) to yield comparisons between P rates in a given year with a given roughness treatment. For other comparisons see Powell et al., 1980.

POTENTIAL PROBLEMS ASSOCIATED WITH RECLAMATION OF ACID SULFATE MINE SPOILS

Dissolution of Clay Minerals. Several reports have been published on weathering of clay minerals by acid solutions (Osthaus, 1954, 1956; Gastuche, 1963; Miller, 1965, 1968; and Gilkes and Young, 1974). Most of these studies were conducted for monomineralic systems, whereas mine spoils contain a mixture of minerals as illustrated in Table 1. However, the same trends in weathering of individual minerals was reported for mixtures of mica and kaolinite by Barnhisel and Rotramel (1974). They observed that both mica and kaolinite weathered as the result of an edge-attack when these clays were subjected to sulfuric acid. The rate of weathering for both minerals was similar and it was concluded that the bulk clay mineralogy of spoils subjected to natural acid weathering would not change significantly with time, although the average particle size might be decreased. Barnhisel and Rotramel also reported a linear relationship between the rate at which K and Al ions were released over a wide range of H_2SO_4 concentrations.

Lime Recommendations. Several potential problems exist in making lime recommendations for neutralization of acid mine spoils (Barnhisel, 1976a). This discussion is with respect to the unusual nature of mine spoils and not with regard to the test employed. It is also assumed that the sample from which the recommendation is made represents the area. Mine land in which the overburden consists of more than one rock stratum, may be very heterogeneous. Contrasts in lime requirements within a short distance may exceed those represented by the three samples in Table 1. The most critical phase of a testing program is that of sampling (Barnhisel, 1976b).

Most lime recommendations for agricultural soils are based on a plow layer depth of about 17 cm that represents 2.24×10^3 kg/ha (2×10^6

lb/acre). When large rates are required for mined land, >30 metric tons per hectare, greater quantities are needed in order to react with upward movement of acid water from below the zone of lime incorporation.

Lime recommendations that do not reflect the level of potential acidity that can be released upon the oxidation of sulfide minerals such as pyrite are subject to failure. The potential for such failures is illustrated in Table 1, since some spoils may contain significant amounts of sulfide minerals.

The lime recommendations that are the most difficult to make are for spoils that contain carbonates. This problem is due to the uncertainty of the particle size distribution of the natural carbonate materials. In the process of sample preparation, large natural fragments of limestone rock may be finely ground and the data collected by either SMP or P.A. tests no longer represent the field condition. Lime recommendations for spoils that have been limed prior to sample collection may also result in erroneous values if the applied lime has become coated with a thin iron oxide layer.

Incorporation of Lime. Improper or inadequate mixing may be a serious limitation in obtaining reliable revegetation of acid mine spoils. Lime can be mixed by disking and rapid neutralization reaction rates may be obtained; however, a heavy-duty disk is needed to mix the lime and under the best conditions, depths of only 15 to 20 cm may be obtained. Such operations are more easily accomplished if the spoils have been freshly graded or disked. Adequate mixing is more frequently achieved when the spoils are dry, provided the disk penetrates the needed depth. If the spoils are damp when the lime is applied, repeated disking treatments are needed to insure adequate incorporation, and the second disking opeartion should be done after the spoils have dried somewhat and prior to the next rainfall event.

Droughty Characteristics of Acid Spoil. Regardless of the amount of lime applied and even with effective incorporation, any spoil which requires a large lime rate (greater than 30 metric tons/ha) will tend to be droughty, even in humid climates. This droughtiness is not a deficiency of rainfall but rather limited root interception. Most plants require a rooting depth of more than 15 to 20 cm; while highly acidic spoils may have pH's less than 4 immediately below the zone of lime incorporation (see Table 4).

Toxicities and Deficiencies of Plant Nutrients. The rooting zone of plants in acidic spoils may contain toxic levels of ions such as Cu, Ni, Zn, Mn, or Fe (Berg and Vogel, 1968; Massey and Barnhisel, 1972) although some of these elements are required for plant growth at lower levels. As the result of a potentially wide range in pH of mined land, deficiencies of Mn, Mo, B may exist for some plants, as well as other essential elements.

Acid spoils may contain large amounts of sulfate which may exist with several cations such as Fe^{2+}, Ca^{2+}, Mg^{2+}, etc. As suggested earlier, potassium may be precipitated as jarosite $[KFe_3(SO_4)_2(OH)_6]$. In such cases, the spoils may not supply enough K for adequate plant growth.

LITERATURE CITED

1. Barnhisel, R. I. 1976a. Lime and fertilizer recommendations for reclamation of surface-mined spoils. Ky. Agric. Exp. Stn. Pub. AGR 40. Univ. of Kentucky, Lexington. 4 p.
2. ———. 1976b. Sampling surface-mined coal spoils. Ky. Agric. Exp. Stn. Pub. AGR 41. Univ. of Kentucky, Lexington. 4 p.
3. ———. 1977. Reclamation of surface-mined coal spoils. EPA-600/7-77-093. EPA Indust. Environ. Res. Lab., Cincinnati, Ohio. 57 p.
4. ———, J. L. Powell, and G. W. Akin. 1975. Keys to successful reclamation in western Kentucky. p. 140–151. In 3rd Symp. on Surface Mining and Reclamation. Natl. Coal Assoc., 1130 17th St., N.W. Washington, DC.
5. ———, and A. R. rotramel. 1974. Weathering of clay minerals by simulated acid coal spoil-bank solutions. Soil Sci. 118:22–27.
6. Bauer, A., W. A. Berg, and W. L. Gould. 1978. Correction of nutrient deficiencies and toxicities in strip-mined lands in semiarid and arid regions. p. 451–466. In Schaller and Sutton (ed.) Reclamation of drastically disturbed lands. Am. Soc. Agron. Publishers.
7. Berg, W. A. 1969. Evaluation of P and K soil fertility tests on coal-mined spoils. p. 93–104. In R. J. Hutnik and G. Davis (ed.) Ecology and reclamation of devastated lands. Gordon and Breach, Publishers, NY. Vol. I.
8. ———, and R. F. May. 1969. Acidity and plant-available phosphorus in strata overlying coal seams. Min. Congr. J. 55:31–34.
9. ———, and W. G. Vogel. 1968. Manganese toxicity of legumes seeded in Kentucky strip-mine spoils. USDA Forest Serv. Res. Paper NE-119.
10. Czapowskyj, M. M. 1976. Annotated bibliography on the ecology and reclamation of drastically disturbed areas. USDA For. Serv. Gen. Tech. Rep. NE-21. 98 p.
11. Funk, D. T. 1962. A revised bibliography of strip-mine reclamation. USDA Forest Service. Central States For. Exp. Stn. Misc. Pub. 35. 20 p.
12. Gastuche, M. C. 1963. Kinetics of acid dissolution of biotite: I. p. 67–76. In Proc. Int. Clay Conf. (Stockholm) Pergamon Press, London.
13. Gilkes, R. J., and R. C. Young. 1974. Artificial weathering of oxidized biotite: IV. The inhibiting effect of potassium on dissolution rate. Soil Sci. Soc. Am. Proc. 38:529–532.
14. Gleason, V., and H. H. Russell. 1976. Coal and the environment abstract series: Mine drainage bibliography 1910–1976. Bituminous Coal Res. Inc., Monroeville, Pa. 288 p.
15. Massey, F. H., and R. I. Barnhisel. 1972. Copper, nickel, and zinc released from acid coal mine spoil materials for eastern Kentucky. Soil Sci. 113:207–212.
16. Miller, R. J. 1965. Mechanisms for hydrogen to aluminum transformations in clays. Soil Sci. Soc. Am. Proc. 29:36–39.
17. ———. 1968. Electron micrographs of acid-edge attack of kaolinite. Soil Sci. 105: 166–171.
18. MCA/BCR. 1975. Reclamation of Coal-mined Land—A Bibliography with Abstracts. Pub. by Natl. Coal Assoc. Bituminous Coal Res., Inc., 350 Hachberg Road, Monroeville, Pa. 188 p.
19. Osthaus, B. B. 1954. Chemical determination of tetrahedral ions in nontronite and montmorillonite. p. 404–417. In Proc. 2nd Natl. Conf. Clays and Clay Minerals, Natl. Acad. Sci.–Natl. Acad. Sci.–Natl. Res. Coun. Pub. 327.
20. ———. 1956. Kinetic studies on montmorillonite and nontronite by acid-dissolution technique. p. 301–321. In Proc. 4th Natl. Conf. Clays and Clay Minerals. Natl. Acad. Sci.–Natl. Research Council Publ. 456.
21. Pearson, R. W., and F. Adams. 1967. Soil acidity and liming. Monograph 12. Am. Soc. Agron., Madison, Wis.
22. Plass, W. T., and W. G. Vogel. 1973. Chemical properties and particle size distribution of 39 surface-mined spoils in southern West Virginia. USDA For. Serv. Res. Pap. NE-276. 8 p.
23. Powell, J. L., R. I. Barnhisel, and G. W. Akin. 1980. Reclamation of surface-mined coal spoils in western Kentucky. Agron. J. 72:597–600.

24. Shoemaker, H. E., E. O. McLean, and P. F. Pratt. 1961. Buffer methods for determining lime requirement of soils with appreciable amounts of exchangeable aluminum. Soil Sci. Soc. Am. Proc. 25:274–277.
25. Smith, R. M., W. E. Grube, Jr., T. Arkle, Jr., and A. Sobek. 1974. Mine spoil potentials for soil and water quality. EPA-670/2-74-070. EPA Indust. Environ. Res. Lab., Cincinnati, Ohio. 303 p.
26. Weiss, N. E., A. A. Sobek, and D. L. Streib. 1977. A selected bibliography of surface coal mining and reclamation literature. Vol. 1. Eastern Coal Province. Argonne Natl. Lab., Argonne, Ill. 158 p.